Mathematics of Linear
and Nonlinear Systems

Mathematics of Linear and Nonlinear Systems

For Engineers and Applied Scientists

D. J. BELL

Department of Mathematics
University of Manchester Institute
of Science and Technology

CLARENDON PRESS · OXFORD
1990

Oxford University Press, Walton Street, Oxford OX2 6DP
Oxford New York Toronto
Delhi Bombay Calcutta Madras Karachi
Petaling Jaya Singapore Hong Kong Tokyo
Nairobi Dar es Salaam Cape Town
Melbourne Auckland
and associated companies in
Berlin Ibadan

Oxford is a trade mark of Oxford University Press

Published in the United States
by Oxford University Press, New York

British Library Cataloguing in Publication Data
Bell, D. J. (David John) 1933-
Mathematics of linear and nonlinear systems.
1. Dynamical systems. Mathematics
I. Title
515.3'5
ISBN 0-19-856332-9

Library of Congress Cataloging in Publication Data
Bell, D. J. (David John), 1933-
Mathematics of linear and nonlinear systems for engineers and applied scientists/D. J. Bell.
p. cm.
Includes bibliographical references.
ISBN 0-19-856332-9
1. Science–Mathematics. 2. Technology–Mathematics.
3. Numerical calculations. I. Title.
Q158.5.B45 1990
510'.02451-dc20
89-37022

Typeset by Macmillan India Ltd., Bangalore 25.
Printed in Great Britain by
Courier International Ltd
Tiptree, Essex

To my mother
CHRISTINE LOUISE
and to my wife
MARJORIE LOUISE

PREFACE

Dynamical systems are evident in many aspects of daily life, whether they be natural, as in the case of the solar system, or those directly controlled by human influence, such as a piloted aircraft. The mathematical theory of uncontrolled systems has a very long history (certainly since the time of Sir Isaac Newton in the 17th century), whereas that for controlled systems is much more recent (e.g. stability of the flyball governor studied by James Clerk Maxwell in 1868).

The design and control of dynamical systems is now a well-established branch of engineering, and engineers were amongst the first to recognize the importance of linear algebra in the study of linear control systems. Yet, at one time, matrix analysis was taught only in mathematics courses. Only much later was this subject introduced into engineering education. Nowadays it would be unthinkable that any qualified engineer involved in linear systems should not have a working knowledge of matrices. Linear algebra has become essential for the analysis of such systems. More recently, other branches of mathematics have proved relevant in the study of both linear and nonlinear system theory. Functional analysis in particular has become important in many areas of engineering (Nowinski 1981). Differential geometry and, very recently, differential algebra are other topics that have gained prominence in the technical and scientific literature. Elements of Galois theory can now be found in the syllabus for undergraduate courses in some departments of electrical engineering.

In recent years, workers in disciplines other than engineering, such as economics, biology, and management science, have also felt the need for some of this more advanced mathematical knowledge. However, a major disadvantage for nonmathematicians who wish to study these branches of mathematics is that the available texts have been written mainly for, and by, mathematicians. In most cases the method of presentation in these texts has been unsuitable for graduates in subjects other than mathematics. After all, it can be somewhat distressing to find that the first line of a research paper or textbook is 'Let M be a paracompact manifold'. Many will not know what a manifold is, let alone a paracompact one! And yet it is not a trivial task to acquire the necessary mathematical background knowledge. On the one hand, there is an abundance of books for the mathematician but with little motivation for the application of the material, while on the other hand, many of the books on dynamical systems written for the nonmathematician avoid discussing the background mathematics in any depth. Thus, at the present time, there is a gap between these two extremes, and the present book is a modest attempt to help bridge that gap. It is true that in the last few years some books have appeared which combine the mathematical and dynamical system theories (Isidori 1985, Banks 1988) but these also demand a high level of mathematical maturity as a prerequisite. Such a level is not usually attained by undergraduates in a department other than one of mathematics.

Students frequently question the need for the mathematics they are taught and they are right to do so. One of the aims of this book is to illustrate how the need for quite sophisticated mathematical concepts arises in the study of mathematical models representing dynamical systems. A further aim is to enable the reader to become familiar with terms and notation currently used in systems theory. Having studied the book, it is hoped that comprehension of the early literature on the subject (some of which is included in the references) will then be in reach of the reader.

Prerequisites required in order to read this book are mainly those obtained from a first year engineering course in mathematics at a university or polytechnic in the United Kingdom. It is assumed that the reader is familiar with the idea of a complex number $z=a+jb$ and its complex conjugate $\bar{z}=a-jb$. Notice that in the time-honoured engineering tradition, the complex number $\sqrt{(-1)}$ is denoted by j in this book, whereas i is used in the mathematical literature. Some knowledge of the differential and integral calculus is essential, together with the expansion of a function in a Taylor series. Although continuity and differentiability are not discussed formally until Chapter 10, these concepts are used in an elementary and intuitive way throughout the book. In Chapter 10 they are defined in a way that allows for generalization. It is useful if the reader has a working knowledge of matrices (addition and multiplication), an understanding of the meaning of symmetric and inverse matrices, and the rank and determinant of a matrix. Right at the end of the book, the Cayley–Hamilton theorem is assumed known (see Strang 1980). A knowledge of methods for solving simple ordinary differential equations is also an advantage.

The book has been written primarily for postgraduate engineers and applied scientists working in the areas of linear and nonlinear dynamical systems. Nevertheless, applied and pure mathematicians in the final year of their undergraduate studies may well be interested in the role of certain mathematical topics in dynamical systems theory. Some mature A-level pupils studying mathematics at school or sixth form college may find the first four chapters of some interest. The book should also be useful to academics in a wide range of departments at universities, polytechnics, and colleges of further education, and to research staff in industry, particularly those working in the fields of aerospace and robotics. As a general introduction to the mathematical content of the present book, the reader could refer to one of the many excellent introductory texts on the foundations of mathematics. A superb example of such a book is that by Courant and Robbins (1941).

This present book can be used both as a reference book and as a textbook. As a reference, many of the technical terms used in the literature are defined without going into too much detail. However, in such cases, sources are given where further information can be found. The book can be used as a textbook, forming the basis of a two-term course (40 hours over 20 weeks) on dynamical systems. Alternatively, the first six chapters on algebraic topics contain the background mathematics for linear system theory and would be suitable for a one-term course of 20 hours for ten weeks. The remaining five chapters contain topics from mathematical analysis, leading to the idea of a Lie algebra, important in the study of nonlinear systems. This latter half of the book could also be used as a self-contained one-term course of 20 hours for ten weeks.

Notation is something of a problem in a book of this sort since there is no uniformity between disciplines. As a general rule, engineering practice has taken precedence over mathematical tradition in this book. For example, it has become acceptable in the engineering literature to use $G\ell(n, \mathbb{R})$ as the set of all nonsingular matrices, whereas in mathematics $G\ell$ is written as GL and $GL(n, \mathbb{R})$ denotes the (isomorphic) group of all general linear nonsingular transformations from $\mathbb{R}^n$ to $\mathbb{R}^n$ under composition. Brockett (1972) refers to the state space $G\ell(n)$ as the *space* of real nonsingular $n \times n$ matrices. Some of the notation used in this book will appear strange to professional mathematicians unfamiliar with the engineering literature. For example, the time profile of a control function u is denoted by $u(\cdot)$, with the dot indicating where the independent variable, time t, should appear in any formulae. Vectors are usually assumed to be column vectors and are denoted either by $(x_1, \ldots, x_n)$ or by $\begin{bmatrix} x_1 \\ \vdots \\ x_n \end{bmatrix}$. A row vector is denoted by $[x_1, \ldots, x_n]$. The words 'function' and 'mapping' are used interchangeably.

Notes at the end of each chapter are referenced by superscript numbers in the text. These notes contain important additions to the material in the chapters, such as definitions, explanations on notation, and comments. Some mathematical proofs, not essential to the main text, are relegated to the notes. These proofs can generally be ignored at a first reading. However, the reader is strongly advised to study them at some stage since insight can be gained into the methods of proof used by mathematicians (e.g. proof of uniqueness of identity in a monoid, Note 4.2, p. 89). Where proofs have been omitted, a reference has usually been included (e.g. Note 4.9, p. 89).

The text is liberally sprinkled with examples and there are exercises at the end of each chapter. A full solution to every exercise is given at the end of the book. The exercises are very important because many contain further results not given anywhere else in the book. They should be attempted immediately after reading the relevant chapter. If an exercise proves difficult, then the reader should study the solution carefully, making sure that each step in the argument is well understood.

Sections, examples, notes, exercises, and solutions are all referenced according to chapter. Examples, sections, and equation numbers are always given in full. But a note in a current chapter is generally referred to by its number only. For example, a reference in Chapter 3 to Note 4 would mean Note 3.4 whereas a reference to that same note in Chapter 5 would be given as Note 3.4. Again, Solution 7.5 refers to the solution of Exercise 5 of Chapter 7, etc. A reference to a note in the solutions to the exercises refers to the notes for the chapter containing those exercises. Thus, Note 18 in the solutions to Chapter 4 (p. 266) refers to Note 18 in the notes at the end of Chapter 4. The reader should be aware that some facts and definitions that appear in a note at the end of a chapter are then assumed known from the point in the chapter where the note is referenced. For example, the definition of $\mathbb{C}$ is given in Note 12 (p. 20 and referenced on p. 7) of Chapter 1 and is assumed known on p. 11.

References include books, which give further details on the various topics, together with journal papers, some of which contain the early work on nonlinear systems. No apology is made for quoting vintage references (e.g. Courant 1936). Some of the older books are classics and still well worth reading. Naturally, I have been influenced by a large number of existing text books during the preparation of this present book, but two in particular have been of outstanding value. One is the introductory text on differential topology by Chillingworth (1976). The other is the book by Hartley and Hawkes (1970) which provides an excellent introduction to the algebraic systems known as rings and modules.

I am indebted to a very large number of colleagues for help and encouragement over a number of years during the preparation of this book. In particular, my grateful thanks go to Dr J. A. H. Shepperd who read the whole of a first draft of the manuscript and made invaluable comments which greatly improved the presentation; also to Mr J. Bentin who did the same with a later draft. I am also very grateful to my UMIST colleagues: Drs I. M. Bride, P. J. Caudrey and P. J. Rowley, and Mr R. Herdan of the Department of Mathematics, and to Prof. H. H. Rosenbrock and Dr P. A. Cook of the Control Systems Centre, all of whom gave excellent advice on various aspects of the book. To Prof. R. W. Brockett I am indebted for permission to use some illustrative examples from his own work and to all my students, past and present, from whom I have learnt so much. Finally, my thanks to the staff of Oxford University Press for their help and guidance during the final preparation of the book for publication.

Manchester D.J.B.
1989

CONTENTS

1

INTRODUCTION TO DYNAMICAL SYSTEMS

When the engines of an aircraft in flight are all switched off, the aircraft will lose altitude. The rate at which heat is dissipated when chemicals are stirred in a reactor determines the concentrations of the constituents in the final mix. If sterling is allowed to fall in value, the price of imports into the United Kingdom will rise. Aircraft, chemical plants, and economies are all examples of dynamical systems. To investigate cause and effect in such systems, it is not always possible to experiment on the real system. It is therefore useful to have a mathematical model (usually a set of differential or discrete equations) the solution of which has a time behaviour approximately the same as that for the real system. Investigations can then be carried out on this mathematical model, and the results tested on the real system. It is only fair to point out that the production of a mathematical model that is a good representation of a physical system can sometimes be a very difficult task (Nicholson 1980).

This first chapter gives a number of examples which illustrate the way in which certain mathematical objects arise in the formation and analysis of mathematical models. The mathematical concepts associated with these objects are then discussed. The examples of this chapter and their generalizations will be used throughout this book in order to motivate the reader when studying the pure-mathematical topics necessary for a full understanding of dynamical systems.

1.1 Mathematical models

In order to describe the basic elements of a dynamical system, consider the example of a space vehicle launched vertically from the surface of the Earth's moon. The *system* is the whole vehicle complete with rocket motors, fuel, and *controller*, the last named being an astronaut sat at the controls in the vehicle. For the sake of simplicity, assume that gravity may be neglected, the change in mass of the vehicle caused by consumption of fuel is negligible, and that the flight of the rocket in a vertical straight line is governed only by one *control*, namely the magnitude T of the thrust from the rocket motors. The altitude and speed of the rocket at any time t' are obviously determined by the controller's choice of T from the initial time of launch at $t=0$ until time t'. Suppose that the astronaut can adjust the magnitude T of the thrust to any value lying between zero and some maximum value $\bar{T}$, so that $0 \leqslant T \leqslant \bar{T}$. The thrust magnitude $T(t)$ in general will depend upon the time t, and can be considered as a control variable or *input* to the system. Input or control variables are usually denoted by the letter u so that, in this case, $u(t) = T(t)$ at time t. The chosen profile of thrust magnitude over a given interval of time then determines the speed of the rocket and its height above the moon's surface over that time interval. The altitude h and speed $\dot{h}$ [1] can be considered as *output* variables. Notice that the input is chosen but the outputs are measured and observed. Output variables are usually denoted by the letter y. Thus, keeping the thrust magnitude constant, typical outputs of altitude $y_1 = h$ and speed $y_2 = \dot{h}$, measured against time t, would be as shown in Fig. 1.1. Different modes of operation of the input give different output behaviour.

The above approach is what engineers call a *black-box* description of a dynamical system, because the *way* in which the system yields an output from a given input is not considered (very often it is not known). In this approach, a dynamical system Σ is thought of as a black box into which is fed an input u and out of which comes a response to this input, namely an output variable y (Fig. 1.2). This description of a system is also known as an *input–output model.* In general, u and y will vary with time, and there may be several inputs to the system and more than one output from it (called a multivariable system, Fig. 1.3).

If the behaviour of the physical system Σ is well understood, then it is often possible to describe it in terms of a set of time-varying *state* variables

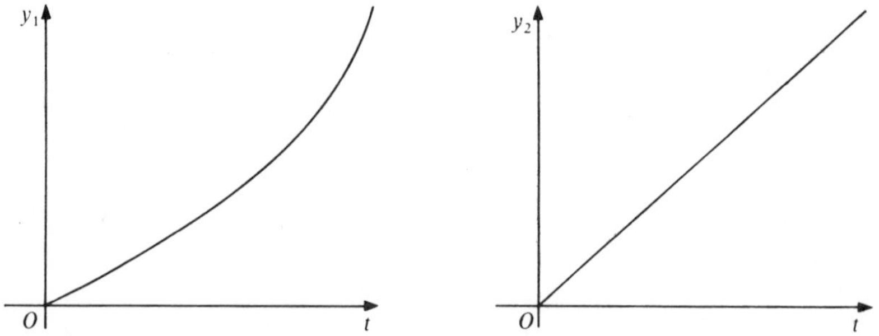

Fig. 1.1 Output variables

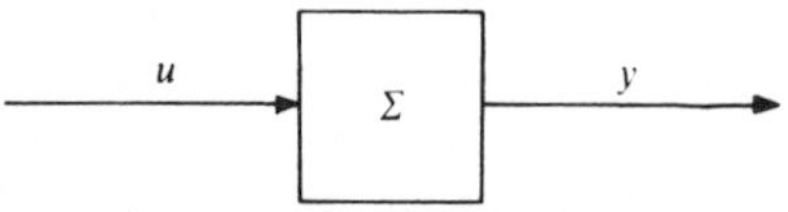

Fig. 1.2 System as a black box

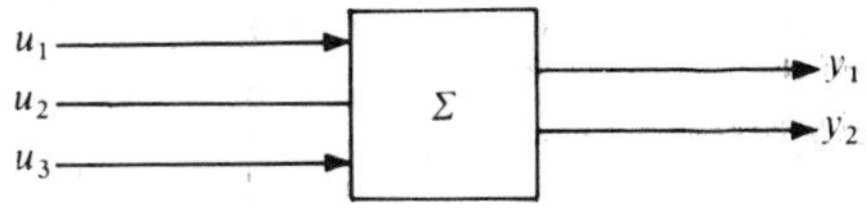

Fig. 1.3 Multivariable system

$x_1(t), \ldots, x_n(t)$. Roughly speaking, Σ is then called a system of *order n* (but see Rosenbrock 1970a). In the example of the rocket vehicle described above, Newton's law of motion

$$\ddot{h}(t) = T(t)/M \tag{1.1}$$

gives the equation of motion, or *system equation*, for a vehicle of mass M. If state variables x_1 and x_2 are chosen as $x_1(t) = h(t)$ and $x_2(t) = \dot{h}(t)$, and the control variable u as $u(t) = T(t)/M$, then (1.1) can be expressed as a system of order 2 in the form of two first-order differential equations

$$\dot{x}_1 = x_2, \qquad \dot{x}_2 = u, \tag{1.2}$$

in which the argument t has been dropped for convenience. If the outputs of the system are chosen to be altitude and speed, then

$$y_1 = x_1, \qquad y_2 = x_2,$$

so that, in this particular case, the output variables are the same as the state variables; this need not always be so. There are two basic reasons why it is useful to write equations such as (1.1) in the form (1.2). First, as will be seen later, techniques of linear algebra are more easily applied to such forms. Secondly, algorithms for solving differential equations on a computer are normally written in a form suitable for sets of first-order equations (e.g. Runge–Kutta methods, Lambert 1973).

A mathematical description of a dynamical system in terms of state variables is called a *state-space model*. The reason for the word 'space' will be explained in Section 1.3. If the model for Σ takes the form of one or more differential equations in the state and control variables, as for example in (1.2), then Σ is called a *continuous* system because the states are defined for all instants of time. On the other hand, as in sampled-data processes, if Σ is described by a set of *discrete* (or *difference*) equations such as

$$x_1(t+1) = x_2(t),$$
$$x_2(t+1) = x_1(t) + u(t) \quad (t = 0, 1, \ldots),$$

then Σ is called a *discrete* system. In both cases, the outputs are expressed in terms of the state variables. Of course, the behaviour of the real system and that of its mathematical model will differ to an extent determined by the assumptions and degree of approximation used in forming the model. For example, in the rocket problem discussed earlier, the mathematical models (1.1) and (1.2) were obtained assuming that gravity could be neglected and that the mass of the vehicle remained constant during flight. A more realistic model would include the acceleration due to the moon's gravitational attraction together with a differential equation for M.

As another example, when the theory of a simple pendulum is based upon the differential equation

$$\ddot{x} + \omega^2 \sin x = 0 \tag{1.3}$$

(ω a constant), the solution $x(t)$ is only an approximation to the angular displacement of a real pendulum, not only because the differential equation has to be solved numerically but also because external effects such as air resistance to the motion have been neglected in forming the mathematical model (1.3). In this book, little reference will be made to the real systems that the mathematical descriptions are supposed to approximate. This text is concerned solely with the analysis of the mathematical model, also referred to as Σ, and a dynamical system in this book is to be interpreted as a given set of differential or discrete equations. However, the reader should not lose sight of the fact that in practice it is the *real* system which is of greatest importance, not its mathematical model.

An alternative to the state-space procedure for describing a dynamical system is the *frequency-domain* approach. This can be used when the differential or difference equations have constant coefficients (time-invariant system) and are linear in the control, state, and output variables. Laplace transforms for continuous systems and z-transforms [2] for discrete systems are used to obtain a frequency-domain model (Examples 1.2 and 1.5 below).

The analysis of mathematical models using either state space or frequency domain quickly give rise to many different mathematical objects: real numbers, complex numbers, matrices, functions, and polynomials, to name but a few. It is well known that calculations using real numbers are subject to the usual rules of arithmetic (e.g. $3+7=7+3$ is an example of the *commutative* law of addition). The reader will also know that it is possible to add and multiply matrices, polynomials, and so on, and it is not too surprising to find that such operations on these more general mathematical objects also have to obey certain rules of their own. Many of these rules differ from those of arithmetic with real numbers, and this is why such sets of mathematical objects are of great interest. The fact that these objects are also very important in the study of dynamical systems is an added bonus.

Investigation into sets of mathematical objects and the rules that operations on them must satisfy form a major part of this book. Ways in which such objects arise in dynamical systems are illustrated in the next section.

1.2 Sets of mathematical objects

The state-space and frequency-domain approaches are illustrated in the following examples. Mathematical objects such as vectors, matrices, polynomials, and rational functions arise from both continuous and discrete systems.

Example 1.1

A rocket vehicle launched vertically from rest on the Moon's surface may be described by the system equations (1.2) and initial conditions

$$\begin{aligned} \dot{x}_1 &= x_2, \qquad x_1(0) = 0, \\ \dot{x}_2 &= u, \qquad x_2(0) = 0. \end{aligned} \tag{1.4}$$

Suppose the input $u(t)$ is chosen as $1-\mathrm{e}^{-t}$, and a single output variable $y(t)$ as $x_1(t)$. Then (1.4) can be written in vector–matrix form as

$$\begin{bmatrix} \dot{x}_1(t) \\ \dot{x}_2(t) \end{bmatrix} = \begin{bmatrix} 0 & 1 \\ 0 & 0 \end{bmatrix} \begin{bmatrix} x_1(t) \\ x_2(t) \end{bmatrix} + \begin{bmatrix} 0 \\ 1 \end{bmatrix} (1-\mathrm{e}^{-t}), \qquad \begin{bmatrix} x_1(0) \\ x_2(0) \end{bmatrix} = \begin{bmatrix} 0 \\ 0 \end{bmatrix},$$

or as

$$\dot{\boldsymbol{x}}(t) = \boldsymbol{A}\boldsymbol{x}(t) + \boldsymbol{b}u(t), \qquad \boldsymbol{x}(0) = \boldsymbol{0},^{3} \tag{1.5}$$

where

$$\boldsymbol{x}(t) = \begin{bmatrix} x_1(t) \\ x_2(t) \end{bmatrix}, \qquad \boldsymbol{b} = \begin{bmatrix} 0 \\ 1 \end{bmatrix}, \qquad \boldsymbol{0} = \begin{bmatrix} 0 \\ 0 \end{bmatrix}, \qquad \boldsymbol{A} = \begin{bmatrix} 0 & 1 \\ 0 & 0 \end{bmatrix}.$$

In order to express the output $y = x_1$ in terms of the *state vector* $\boldsymbol{x}$, the output variable y may be written in the form $y(t) = \boldsymbol{c}^{\mathrm{T}}\boldsymbol{x}(t)$ where $\boldsymbol{c}^{\mathrm{T}} = [1, 0]$. The magnitudes of the control and output at time t are given respectively by

$$u(t) = 1 - \mathrm{e}^{-t}, \qquad y(t) = x_1(t).$$

In Example 1.1, the value of the time variable t is assumed to increase from its initial value of zero through all positive real numbers. In other applications, it is sometimes convenient or necessary to allow the value of t to run through all negative real numbers. At any time t, the control, state, and output variables have values $u(t)$, $x_1(t)$, $x_2(t)$, and $y(t)$, all of which are also real numbers. Indeed, in engineering and other disciplines where mathematics is applied, solutions to practical problems have to be obtained by ordinary arithmetic using real numbers no matter how abstract the underlying theory. It is convenient to think of the real numbers as a set of points (or elements) distributed along a straight line (called the *real line*) as shown in

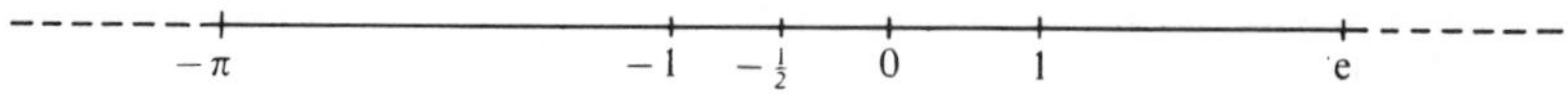

Fig. 1.4 The real line

Fig. 1.4 and on which a few of the real numbers have been declared. The set of all real numbers is denoted by $\mathbb{R}$. Some of the tools for arithmetical work on the real numbers are the operations of addition ($+$), multiplication ($\times$ or $\cdot$), and relations between real numbers such as 'equality' ($=$), 'greater than' ($>$) and 'less than' ($<$).

Returning to Example 1.1: at any time t, the components of the vector $\boldsymbol{x}(t)$ will be the real numbers $x_1(t)$ and $x_2(t)$; and each such vector $\boldsymbol{x}(t)$ will be a member of the set of all vectors like (a, b), where a and b are arbitrary real numbers. This set is denoted by $\mathbb{R}^2$, and so in particular the vectors $\dot{\boldsymbol{x}}(t)$ [4, 5], $\boldsymbol{b}$, and $\boldsymbol{x}(0)$ in (1.5) belong to $\mathbb{R}^2$. Similarly the matrix $\boldsymbol{A}$ in (1.5) is a member of the set of all 2×2 matrices with real components, a set denoted by $\mathbb{R}^{2 \times 2}$ (or sometimes by $M_2(\mathbb{R})$).

The input u in Example 1.1 is the function given by $u(t) = 1 - e^{-t}$. The graph of this function is said to be continuous, by which is meant there are no breaks in the curve. This function is a member or element of the set $C^0(\mathbb{R})$ of all functions that are continuous on $\mathbb{R}$, and it could (if necessary) be stipulated that the input u to a dynamical system should always be a member of $C^0(\mathbb{R})$.

Example 1.2

The discrete analogue of (1.5) is the vector–matrix equation

$$\boldsymbol{x}(t+1) = \boldsymbol{A}\boldsymbol{x}(t) + \boldsymbol{b}u(t), \tag{1.6}$$

where now the time t is restricted to integral values only, i.e. $t = \dots, -3, -2, -1, 0, 1, 2, 3, \dots$. In this case, the input will be a sequence of real values. For example, $u(-4) = 3$, $u(-3) = -\frac{1}{2}$, $u(-2) = -2$, $u(-1) = 0$, and $u(0) = 1$, giving the finite sequence $(3, -\frac{1}{2}, -2, 0, 1)$ as input. The z-transform of this sequence is defined as

$$\begin{aligned}\bar{u}(z) &= \sum_{t=-4}^{0} u(t)z^{-t}\\ &= u(-4)z^4 + u(-3)z^3 + u(-2)z^2 + u(-1)z + u(0)\\ &= 3z^4 - \tfrac{1}{2}z^3 - 2z^2 + 1.\end{aligned}$$

Unlike Example 1.1, the time t in Example 1.2 takes integral values only, i.e. values of t belong to the set of integers $\{0, \pm 1, \pm 2, \ldots\}$.[6] Denoted by $\mathbb{Z}$, this set is one of a number of important *subsets* of the set $\mathbb{R}$. The set $\mathbb{Z}$ is called a subset of $\mathbb{R}$ because every element in $\mathbb{Z}$ is also an element of $\mathbb{R}$. Other subsets of $\mathbb{R}$ are the set of all positive integers (the *natural numbers*) $\mathbb{N} = \{1, 2, \ldots\}$,[7], the set of all rational numbers $\mathbb{Q} = \{x : x = p/q, p, q \in \mathbb{Z}, q \neq 0\}$,[8,9,10] and the set of all irrational numbers $\{x : x \neq p/q$ for any $p \in \mathbb{Z}$, $q \in \mathbb{N}\}$, this last set containing real numbers such as $\sqrt{2}$, e, and π which are expressible only as nonterminating decimals. The fact that the subsets $\mathbb{Z}$, $\mathbb{N}$, $\mathbb{Q}$, ... are contained in the set $\mathbb{R}$ is denoted by $\mathbb{Z} \subset \mathbb{R}$, $\mathbb{N} \subset \mathbb{R}$, $\mathbb{Q} \subset \mathbb{R}$, ... ($\subset$ denotes *inclusion*). These subsets are what may be called *proper* subsets because they do not contain all elements of $\mathbb{R}$. When a subset S of real numbers may possibly contain all the real numbers, the notation $S \subseteq \mathbb{R}$ will be used in this book (similar to $a \leqslant b$ indicating that the real number a is less than or equal to b). Some authors use $\subset$ to cover both $\subset$ and $\subseteq$.

To illustrate how the above notation is used, consider the subset of $\mathbb{R}$ given by the interval $-\varepsilon < t < \varepsilon$, where ε is a positive real number. When studying a dynamical system, its time behaviour is sometimes of interest over such an interval I_ε of time. This interval is a proper subset of the real line $\mathbb{R}$ (i.e. $I_\varepsilon \subset \mathbb{R}$) and is usually written as

$$I_\varepsilon = \{t \in \mathbb{R} \colon |t| < \varepsilon\}.$$ [11]

Example 1.2 also illustrates how an input sequence applied to a system Σ over a given interval of time may be represented by a polynomial in an *indeterminate* z [12] and with real coefficients (i.e. coefficients of the polynomial are real numbers). Thus, the set of all such polynomials, denoted by $\mathbb{R}[z]$, arises quite naturally in the study of discrete systems.

If a discrete system has m inputs $u_1, \ldots, u_m$ (with $m \geqslant 2$), then the vector $\boldsymbol{b}$ in (1.6) becomes an $n \times m$ matrix $\boldsymbol{B}$ and $\boldsymbol{u}(t) = (u_1(t), \ldots, u_m(t))$ is the input vector at time t. The next example shows that, in this case, another mathematical object arises when each individual input is a finite sequence of real values, just as in Example 1.2.

Example 1.3

A multi-input sequence of real numbers is given by

$$\boldsymbol{u} = (\boldsymbol{u}(-q), \boldsymbol{u}(-q+1), \ldots, \boldsymbol{u}(-1), \boldsymbol{u}(0)),$$

in which each $\boldsymbol{u}(t)$ is a column vector $(u_1(t), \ldots, u_m(t))$, with $u_j(t) \in \mathbb{R}$, and $t = -q, -q+1, \ldots, -1, 0$. The z-transform of this sequence $\boldsymbol{u}$ is

$$\bar{\boldsymbol{u}}(z) = \sum_{t=-q}^{0} \boldsymbol{u}(t) z^{-t}$$

$$= \begin{bmatrix} u_1(-q) \\ u_2(-q) \\ \vdots \\ u_m(-q) \end{bmatrix} z^q + \begin{bmatrix} u_1(-q+1) \\ u_2(-q+1) \\ \vdots \\ u_m(-q+1) \end{bmatrix} z^{q-1} + \cdots + \begin{bmatrix} u_1(-1) \\ u_2(-1) \\ \vdots \\ u_m(-1) \end{bmatrix} z$$

$$+ \begin{bmatrix} u_1(0) \\ u_2(0) \\ \vdots \\ u_m(0) \end{bmatrix} = \begin{bmatrix} \bar{u}_1(z) \\ \bar{u}_2(z) \\ \vdots \\ \bar{u}_m(z) \end{bmatrix}, \tag{1.7}$$

where

$$\bar{u}_j(z) = u_j(-q)z^q + u_j(-q+1)z^{q-1} + \cdots + u_j(-1)z + u_j(0)$$
$$(j = 1, \ldots, m).$$

Thus, a multi-input sequence can be represented as a vector, each element of this vector being a polynomial $\bar{u}_j(z)$. Every $\bar{u}_j(z)$ is a polynomial in the indeterminate z with coefficients which are real numbers. Therefore, each polynomial belongs to $\mathbb{R}[z]$.

Polynomials arising from dynamical systems may not always have their coefficients in $\mathbb{R}$, as the next example shows.

Example 1.4

The solution of the time-varying vector differential equation

$$\dot{\boldsymbol{x}}(t) = \boldsymbol{A}(t)\boldsymbol{x}(t), \qquad \boldsymbol{x}(t_0) = \boldsymbol{x}_0,$$

is given by $\boldsymbol{x}(t) = \boldsymbol{\Phi}(t, t_0)\boldsymbol{x}_0$, where $\boldsymbol{\Phi}(t, t_0)$ is called the *state transition matrix* (Barnett and Cameron 1985; see also Note 2.6) and is a time-varying nonsingular matrix. Suppose that

$$\boldsymbol{A}(t) = \begin{bmatrix} t^2 & 1+t \\ 1-t^2 & t \end{bmatrix}.$$

Then $\boldsymbol{A}(t)$ can be written as

$$\begin{bmatrix} 1 & 0 \\ -1 & 0 \end{bmatrix} t^2 + \begin{bmatrix} 0 & 1 \\ 0 & 1 \end{bmatrix} t + \begin{bmatrix} 0 & 1 \\ 1 & 0 \end{bmatrix},$$

and this may be considered as a polynomial in the indeterminate t with coefficients in $\mathbb{R}^{2\times 2}$, i.e. as an element of the set $\mathbb{R}^{2\times 2}[t]$ of all such polynomials.[13]

Example 1.5

Consider a damped oscillator with forcing term. Let a particle of unit mass be suspended under gravity from a fixed point O by a spring of stiffness k (Fig. 1.5). A dash-pot in the system produces a resistive force to motion of $a\dot{x}$ when the displacement of the mass from its equilibrium position is x. Both k and a are assumed to be known constants. The mass is at rest in its equilibrium position when an external time-varying force (the input), of magnitude $\cos t$ at time t, is applied to the mass in a downward vertical direction at time $t = 0$. The differential equation governing the subsequent displacement $x(t)$ at time t, together with initial conditions, are given by

$$\begin{gathered} \ddot{x}(t) + a\dot{x}(t) + kx(t) = \cos t, \\ x(0) = 0, \qquad \dot{x}(0) = 0, \end{gathered} \tag{1.8}$$

where time is measured from the instant $t = 0$ when the input is first applied. In this case, the magnitude of the input u at any time t is $\cos t$ and an obvious choice of output is the displacement x.

System (1.8) can be investigated using Laplace transforms. The transformed equation is

$$(s^2 + as + k)\bar{y}(s) = \bar{u}(s), \tag{1.9}$$

where $\bar{y}(s)$ $(= \bar{x}(s))$ and $\bar{u}(s)$ $(= s/(s^2+1))$ are the Laplace transforms of the output and input respectively. Then (1.9) can be written as

$$\bar{y}(s) = \frac{s}{(s^2+1)(s^2+as+k)} = \frac{s}{s^4 + as^3 + (k+1)s^2 + as + k},$$

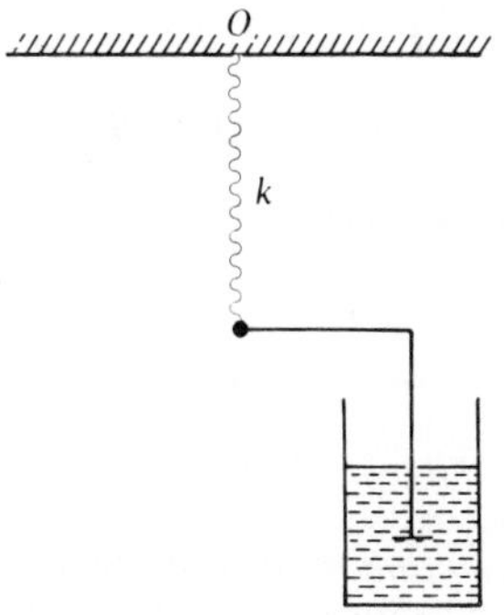

Fig. 1.5 Damped oscillator

which is a quotient of two polynomials in the indeterminate s. Such a quotient is called a *rational function* in s. The ratio $\bar{y}(s)/\bar{u}(s)$ obtained from (1.9) is called the *transfer function* of the system (1.8) and is denoted by $g(s)$. Thus, $g(s) = 1/(s^2 + as + k)$ is also a rational function in s (since the numerator is the polynomial s^0) and

$$\bar{y}(s) = g(s)\bar{u}(s).$$

Example 1.5 demonstrates how real polynomials arise from continuous systems, but in this case they appear as numerator and denominator of rational functions. Both numerator and denominator are elements of $\mathbb{R}[s]$ when considered separately; but the rational function formed by their quotient is considered as a single element of the set of all rational functions in s. This set is denoted by $\mathbb{R}(s)$ and can be expressed as

$$\mathbb{R}(s) = \{p(s)/q(s): p(s), q(s) \in \mathbb{R}[s], q(s) \neq 0\}.$$ [3]

Notice that $\mathbb{R}(s)$ is obtained from $\mathbb{R}[s]$ in the same way as the rational numbers are obtained from $\mathbb{Z}$.

The damped oscillator in Example 1.5 is a single-input single-output (SISO) system. A system with m inputs and l outputs (where $m, l > 1$) is called a multiple-input multiple-output (MIMO) or multivariable system. A linear time-invariant continuous multivariable system of order n can be expressed as

$$\dot{\boldsymbol{x}} = \boldsymbol{Ax} + \boldsymbol{Bu}, \qquad \boldsymbol{y} = \boldsymbol{Cx}, \tag{1.10}$$

where $\boldsymbol{A} \in \mathbb{R}^{n \times n}$, $\boldsymbol{B} \in \mathbb{R}^{n \times m}$, and $\boldsymbol{C} \in \mathbb{R}^{l \times n}$. Here $\mathbb{R}^{n \times m}$ is the set of all $n \times m$ matrices with real entries and is a generalization of $\mathbb{R}^{2 \times 2}$—similarly with $\mathbb{R}^{n \times n}$ and $\mathbb{R}^{l \times n}$. Taking Laplace transforms in (1.10) with zero initial conditions, we obtain

$$s\bar{\boldsymbol{x}} = \boldsymbol{A}\bar{\boldsymbol{x}} + \boldsymbol{B}\bar{\boldsymbol{u}}, \tag{1.11}$$

$$\bar{\boldsymbol{y}} = \boldsymbol{C}\bar{\boldsymbol{x}}. \tag{1.12}$$

From (1.11), it follows that $(s\boldsymbol{I} - \boldsymbol{A})\bar{\boldsymbol{x}} = \boldsymbol{B}\bar{\boldsymbol{u}}$; whence

$$\bar{\boldsymbol{x}} = (s\boldsymbol{I} - \boldsymbol{A})^{-1}\boldsymbol{B}\bar{\boldsymbol{u}}.$$

Substituting for $\bar{\boldsymbol{x}}$ in (1.12), we have

$$\bar{\boldsymbol{y}} = \boldsymbol{G}(s)\bar{\boldsymbol{u}}, \tag{1.13}$$

where the $l \times m$ matrix $\boldsymbol{G}(s)$ is given by

$$\boldsymbol{G}(s) = \boldsymbol{C}(s\boldsymbol{I} - \boldsymbol{A})^{-1}\boldsymbol{B}. \tag{1.14}$$

$\boldsymbol{G}(s)$ is called the *transfer-function matrix* of the system (1.10) and its elements are rational functions of s. The next example illustrates the form of $\boldsymbol{G}(s)$ for a second-order system with two inputs and two outputs.

Example 1.6

$$\dot{x}_1 = -x_1 + x_2 + u_1, \qquad \dot{x}_2 = x_1 + x_2 + u_1 + u_2,$$

$$y_1 = x_1, \qquad y_2 = x_2.$$

Here $n = 2$, $m = 2$, and $l = 2$. Writing these equations in vector–matrix form (1.10),

$$\boldsymbol{A} = \begin{bmatrix} -1 & 1 \\ 1 & 1 \end{bmatrix}, \qquad \boldsymbol{B} = \begin{bmatrix} 1 & 0 \\ 1 & 1 \end{bmatrix}, \qquad \boldsymbol{C} = \begin{bmatrix} 1 & 0 \\ 0 & 1 \end{bmatrix}.$$

Then

$$s\boldsymbol{I} - \boldsymbol{A} = \begin{bmatrix} s+1 & -1 \\ -1 & s-1 \end{bmatrix}$$

and $\det(s\boldsymbol{I} - \boldsymbol{A}) = s^2 - 2$. Using the adjoint matrix of $s\boldsymbol{I} - \boldsymbol{A}$, we have

$$(s\boldsymbol{I} - \boldsymbol{A})^{-1} = \frac{1}{s^2 - 2}\begin{bmatrix} s-1 & 1 \\ 1 & s+1 \end{bmatrix};$$

whence the transfer function matrix $\boldsymbol{G}(s)$ is obtained from (1.14) as

$$\boldsymbol{G}(s) = \begin{bmatrix} \dfrac{s}{s^2-2} & \dfrac{1}{s^2-2} \\ \dfrac{s+2}{s^2-2} & \dfrac{s+1}{s^2-2} \end{bmatrix}.$$

Each element of $\boldsymbol{G}(s)$ is a rational function of s and therefore belongs to $\mathbb{R}(s)$.

In exactly the same way, a multi-input multi-output discrete system

$$\boldsymbol{x}(t+1) = \boldsymbol{A}\boldsymbol{x}(t) + \boldsymbol{B}\boldsymbol{u}(t), \qquad \boldsymbol{y}(t) = \boldsymbol{C}\boldsymbol{x}(t),$$

analogous to the continuous system (1.10), gives rise to a transfer-function matrix (Ogata 1987)

$$\boldsymbol{G}(z) = \boldsymbol{C}(z\boldsymbol{I} - \boldsymbol{A})^{-1}\boldsymbol{B}, \tag{1.15}$$

which relates the (transformed) input $\bar{\boldsymbol{u}}(z)$ to the (transformed) output $\bar{\boldsymbol{y}}(z)$ through the equation

$$\bar{\boldsymbol{y}}(z) = \boldsymbol{G}(z)\bar{\boldsymbol{u}}(z).$$

From the above examples and remarks, it will be clear to the reader that a study of dynamical systems will necessarily involve sets of mathematical objects such as $\mathbb{R}$, $\mathbb{C}$, $\mathbb{R}^2$, $\mathbb{R}^{2\times 2}$, $\mathbb{R}[z]$, $\mathbb{R}(s)$, and $C^0(\mathbb{R})$. A set is determined by the elements in it. In many cases there are rules associated with a set which enables one to determine whether a particular element belongs to that set or not.

Example 1.7

The set of all nonsingular 2×2 matrices with real entries is usually denoted by $G\ell(2, \mathbb{R})$.[14] The rules associated with this set are (i) the order of matrices must be 2×2, (ii) matrices must be nonsingular (i.e. the determinant of each matrix must be nonzero), and (iii) all entries in each matrix must be real numbers. Thus:

$$\begin{bmatrix} 2 & 1 \\ -1 & 1 \end{bmatrix} \in G\ell(2, \mathbb{R}) \quad \text{but} \quad \begin{bmatrix} 1 & 1 \\ 1 & 1 \end{bmatrix} \notin G\ell(2, \mathbb{R}),$$

$$\begin{bmatrix} 4 & -\mathrm{j} \\ \mathrm{j} & -2 \end{bmatrix} \notin G\ell(2, \mathbb{R}) \qquad (\mathrm{j} = \sqrt{(-1)}).$$

1.3 The space $\mathbb{R}^2$

The vector[5] $\boldsymbol{x} = (x_1, x_2) \in \mathbb{R}^2$ may be associated with the geometrical vector $\overrightarrow{OP}$ in which x_1 and x_2 are the cartesian coordinates of the point P relative to rectangular axes OX_1 and OX_2 as shown in Fig. 1.6. No confusion arises from the use of the same notation to denote both the vector $\boldsymbol{x}$ and the point P. Then $\mathbb{R}^2$ may be thought of as the set of all points in the plane of these axes (called the *real plane*). Since each point P with coordinates (x_1, x_2) can be associated with one and only one vector (x_1, x_2) an alternative interpretation of $\mathbb{R}^2$ is as the set of all vectors from the origin O to points P. Mathematicians and physicists sometimes refer to the state variables x_1 and x_2 of a dynamical system, such as that described in Example 1.1, as the phase coordinates, and the real plane $\mathbb{R}^2$ as the phase plane, unlike engineers who generally call the latter the state space. Any set of elements is called a 'space' rather than a 'set'

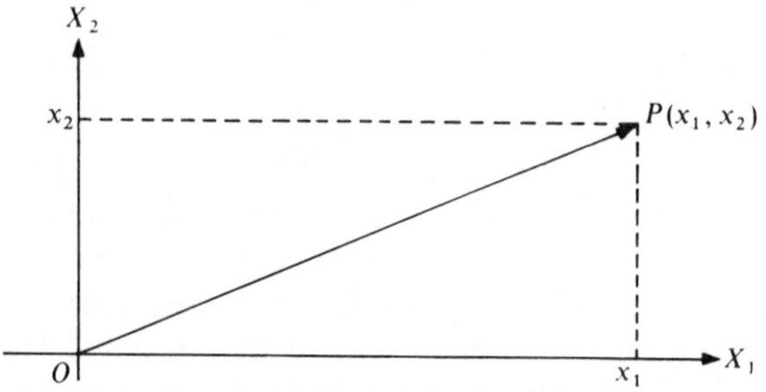

Fig. 1.6 Geometrical vector $\overrightarrow{OP}$

when some operations on the elements of the set, such as addition, are defined (e.g. groups in Chapter 4). In other words a set which is endowed with some *structure* is called a space. In order to carry out analysis on a set of states for a dynamical system, some operations are essential; hence the term 'state space'.

The behaviour of a second order system in time can be represented by a path or trajectory in the state space $\mathbb{R}^2$. In general, the magnitudes of the states $x_1(t)$ and $x_2(t)$ will vary with time, and so the state vector $\boldsymbol{x}(t) = (x_1(t), x_2(t))$ represented by a point P in the state space will move about in $\mathbb{R}^2$ as time increases. The locus of P will generally be a continuous curve in $\mathbb{R}^2$. This is because the states of a dynamical system are usually continuous functions of time (continuity of functions is discussed in Chapter 10). For example, one of the state variables of an aircraft may be altitude h with a value $h(t_1) = h_1$ at time t_1. It is impossible for the aircraft to change from this altitude h_1 to some different altitude h_2 instantaneously. It must fly through all altitudes between h_1 and h_2. On the other hand, some states do experience sudden changes in their magnitudes (these are called discontinuous states). The launch of space vehicles is often achieved by booster rockets which fall away when their fuel is all consumed. The mass of the whole vehicle is a state which experiences a sudden change of magnitude at the instant the rocket boosters are jettisoned.

A particular example will help to illustrate the usefulness of the state space $\mathbb{R}^2$ in representing the behaviour of a dynamical system.

Example 1.8

Consider the lightly damped oscillator

$$\ddot{x} + 2\dot{x} + 2x = 0, \qquad x(0) = 1, \quad \dot{x}(0) = -1. \tag{1.16}$$

This is the damped oscillator of Example 1.5 with $a = 2$ and $k = 2$, but without the forcing term $\cos t$ and with nonzero initial conditions. It can be shown (check!) that the solution of (1.16) is

$$x(t) = e^{-t}\cos t, \qquad \dot{x}(t) = -\sqrt{2}e^{-t}\cos(t - \tfrac{1}{4}\pi).$$

With $x_1 = x$ and $x_2 = \dot{x}$ as the states of the system, this solution is represented by a spiral in the (x_1, x_2)-state space, starting from the initial point P with coordinates $(1, -1)$, as shown in Fig. 1.7. For (1.16) may be written in state-space form as

$$\dot{x}_1 = x_2, \qquad x_1(0) = 1,$$

$$\dot{x}_2 = -2x_1 - 2x_2, \qquad x_2(0) = -1,$$

and the equilibrium point of the system (i.e. where $\dot{x}_1$ and $\dot{x}_2$ are both zero) is the origin. Then the state of the system follows a path which spirals inwards towards the origin of $\mathbb{R}^2$.

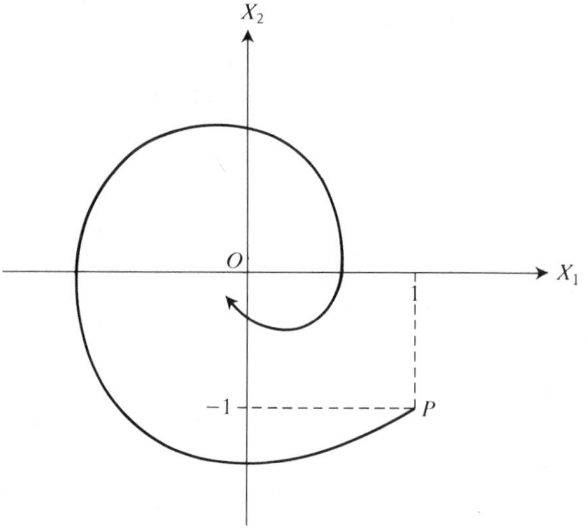

Fig. 1.7 State space trajectory

1.4 Optimal control problems

The general problem associated with a system such as (1.10) is to choose the control vector $\boldsymbol{u}(\cdot)$ [15] in such a way that the system behaviour is in some sense satisfactory.

Example 1.9

Consider the system model defined by the equation

$$\dot{x} = xu, \qquad x(0) = 1,$$

where $x(t) \in \mathbb{R}$ and $u(t) \in \mathbb{R}$. For all $t \geqslant 0$, choose the control $u(t) = 1$. Then the system behaviour is $x(t) = e^t$, and the value of the state x becomes very large when t is large, behaviour which is often unacceptable in practice. Now, for all $t \geqslant 0$, choose $u(t) = -1$; the solution becomes $x(t) = e^{-t}$, which decreases towards zero as time increases, sometimes a much more acceptable response.

Instead of choosing the control u so that the resultant behaviour is acceptable according to some general practical criteria (e.g. stability, no oscillations, etc.), it may be a requirement that the control u should yield the best possible behaviour. Of course, it is necessary to provide a mathematical formula for the performance of the system so that there is some way in which the word 'best' can be interpreted. Once such a performance index (or objective function) has been developed as part of the system modelling process then an *optimal-control* problem is that of determining the control vector $u(\cdot)$ that minimizes (or maximizes) that performance index. An example will illustrate the concept of a performance index.

Example 1.10

Assume that a civil airliner can be represented as a point mass moving in a vertical plane with velocity vector and with forces acting on the aircraft as shown in Fig. 1.8. The speed $V(t)$, flight path angle $\gamma(t)$, range $x(t)$, altitude $h(t)$, and mass $M(t)$ determine the configuration or state of the aircraft at any time t and can be chosen as state variables. The thrust $T(t)$ of the engines and the angle of incidence (or angle of attack) $\alpha(t)$ are variables that the pilot can adjust within certain given limits, and so may be chosen as control variables. Outputs are generally chosen to be h, $\dot{h}$, θ, and $\dot{\theta}$, where the pitch angle θ is given by $\gamma + \alpha$. Suppose the lift $L(V, h, \alpha)$, drag $D(V, h, \alpha)$, and weight Mg are the only external forces acting on the aircraft, g being the acceleration due to gravity. The thrust T can be approximated as $T = cm$, where c is a constant and m the rate of fuel consumption, so that m may be taken as a control variable instead of T. Because fuel is consumed during flight, the mass of the aircraft decreases and $\dot{M} = -m$. The equations of motion are given by

$$\dot{V} = \frac{cm\cos\alpha - D(V, h, \alpha)}{M} - g\sin\gamma,$$

$$\dot{\gamma} = \frac{1}{V}\left(\frac{cm\sin\alpha + L(V, h, \alpha)}{M} - g\cos\gamma\right), \tag{1.17}$$

$$\dot{x} = V\cos\gamma, \qquad \dot{h} = V\sin\gamma, \qquad \dot{M} = -m.$$

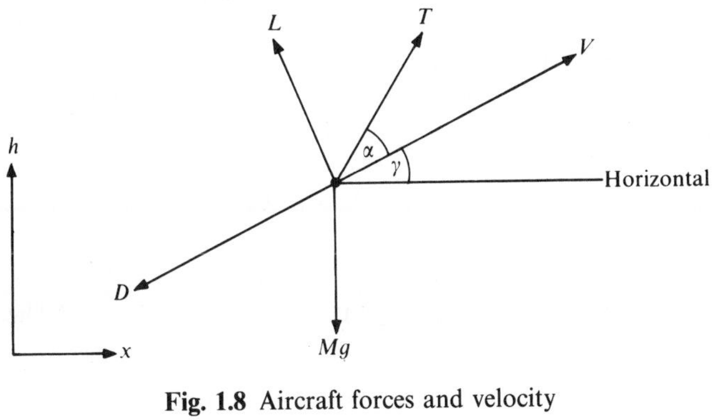

Fig. 1.8 Aircraft forces and velocity

If initial values of the states V, γ, x, h, and M are known, then a choice of time profiles $m(\bullet)$ and $\alpha(\bullet)$ for the controls would enable the above set of first-order nonlinear differential equations (1.17) to be solved, giving the flight trajectory of the aircraft over any interval of time.

Suppose that the aircraft is about to make its final descent to an airport. In controlling the airliner from the start of the final approach at $t = 0$ to touch-down at $t = t_f$, the pilot must adjust the controls α and m so that certain end conditions are satisfied at time t_f. Obviously $h(t_f) = 0$ and $\gamma(t_f) = 0$, but speed at touch-down may also have to be a specified value. Apart from satisfying these final end conditions, the pilot will wish to carry out the descent in a manner that causes no discomfort to the passengers and that will also use no more fuel than is necessary. These requirements can be achieved by keeping the flight-path angle and fuel consumption as small as possible during the descent. A suitable performance index which, if minimized over the total descent, would achieve these two requirements is

$$\int_0^{t_f} (q\gamma^2 + rm^2)\,\mathrm{d}t. \tag{1.18}$$

Here the final time t_f is assumed known, although this would not normally be the case. Constants q and r are weighting factors which may be chosen according to the importance of keeping γ and m small. If it is more important to keep the flight-path angle γ (rather than m) small, then q would be chosen larger than r, say $q = 100$ and $r = 1$. The optimal-control problem is then to choose $m(t)$ and $\alpha(t)$ for $0 \leqslant t \leqslant t_f$ so that the corresponding solution of (1.17) from the given initial conditions yields a trajectory that satisfies the specified end conditions and also gives the minimum value of the performance index (1.18).

In Example 1.10 the system equations (1.17) are nonlinear in the states and the control α. The performance index (1.18) is quadratic in the state γ and the control m. Many nonlinear systems can be approximated by differential equations that are linear in the states and controls together with outputs that are linear in the states. The equations are then of the form (1.10). If a performance index is chosen quadratic in the states and the controls, then such an index J can be expressed in the form

$$J = \int_0^{t_f} (\boldsymbol{x}^{\mathrm{T}}\boldsymbol{Q}\boldsymbol{x} + \boldsymbol{u}^{\mathrm{T}}\boldsymbol{R}\boldsymbol{u})\,\mathrm{d}t, \tag{1.19}$$

where $\boldsymbol{Q}$ is a positive semidefinite $n \times n$ weighting matrix on the states and $\boldsymbol{R}$ is a positive definite $m \times m$ weighting matrix on the controls. [16] The optimal-control problem with system equation

$$\dot{\boldsymbol{x}} = \boldsymbol{A}\boldsymbol{x} + \boldsymbol{B}\boldsymbol{u}$$

and performance index (1.19) is known as an LQP (linear–quadratic performance) or regulator problem.

If the system equations are nonlinear, given by

$$\dot{\boldsymbol{x}} = \boldsymbol{f}(\boldsymbol{x}, \boldsymbol{u}), \qquad \boldsymbol{x}(t_0) = \boldsymbol{x}_0, \tag{1.20}$$

and if the performance index is of a general nature

$$J = \int_{t_0}^{t_f} L(\boldsymbol{x}, \boldsymbol{u})\,\mathrm{d}t, \tag{1.21}$$

then again an optimal-control problem is to minimize J with respect to the control vector $\boldsymbol{u}$ and subject to the system equation and initial condition (1.20). Of course, in some optimal-control problems, a performance index may be chosen which is to be maximized rather than minimized. This case is really incorporated in the minimization problem described above. For example, maximizing J in (1.21) is equivalent to minimizing the performance index

$$-\int_{t_0}^{t_f} L(\boldsymbol{x}, \boldsymbol{u})\,\mathrm{d}t.$$

1.5 The space $\mathbb{R}^n$

The space $\mathbb{R}^2$ considered as a set of vectors $\{\boldsymbol{x} : \boldsymbol{x} \in \mathbb{R}^2\}$ can obviously be generalized to the set of vectors $\{\boldsymbol{x} : \boldsymbol{x} \in \mathbb{R}^n\}$ in which each vector $\boldsymbol{x}$ has n components: $\boldsymbol{x} = (x_1, \dots, x_n)$. If these components are real numbers, then all such vectors $\boldsymbol{x}$ (called *n-tuples* of real numbers) form a set which is denoted by $\mathbb{R}^n$. Following this generalization to its logical conclusion an infinite sequence $(x_1, x_2, \dots)$ of real numbers may be considered as the real components of an infinite vector. The set of all such infinite vectors (with some additional conditions) is denoted by $\mathbb{R}^\infty$. Note that $\mathbb{R}^1$ is just the set of all real numbers $\mathbb{R}$.

The space $\mathbb{R}^n$ plays an important role in the study of dynamical systems of order n. Suppose there are m control variables $u_1, \dots, u_m$ (with $m \leqslant n$) available with which to control a system governed by a set of n first-order nonlinear differential equations in n state variables $x_1, \dots, x_n$ which take the form

$$\begin{aligned}
\dot{x}_1(t) &= f_1(x_1(t), \dots, x_n(t), u_1(t), \dots, u_m(t)),\\
\dot{x}_2(t) &= f_2(x_1(t), \dots, x_n(t), u_1(t), \dots, u_m(t)),\\
&\ \ \vdots \\
\dot{x}_n(t) &= f_n(x_1(t), \dots, x_n(t), u_1(t), \dots, u_m(t)),
\end{aligned} \tag{1.22}$$

with initial conditions

$$x_1(t_0) = x_{10}, \quad x_2(t_0) = x_{20}, \ldots, x_n(t_0) = x_{n0}. \tag{1.23}$$

Forming the vectors

$$\boldsymbol{x}(t) = \big(x_1(t), \ldots, x_n(t)\big) \in \mathbb{R}^n, \qquad u(t) = \big(u_1(t), \ldots, u_m(t)\big) \in \mathbb{R}^m,$$
$$\boldsymbol{f}\big(\boldsymbol{x}(t), \boldsymbol{u}(t)\big) = \big(f_1(\boldsymbol{x}(t), \boldsymbol{u}(t)), \ldots, f_n(\boldsymbol{x}(t), \boldsymbol{u}(t))\big) \in \mathbb{R}^n,$$
$$\boldsymbol{x}_0 = (x_{10}, \ldots, x_{n0}) \in \mathbb{R}^n,$$

the set of n differential equations (1.22) can be expressed in vector form either as

$$\dot{\boldsymbol{x}}(t) = \boldsymbol{f}\big(\boldsymbol{x}(t), \boldsymbol{u}(t)\big) \tag{1.24}$$

or, when there is no need to emphasize the dependence of $\boldsymbol{x}$ and $\boldsymbol{u}$ on the time variable, as

$$\dot{\boldsymbol{x}} = \boldsymbol{f}(\boldsymbol{x}, \boldsymbol{u}), \tag{1.25}$$

with initial condition

$$\boldsymbol{x}(t_0) = \boldsymbol{x}_0 \tag{1.26}$$

as in (1.20). Of course, for an uncontrolled system, (1.25) reduces to

$$\dot{\boldsymbol{x}} = \boldsymbol{f}(\boldsymbol{x}). \tag{1.27}$$

Equation (1.25) is called *autonomous* since $\boldsymbol{f}$ (and hence $f_1, \ldots, f_n$) does not contain the time t explicitly. If t does appear explicitly in $\boldsymbol{f}$, then (1.25) becomes

$$\dot{\boldsymbol{x}} = \boldsymbol{f}(\boldsymbol{x}, \boldsymbol{u}, t), \tag{1.28}$$

an equation which is called *nonautonomous*. The explicit dependence on t in (1.28) can be eliminated to yield an autonomous differential equation by introducing an extra state variable $x_{n+1}(t)$ which is actually time t itself written under a different name! So $x_{n+1}(t) = t$, from which it follows that

$$\dot{x}_{n+1} = 1, \qquad x_{n+1}(t_0) = t_0. \tag{1.29}$$

Augmenting (1.28) (with t replaced by x_{n+1}) and (1.26) with (1.29) the system again can be represented by a vector differential equation and initial condition

$$\dot{\boldsymbol{x}} = \boldsymbol{f}(\boldsymbol{x}, \boldsymbol{u}), \qquad \boldsymbol{x}(t_0) = \boldsymbol{x}_0,$$

but now

$$\boldsymbol{x}(t) = \big(x_1(t), \ldots, x_n(t), x_{n+1}(t)\big) \in \mathbb{R}^{n+1},$$
$$\boldsymbol{f}(\boldsymbol{x}(t), \boldsymbol{u}(t)) = \big(f_1(\boldsymbol{x}(t), \boldsymbol{u}(t)), \ldots, f_n(\boldsymbol{x}(t), \boldsymbol{u}(t)), 1\big) \in \mathbb{R}^{n+1},$$
$$\boldsymbol{x}_0 = (x_{10}, \ldots, x_{n0}, t_0) \in \mathbb{R}^{n+1}.$$

Exercises

1. Write down a state-space form $\dot{\boldsymbol{x}} = \boldsymbol{A}\boldsymbol{x} + \boldsymbol{b}u$, $\boldsymbol{y} = \boldsymbol{C}\boldsymbol{x}$, for the system

$$\ddot{p} + 3\dot{p} + 9q = 0, \qquad \ddot{q} + 2\dot{q} + p = \cos t,$$

where $\boldsymbol{x} = (p, \dot{p}, q, \dot{q})$, $u = \cos t$, and $\boldsymbol{y} = (p+q, \dot{p}-\dot{q})$.

2. Find the transfer function of the system

$$\ddot{y} + 4\dot{y} + 2y = \dot{u} + 2u.$$

3. The input values $u(t)$ $(t \in \mathbb{Z})$ for a discrete system are all zero except for the values $u(t) = t-1$ $(t = -5, -4, -3, -2, -1, 0)$. By using z-transforms, find the polynomial in z which represents this input sequence.

4. Find the state-space ($\mathbb{R}^2$) trajectory for the system

$$\dot{x}_1 = x_2, \qquad x_1(0) = 0,$$

$$\dot{x}_2 = -x_1, \qquad x_2(0) = 1.$$

5. By introducing an extra state variable x_{n+1}, show that the performance index

$$\int_{t_0}^{t_f} L(\boldsymbol{x}, \boldsymbol{u})\,\mathrm{d}t$$

can be replaced by $x_{n+1}(t_f)$.

6. By treating the time t as a state variable, express the system

$$\dot{x}_1 = 3x_2 - tx_1, \qquad x_1(0) = 1,$$

$$\dot{x}_2 = -x_1 + \mathrm{e}^t, \qquad x_2(0) = -1,$$

as an autonomous system.

Notes for Chapter 1

1. Dots above a variable denote differentiation with respect to time t. Thus, $\dot{h} = \mathrm{d}h/\mathrm{d}t$, $\ddot{x} = \mathrm{d}^2x/\mathrm{d}t^2$, etc.

2. It is assumed that the reader is familiar with Laplace and z-transforms. If this is not the case then it is worth reading about these very useful techniques because they are used later in this book. A good reference is (Ogata 1987).

3. Throughout this book, $\mathbf{0}$ will denote vectors and matrices in which all entries are zero, and 0 for polynomials in which all coefficients are zero. In general, 0 is used to denote the *zero* or *null* element of the particular set of objects under discussion. Equality between vectors or matrices means equality between corresponding entries so that in (1.5)

$$\boldsymbol{x}(0) = \mathbf{0} \quad \text{means} \quad \begin{bmatrix} x_1(0) \\ x_2(0) \end{bmatrix} = \begin{bmatrix} 0 \\ 0 \end{bmatrix},$$

which in turn implies $x_1(0) = 0$ and $x_2(0) = 0$.

4. $\dot{\boldsymbol{x}}(t)$ is the vector $\begin{bmatrix} \dot{x}_1(t) \\ \dot{x}_2(t) \end{bmatrix}$ and thus belongs to $\mathbb{R}^2$ for a given t, because $\dot{x}_1(t)$ and $\dot{x}_2(t)$ are real numbers.
5. It is usual in dynamical system theory to assume that all vectors are column vectors unless otherwise stated. A column vector such as $\boldsymbol{x}(t) = \begin{bmatrix} x_1(t) \\ x_2(t) \end{bmatrix}$ is conveniently written as $(x_1(t), x_2(t))$ in order to save space, and the argument t in $\boldsymbol{x}(t)$, $x_1(t)$, $x_2(t)$ is often omitted for the same reason. Whether a given vector is a column vector or a row vector is not usually important unless operations such as multiplication of vectors and matrices are being considered (e.g. in Example 1.1, output $y(t) = \boldsymbol{c}^{\mathrm{T}}\boldsymbol{x}(t)$ must be interpreted as the row vector $[1, 0]$ pre-multiplying the column vector $\boldsymbol{x}(t)$ in order for $\boldsymbol{c}^{\mathrm{T}}\boldsymbol{x}(t)$ to be the *scalar* output $y(t)$). In this book, we follow the convention that a row vector such as $\boldsymbol{c}^{\mathrm{T}}$ is denoted by the transpose of a column vector.
6. Curly brackets are always used when declaring elements of a set. The order in which the elements are declared has no relevance here, unlike elements in a sequence.
7. Some writers, particularly those on the continent of Europe, use $\mathbb{N}$ to denote the set of non-negative integers $\{0, 1, ...\}$ including zero. In this case the set of positive integers $\{1, 2, ...\}$, which in this book is denoted by $\mathbb{N}$, is written as $\mathbb{N}^*$.
8. When it is impossible or inconvenient to write out all the individual elements of a set, curly brackets are still used but a general element (say x) of the set is first declared followed by the sign : (or sometimes a vertical line |) meaning 'such that'. Then follows the properties associated with x. The phrase 'such that' is often abbreviated to s.t.
9. If x is an element of a set S, then one writes $x \in S$ (read as 'x belongs to s'), whereas $x \notin S$ denotes the fact that x does not belong to S. For example, $3 \in \mathbb{N}$ and $\frac{3}{2} \in \mathbb{Q}$, but $\frac{3}{2} \notin \mathbb{Z}$.
10. The sign $\neq$ means 'not equal to'.
11. The modulus of a real number t is defined as

$$|t| \triangleq \begin{cases} t & \text{if } t \geqslant 0, \\ -t & \text{if } t < 0. \end{cases}$$

Note that the sign $\triangleq$ (or sometimes $:=$) is used to denote a definition. The condition $|t| < \varepsilon$ is equivalent to $-\varepsilon < t < \varepsilon$: if t is positive, then $|t| < \varepsilon$ means $t < \varepsilon$; if t is negative—say $t = -\beta$ with $\beta > 0$—then $|t| < \varepsilon$ means $|-\beta| = \beta < \varepsilon$, i.e. $\beta = -t < \varepsilon$, whence $-t < \varepsilon$ and so $t > -\varepsilon$.

Similarly, $|3t+2| < 3$ is equivalent to $-\frac{5}{3} < t < \frac{1}{3}$ because $-3 < 3t + 2 < 3$ is equivalent to the two inequalities $3t + 2 > -3$ and $3t + 2 < 3$, i.e. $t > -\frac{5}{3}$ and $t < \frac{1}{3}$. Note that

$$\frac{t}{|t|} = \begin{cases} 1 & \text{if } t > 0, \\ -1 & \text{if } t < 0. \end{cases}$$

12. A polynomial such as $3z^4 - \frac{1}{2}z^3 - 2z^2 + 1$ in Example 1.2 may be looked at in two different ways. The 'z' of the z-transform may be thought of as a complex number, and the polynomial as a function of $z \in \mathbb{C}$, where $\mathbb{C}$ denotes the set of all complex numbers. Alternatively, the polynomial can be considered as a single element in the set of all such polynomials in z. In the latter case, it is the coefficients of the polynomial ($3, -\frac{1}{2}, -2, 0, 1$ for $3z^4 - \frac{1}{2}z^3 - 2z^2 + 0z + 1z^0$) which determine a unique element of the set, and the powers of z play only a minor role. This alternative interpretation avoids the

difficulty of the polynomial $z^2 - z$, for example, vanishing identically over the set of integers $\{0, 1\}$ but not being the zero polynomial, i.e. having all coefficients zero (Jenner 1963). Here, z is not considered as a complex number but simply as an 'indeterminate'. Similar remarks can be made about polynomials in the Laplace transform operator s (see Example 1.5).

13. Thus, $\boldsymbol{A}(t)$ may be considered either as a matrix with real polynomial entries, i.e. as an element belonging to the set $M_2(\mathbb{R}[t])$ of all such matrices or as an element of $\mathbb{R}^{2\times 2}[t]$.

14. $G\ell$ stands for 'general linear group' (see comments in the Preface to this book on this notation and Chapter 4 for groups). $G\ell$ is used in the engineering literature (Brockett 1973) but GL in the mathematical books (MacLane and Birkoff 1979). $G\ell(2, \mathbb{R}) \subset \mathbb{R}^{2\times 2}$ since $G\ell(2, \mathbb{R})$ contains only nonsingular 2×2 matrices whereas $\mathbb{R}^{2\times 2}$ contains all 2×2 singular matrices as well as all the nonsingular ones. The notation $G\ell(2, \mathbb{R})$ is sometimes shortened to $G\ell(2)$ provided that no confusion is likely. The set of all nonsingular $n\times n$ matrices with entries from $\mathbb{C}$ is denoted by $G\ell(n, \mathbb{C})$, the set of all $n\times n$ matrices with entries taken from $\mathbb{C}$ by $\mathbb{C}^{n\times n}$, and so on.

15. When all values of a function such as an input or control $\boldsymbol{u}$ are referred to, rather than one value $\boldsymbol{u}(t)$ at a particular time t, the notation $\boldsymbol{u}(\bullet)$ is often employed (see comments in the Preface to this book).

16. A positive semidefinite matrix $\boldsymbol{Q}$ (written $\boldsymbol{Q} \geqslant \mathbf{0}$) is such that $\boldsymbol{x}^{\mathrm{T}}\boldsymbol{Q}\boldsymbol{x} \geqslant 0$ for all $\boldsymbol{x} \neq \mathbf{0}$. A positive definite matrix $\boldsymbol{R}$ (written $\boldsymbol{R} > \mathbf{0}$) is such that $\boldsymbol{u}^{\mathrm{T}}\boldsymbol{R}\boldsymbol{u} > \mathbf{0}$ for all $\boldsymbol{u} \neq \mathbf{0}$.

2

ASPECTS OF SET THEORY

It will be clear from Chapter 1 that sets of various kinds figure prominently in any mathematical study of dynamical systems. The general theory of sets is enormously rich in ideas and leads to many modern branches of mathematics. However, it is not the purpose of this chapter to present a general discussion of set theory; that is already available in a very large number of texts (of which the book by Halmos (1974) is an interesting one to read). Rather it is those aspects of the theory particularly relevant to dynamical systems which will be highlighted here. Wherever possible, in order to avoid a presentation solely in terms of general abstract sets, particular sets will be used to emphasize the ideas behind the mathematics. Even so, it will be shown in this chapter that a large number of concepts emerge from the basic idea of a set, concepts which prove extremely useful in discussing topics of fundamental importance in dynamical system theory.

2.1 Unions and intersections

Sets can contain either a finite number of elements (a finite set) or an infinite number (an infinite set). If a set S is finite, then the number of different elements in S is called the *order* of the set and denoted by $|S|$. Sets are often subsets of some fixed universal set U. When dealing with real numbers, the obvious choice for U is $\mathbb{R}$; with complex numbers, U could be $\mathbb{C}$ with $\mathbb{R} \subset \mathbb{C}$. A universal set will generally contain very many subsets. For example, the set $\mathbb{R}^{n \times n}$ of all real $n \times n$ matrices contains the set $G\ell(n, \mathbb{R})$ of all real nonsingular $n \times n$ matrices, the set of all real singular matrices, the set $So(n, \mathbb{R})$ of all real orthogonal $n \times n$ matrices, the set of all real symmetric $n \times n$ matrices and so on. Some elements will lie in several subsets just as the real number 2 lies in both $\mathbb{N}$ and $\mathbb{Z}$.

Example 2.1

$$\begin{bmatrix} 0 & 0 \\ 0 & 1 \end{bmatrix} \quad \text{is both symmetric and singular,}$$

$$\begin{bmatrix} 1 & 0 \\ 0 & 2 \end{bmatrix} \quad \text{is both symmetric and nonsingular.}$$

The unit matrix

$$\mathbf{I}_2 = \begin{bmatrix} 1 & 0 \\ 0 & 1 \end{bmatrix}$$

is symmetric, nonsingular, and orthogonal.

The different subsets of $\mathbb{R}^{n \times n}$ can be illustrated in a *Venn diagram* (Fig. 2.1) in which the shaded portion of the universal set $\mathbb{R}^{n \times n}$ represents the set of all real singular $n \times n$ matrices and S the set of all real symmetric $n \times n$ matrices.

If X is a proper nonempty (i.e. contains at least one element) subset of some universal set U, then the elements of U that do not belong to X form another proper subset called the *complement* of X, written X' (sometimes X^c or $\mathscr{C}X$). So the set of all real singular $n \times n$ matrices is the complement of $G\ell(n, \mathbb{R})$ (Fig. 2.1). Again, the set of all irrational numbers in $\mathbb{R}$ is the complement of the set $\mathbb{Q}$. The concept of the complement X' of a set X in a universal set U is illustrated in a Venn diagram (Fig. 2.2) in which X' is the shaded portion of U.

Very often, general sets are discussed without any reference to a universal set U; but in practical applications it is usually fairly obvious what the set U

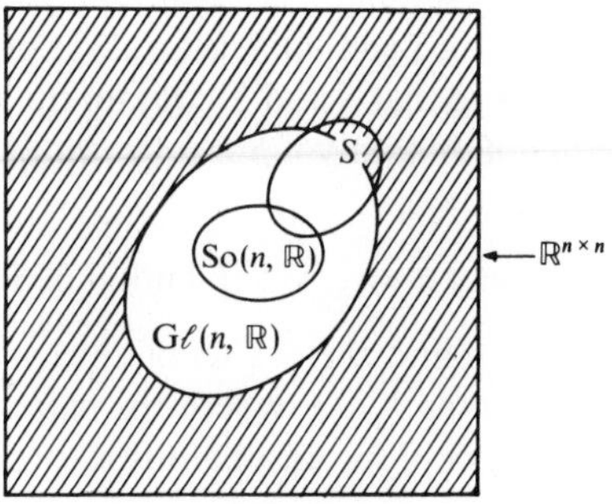

Fig. 2.1 Venn diagram

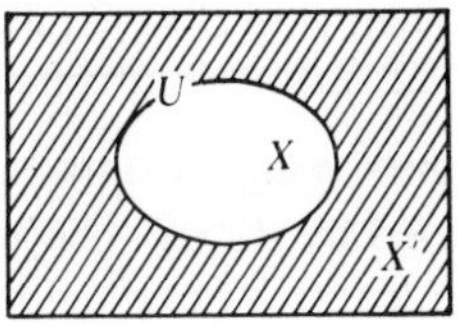

Fig. 2.2 The complement of set X

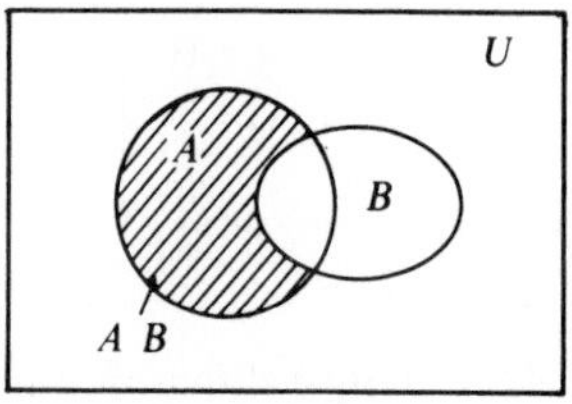

Fig. 2.3 Difference of two sets

is, even if it is not defined explicitly. Given two sets A and B in U, the set of all elements in A that are not in B is denoted by either $A \backslash B$ or $A - B$ (Fig. 2.3).

Example 2.2 [1]

$A = (0, 2)$ and $B = [1, 3)$. Then $A \backslash B = (0, 1)$.

In Example 2.2, an equation such as $A \backslash B = (0, 1)$ expresses equality between two sets $A \backslash B$ and $(0, 1)$ and a formal definition of such an equality is as follows. Two sets are equal if they contain the same elements. To prove that two sets S and T are equal, it is necessary to prove first that $x \in S$ implies $x \in T$, because it then follows that S is a subset of T. Secondly, assuming $x \in T$, it has to be proved that $x \in S$ so that T is a subset of S. The two results $S \subseteq T$ and $T \subseteq S$ together ensure that S and T contain the same elements. Thus, two sets S and T are equal if and only if S is a subset of T and T is a subset of S. Written symbolically

$$S = T \quad \Leftrightarrow \quad S \subseteq T \text{ and } T \subseteq S. \tag{2.1}$$ [2, 3]

The *union* of two sets A and B is the set of all elements that lie in A or B (or both). This set is written as $A \cup B$ where the sign $\cup$ stands for union ($\cup$ is sometimes pronounced 'cup' for an obvious reason). Clearly

$$A \cup B = \{x : x \in A \text{ and/or } x \in B\}.$$

Example 2.3

$A = (0, 2)$ and $B = [1, 3)$. Then $A \cup B = (0, 3)$.

Similarly, the set of elements lying in both set A and set B is called the *intersection* of A and B, denoted by $A \cap B$. The sign $\cap$ stands for intersection ($\cap$ is sometimes pronounced 'cap'). Thus,

$$A \cap B = \{x : x \in A \text{ and } x \in B\}.$$

There are many relationships involving unions and intersections. A typical result is given in the next example.

Example 2.4

$A = (0, 2)$ and $B = [1, 3)$. Then $A \cap B = [1, 2)$. With $\mathbb{R}$ as the universal set, the complement of B is given by

$$B' = (-\infty, 1) \cup [3, \infty),$$

and so $A \cap B' = (0, 2) \cap ((-\infty, 1) \cup [3, \infty)) = (0, 1)$ (draw the real line $\mathbb{R}$ and mark off the intervals to see this). From Example 2.2, $A \backslash B = (0, 1)$ and so, for this example,

$$A \backslash B = A \cap B',$$

a result which is true for all sets A and B (see Fig. 2.3).

To illustrate how the union and intersection of sets can arise in dynamical system theory, look again at the damped oscillator system discussed in Example 1.5. Suppose that $a = 1$ and $k = 1$, and that, instead of $\cos t$, the input u is such that its magnitude $u(t)$ can take values only in the range $-1 \leqslant u(t) \leqslant 1$, but is otherwise unrestricted. If the system starts at zero state, then the dynamical system can be expressed in state-space form (cf. (1.8) and Example 1.1), in which $x_1 = x$ and $x_2 = \dot{x}$, as

$$\dot{x}_1(t) = x_2(t), \qquad x_1(0) = 0,$$

$$\dot{x}_2(t) = -x_1(t) - x_2(t) + u(t), \quad x_2(0) = 0,$$

$$|u(t)| \leqslant 1, \qquad t \in [0, \infty).$$

Let the value of control u be kept constant at 1 for an initial period of time. The equilibrium point will be at (1, 0) and the resulting trajectory Γ_1 in $\mathbb{R}^2$ (cf. Example 1.8) will start to spiral in towards that point (Fig. 2.4). At the instant when $x_2(t)$ first returns to zero, switch the control to the constant value zero and keep it zero thereafter. The model then becomes

$$\dot{x}_1(t) = x_2(t), \qquad \dot{x}_2(t) = -x_1(t) - x_2(t),$$

with the equilibrium point now at the origin (0, 0). The trajectory Γ_1 now spirals in towards this new equilibrium point as shown in Fig. 2.4.

Now suppose that, instead of choosing a value 1 for the initial control from the origin, the value -1 is chosen for an initial period of time. Again the control is switched to zero value when $x_2(t)$ vanishes. The trajectory Γ_2 resulting from this control programme is also shown in Fig. 2.4. Now let P_1 be the set of control programmes such that $u(t) \in [0, 1]$ until the first instant at which $x_2(t)$ vanishes after leaving the origin and $u(t) = 0$ thereafter. Let P_2

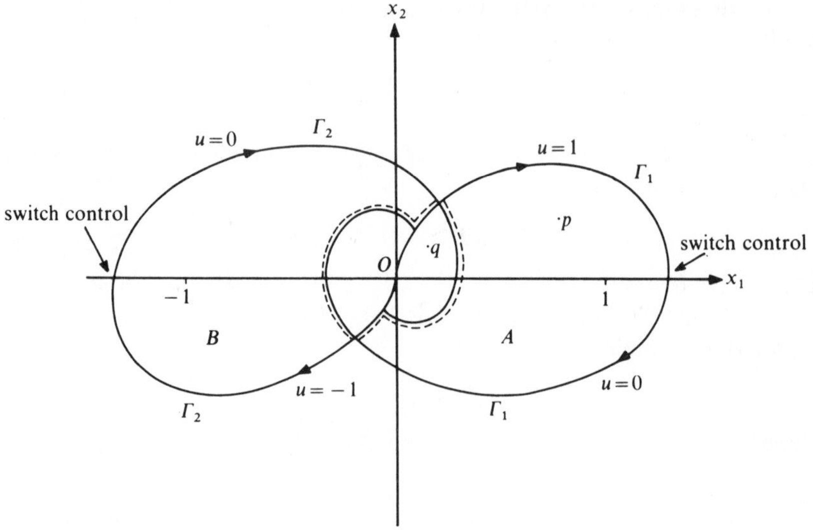

Fig. 2.4 Controlled damped oscillator

be the set of control programmes for which $u(t) \in [-1, 0]$ until the first instant at which $x_2(t)$ vanishes after leaving the origin and $u(t) = 0$ at all subsequent time t. The two paths Γ_1 and Γ_2 enclose all trajectories (including Γ_1 and Γ_2) starting at the origin under either of the programmes P_1 and P_2. Of course, if $u(t) = 0$ for an initial period, then the system just remains at the origin—not a very interesting case! All points inside and on the trajectories Γ_1 and Γ_2 are points that can be reached from the origin using the above admissible control values. This set of points is the union of that set A bounded by Γ_1 and that set B bounded by Γ_2 so that

$$A \cup B = \{(x_1, x_2) \in \mathbb{R}^2 : u \in P_1 \text{ or } u \in P_2\}.$$

Similarly, all points that can be reached by controls from P_1 and can also be reached by controls from P_2 form the intersection of sets A and B, so that

$$A \cap B = \{(x_1, x_2) \in \mathbb{R}^2 \colon u \in P_1 \text{ and } u \in P_2\}.$$

For example, the point p in Fig. 2.4 could be reached by a control from P_1 but not by one from P_2, so that $p \in A \cup B$ but $p \notin A \cap B$. However, the point q could be reached either by using a control from P_1 or by using one from P_2, so that $q \in A \cap B$.

Although it may seem rather strange, the set that contains no elements is a very useful set. It is called the *empty set* and is denoted by $\varnothing$.

Example 2.5

$A = [0, 1)$ and $B = [1, 2]$. Then $A \cap B = \varnothing$.

Be careful not to interpret $\varnothing$ as the set $\{0\}$ consisting of just the null element of some set A. The null element of A, if it exists, is a perfectly valid element of A. Therefore, the set $\{0\}$ contains one element so cannot be the empty set $\varnothing$. A very useful way in which the set $\varnothing$ can be used is illustrated in Example 2.5. If two sets A and B do not have any common element then

$$A \cap B = \varnothing$$

and the sets A and B are said to be *disjoint*. With a universal set U and $A \subset U$ it is useful to note that

$$A \cap A' = \varnothing \quad \text{and} \quad A \cup A' = U.$$

In dynamical system theory, one is often interested in the set of points $S(\boldsymbol{x}_0)$ that can be reached from the point $\boldsymbol{x}_0$ (Fig. 2.5(a)) using a given set of admissible controls (say piecewise continuous, i.e. continuous on finite intervals, with $|u(t)| \leqslant 1, t \in [0, \infty)$). With the same set of controls, let $E(\boldsymbol{x}_f)$ be the set of points from which $\boldsymbol{x}_f$ can be reached (Fig. 2.5(b)). [4] If there exists a point $\boldsymbol{x}$ which belongs to the intersection of sets $S(\boldsymbol{x}_0)$ and $E(\boldsymbol{x}_f)$, i.e. $\boldsymbol{x} \in S(\boldsymbol{x}_0) \cap E(\boldsymbol{x}_f)$, then it follows that $\boldsymbol{x} \in S(\boldsymbol{x}_0)$ and $\boldsymbol{x} \in E(\boldsymbol{x}_f)$. This means that $\boldsymbol{x}$ can be reached from $\boldsymbol{x}_0$ by an admissible control and furthermore $\boldsymbol{x}_f$ can be reached from $\boldsymbol{x}$ also by an admissible control. Thus, if

$$S(\boldsymbol{x}_0) \cap E(\boldsymbol{x}_f) \neq \varnothing,$$

then there exists an admissible control which yields a trajectory joining $\boldsymbol{x}_0$ to $\boldsymbol{x}_f$. Therefore, the existence of an admissible trajectory (i.e. one produced by an admissible control) can be demonstrated by showing that the intersection of sets $S(\boldsymbol{x}_0)$ and $E(\boldsymbol{x}_f)$ is not empty. On the other hand, if the sets $S(\boldsymbol{x}_0)$ and

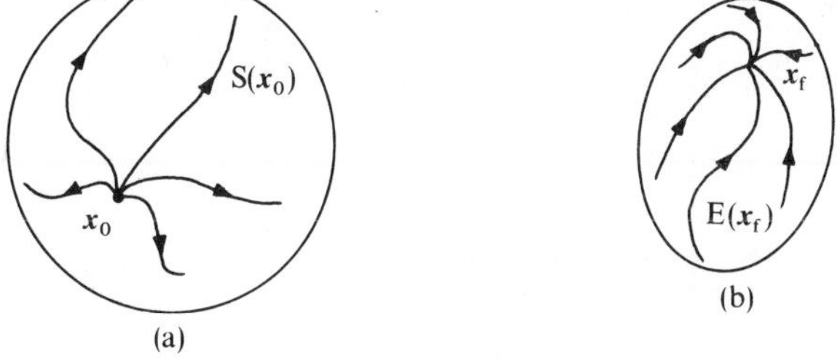

Fig. 2.5 Reachable sets

$E(\boldsymbol{x}_{\mathrm{f}})$ are found to be disjoint, so that

$$S(\boldsymbol{x}_0) \cap E(\boldsymbol{x}_{\mathrm{f}}) = \varnothing,$$

then there can be no trajectory arising from an admissible control which joins $\boldsymbol{x}_0$ to $\boldsymbol{x}_{\mathrm{f}}$.

Suppose two elements $a \in A$ and $b \in B$ are chosen from two sets A and B with a chosen first followed by b. Denote this choice by (a, b) in which the first entry a is the element chosen first (for example, $a \in \mathbb{R}$ and $b \in \mathbb{R}$ might be the cartesian coordinates of a point in $\mathbb{R}^2$ or the real and imaginary parts of a complex number $a + jb$). The pair of elements (a, b) is called an *ordered pair*. The set of all such ordered pairs is called the *cartesian product* of A and B, denoted by $A \times B$. Figure. 2.6 illustrates the set $A \times B$ when A and B are the intervals $[1, 2]$ and $[0, 2]$ respectively on $\mathbb{R}$. The vectors arising in Example 1.1 furnish other examples of ordered pairs. The vector $(x_1(t), x_2(t)) \in \mathbb{R}^2$ has real numbers $x_1(t)$ and $x_2(t)$ as elements and corresponds to the ordered pair $(x_1(t), x_2(t))$ belonging to the cartesian product $\mathbb{R} \times \mathbb{R}$. Thus, no distinction is to be made between the set $\mathbb{R} \times \mathbb{R}$ and the set $\mathbb{R}^2$. Some structure on the set $A \times B$ leads to the concept of a cartesian product space.

An obvious extension to the cartesian product $A \times B$ discussed above is the choice of n elements $x_1 \in X_1, \ldots, x_n \in X_n$, one from each of the n sets $X_1, \ldots, X_n$, to form the ordered n-tuple $(x_1, \ldots, x_n)$. The set of all such n-tuples is the cartesian product $X_1 \times \cdots \times X_n$. As in the case of $\mathbb{R} \times \mathbb{R}$ above, no distinction is to be made between the cartesian product space $\mathbb{R} \times \cdots \times \mathbb{R}$ (n terms) and the space $\mathbb{R}^n$. In general, $\mathbb{R}^{n_1} \times \cdots \times \mathbb{R}^{n_p}$ can be identified with $\mathbb{R}^n$ where $n = n_1 + \cdots + n_p$. This is useful when the total number of variables fall naturally into a number of separate sets.

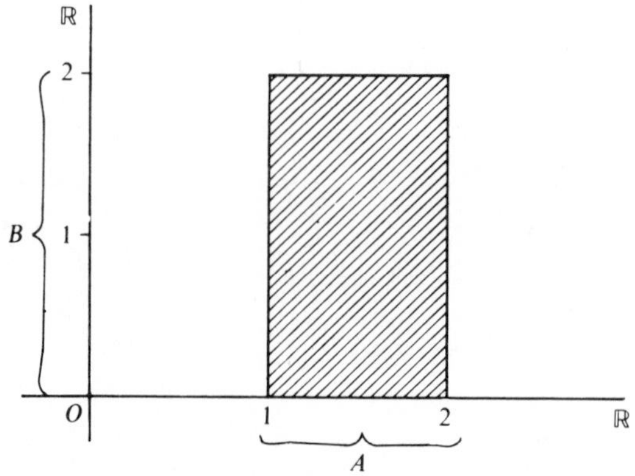

Fig. 2.6 Cartesian product $A \times B$

Example 2.6

In the vector differential equation (1.28), namely $\dot{\boldsymbol{x}} = \boldsymbol{f}(\boldsymbol{x}, \boldsymbol{u}, t)$, suppose that the state vector $\boldsymbol{x}(t)$ lies in X, with $X \subseteq \mathbb{R}^n$, the control vector $\boldsymbol{u}(t)$ satisfies $\boldsymbol{u}(t) \in U \subseteq \mathbb{R}^m$, and $t \in [0, T] \subset \mathbb{R}$. Then the $(n+m+1)$-tuple

$$(x_1, \ldots, x_n; u_1, \ldots, u_m; t)$$

belongs to $X \times U \times [0, T]$.

2.2 Equivalence relations and classes

Very often, interest is focused on sets X and Y between which certain elements are related in some way. A fairly trivial example is the set of all ordered pairs (t, x) of real numbers satisfying the equation $x - t = 2$. This set is clearly a subset of $\mathbb{R} \times \mathbb{R}$ and contains all points on the graph of $x - t = 2$. The subset is a *relation* from $\mathbb{R}$ to $\mathbb{R}$.

Formally, a subset $A \subseteq X \times Y$ is called a *relation from X to Y* (more precisely a *binary* relation, since two elements $x \in X$ and $y \in Y$ are involved). The statement $(x, y) \in A$ is written $x \sim y$ (x twiddles y), [5] where the twiddle $\sim$ stands for the phrase 'is related to'. So $\sim$ denotes a relation on a set A. Of course, in any particular case, $\sim$ will be associated with some definite relation such as 'equals' or 'less than'. In the first case $x \sim y$ would mean $x = y$, whereas in the second it would mean $x < y$.

Example 2.7

Suppose that $X = \{1, 2, 3, 4\}$ and $Y = \{2, 4, 6\}$, and let $\sim$ mean 'less than'. Then, from the set of all ordered pairs $X \times Y$, the subset

$$\{(1,2), (1,4), (1,6), (2,4), (2,6), (3,4), (3,6), (4,6)\}$$

is the relation $\sim$.

In the case when $Y = X$, the relation $\sim$ is from set X to the same set X, and so is a subset of $X \times X$. Then it is usual to speak of $\sim$ as a relation on X.

Example 2.8

Let $X = \{2, 3, 4, 6\}$ and let $\sim$ be the relation on X meaning 'divides'. Then $2 \sim 2$, $2 \sim 4$, $2 \sim 6$, $3 \sim 3$, $3 \sim 6$, $4 \sim 4$, and $6 \sim 6$. Thus, the relation $\sim$ is the subset

$$\{(2,2), (2,4), (2,6), (3,3), (3,6), (4,4), (6,6)\} \subset X \times X.$$

A *partial order* on a set A is a relation (often denoted by $\preccurlyeq$) such that

$$\text{(i)} \quad a \preccurlyeq a \text{ for all } a \in A,$$

$$\text{(ii)} \quad a \preccurlyeq b \text{ and } b \preccurlyeq a \quad \Rightarrow \quad a = b,$$

$$\text{(iii)} \quad a \preccurlyeq b \text{ and } b \preccurlyeq c \quad \Rightarrow \quad a \preccurlyeq c.$$

Example 2.9

Suppose $S = \{1, 2, 3\}$. The *power set* of S, denoted by $\mathscr{P}(S)$, is the set of all subsets of S, i.e.

$$\mathscr{P}(S) = \{\varnothing, S, \{1\}, \{2\}, \{3\}, \{1,2\}, \{1,3\}, \{2,3\}\}.$$

Then a partial order on the set $\mathscr{P}(S)$ is set inclusion (here $\preccurlyeq$ is replaced by $\subseteq$). This is because, for $T,P,Q,R \in \mathscr{P}(S)$, the following properties hold: (i) $T \subseteq T$; (ii) $P \subseteq Q$ and $Q \subseteq P$ together imply that $P = Q$; (iii) $P \subseteq Q$ and $Q \subseteq R$ together imply that $P \subseteq R$.

This partial order can be illustrated diagrammatically and is shown in Fig. 2.7, in which the arrows indicate set inclusion. Thus, $\{3\} \to \{2, 3\}$ means $\{3\} \subset \{2, 3\}$, etc. Not all admissible lines are drawn in Fig. 2.7, because $\{3\} \to \{1, 2, 3\}$ is implied by $\{3\} \to \{2, 3\} \to \{1, 2, 3\}$ and so on. Notice that some pairs of elements in $\mathscr{P}(S)$ do not satisfy any of the relations $\subseteq$, $=$, $\supseteq$. For example, $\{1, 3\} \not\subseteq \{2, 3\}$ and $\{1, 3\} \not\supseteq \{2, 3\}$.

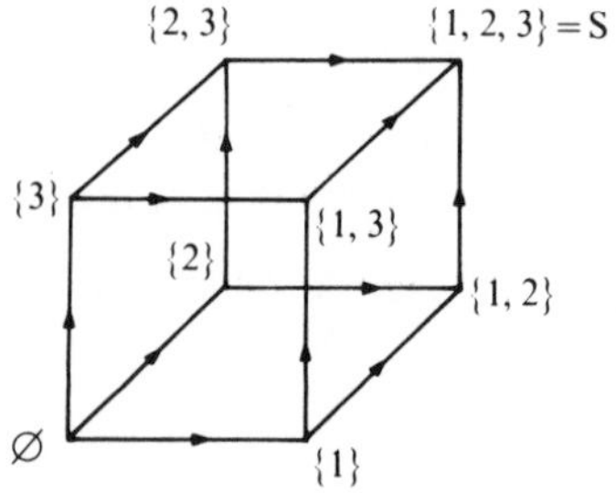

Fig. 2.7 Partial order

A set A and a partial order $\preccurlyeq$ together form a pair $(A, \preccurlyeq)$ which is called a partially ordered set. When $a \preccurlyeq b$, then a is said to be smaller than (or precede) b and b is greater than (or follows or dominates) a. If $a \neq b$, then

$a \preccurlyeq b$ is usually written as $a \prec b$ (as in $a < b$ with $a,b \in \mathbb{R}$). A partially ordered set $(A, \preccurlyeq)$ is *totally ordered* if, for every $a,b \in A$, either $a \preccurlyeq b$ or $b \preccurlyeq a$. So, in Example 2.9, the pair $(S, \leqslant)$ is a totally ordered set, but the pair $(\mathscr{P}(S), \subseteq)$ is not.

Example 2.10

An order relation for matrices can be defined in the following way. A matrix $\boldsymbol{A}$ is said to be *smaller* than a matrix $\boldsymbol{B}$ (written ($\boldsymbol{A} < \boldsymbol{B}$) iff the matrix $\boldsymbol{B} - \boldsymbol{A}$ is positive definite (see Note 1.16), i.e.

$$\boldsymbol{A} < \boldsymbol{B} \quad \Leftrightarrow \quad \boldsymbol{B} - \boldsymbol{A} > \boldsymbol{0}.$$

Alternatively, a dominance ordering $<^*$ can be defined as follows. For any two matrices A and B,

$$\boldsymbol{A} <^* \boldsymbol{B} \quad \Leftrightarrow \quad \begin{cases} \boldsymbol{x}^{\mathrm{T}} \boldsymbol{A} \boldsymbol{x} \leqslant \boldsymbol{x}^{\mathrm{T}} \boldsymbol{B} \boldsymbol{x} & \text{for all } \boldsymbol{x}, \\ \bar{\boldsymbol{x}}^{\mathrm{T}} \boldsymbol{A} \bar{\boldsymbol{x}} < \bar{\boldsymbol{x}}^{\mathrm{T}} \boldsymbol{B} \bar{\boldsymbol{x}} & \text{for some } \bar{\boldsymbol{x}}. \end{cases}$$

Note that neither of the relations in Example 2.10 is totally ordered since it may be that, for two particular matrices $\boldsymbol{A}$ and $\boldsymbol{B}$, neither $\boldsymbol{A} \leqslant \boldsymbol{B}$ nor $\boldsymbol{B} \leqslant \boldsymbol{A}$ is true. For example, the matrices

$$\boldsymbol{A} = \begin{bmatrix} 3 & 1 \\ 0 & -2 \end{bmatrix} \quad \text{and} \quad \boldsymbol{B} = \begin{bmatrix} 2 & 1 \\ 0 & -1 \end{bmatrix}$$

are such that neither $\boldsymbol{A} - \boldsymbol{B}$ nor $\boldsymbol{B} - \boldsymbol{A}$ is a positive definite matrix and so $\boldsymbol{A} \nless \boldsymbol{B}$, $\boldsymbol{B} \nless \boldsymbol{A}$, and certainly $\boldsymbol{A} \neq \boldsymbol{B}$.

An example of the way in which the partial orders of Example 2.10 can arise in system theory is obtained from the so-called output-feedback problem (Allwright 1982). In this problem, a system Σ is represented by the equations (cf. (1.10))

$$\dot{\boldsymbol{x}} = \boldsymbol{A}\boldsymbol{x} + \boldsymbol{B}\boldsymbol{u}, \qquad \boldsymbol{x}(0) = \boldsymbol{x}_0,$$

$$\boldsymbol{y} = \boldsymbol{C}\boldsymbol{x},$$

and a matrix $\boldsymbol{F}$ (called a feedback matrix) is to be found such that the control $\boldsymbol{u} = \boldsymbol{F}\boldsymbol{y}$ minimizes the objective function

$$J(\boldsymbol{x}, \boldsymbol{u}) = \int_0^{\infty} (\boldsymbol{x}^{\mathrm{T}} \boldsymbol{Q} \boldsymbol{x} + \boldsymbol{u}^{\mathrm{T}} \boldsymbol{R} \boldsymbol{u}) \mathrm{d}t,$$

where $\boldsymbol{x}(t) \in \mathbb{R}^n$, $\boldsymbol{u}(t) \in \mathbb{R}^m$, $\boldsymbol{y}(t) \in \mathbb{R}^l$, $\boldsymbol{A}, \boldsymbol{Q} \in \mathbb{R}^{n \times n}$, $\boldsymbol{B} \in \mathbb{R}^{n \times m}$, $\boldsymbol{C} \in \mathbb{R}^{l \times n}$, $\boldsymbol{F} \in \mathbb{R}^{m \times l}$, $\boldsymbol{R} \in \mathbb{R}^{m \times m}$.

Since $\boldsymbol{u} = \boldsymbol{F}\boldsymbol{y} = \boldsymbol{F}\boldsymbol{C}\boldsymbol{x}$ it follows that

$$\dot{\boldsymbol{x}} = (\boldsymbol{A} + \boldsymbol{B}\boldsymbol{F}\boldsymbol{C})\boldsymbol{x}, \qquad \boldsymbol{x}(0) = \boldsymbol{x}_0.$$

Suppose the solution to this vector differential equation is [6]

$$x(t) = X(F, t)x_0, \qquad X \in \mathbb{R}^{n \times n}.$$

Then the objective function J may be expressed as

$$\begin{aligned} J(x_0, F) &= \int_0^{\infty} x_0^{\mathrm{T}} X^{\mathrm{T}}(F, t)(Q + C^{\mathrm{T}} F^{\mathrm{T}} R F C) X(F, t) x_0 \, \mathrm{d}t \\ &= x_0^{\mathrm{T}} K(F) x_0 \quad \text{(say)}, \qquad K(F) \in \mathbb{R}^{n \times n}. \end{aligned}$$

The minimization of $x_0^{\mathrm{T}} K(F) x_0$ requires the definition of an order relationship as in Example 2.10. Suppose F_1 and F_2 are two feedback matrices such that $K(F_2) < K(F_1)$. Then $K(F_1) - K(F_2) > \mathbf{0}$ and $x_0^{\mathrm{T}}[K(F_1) - K(F_2)]x_0 > \mathbf{0}$ for all $x_0 \neq \mathbf{0}$. It follows that

$$x_0^{\mathrm{T}} K(F_2) x_0 < x_0^{\mathrm{T}} K(F_1) x_0$$

i.e. $J(x_0, F_2) < J(x_0, F_1)$. So, if $K(F_2)$ is 'smaller' than $K(F_1)$, then the cost of F_2 is smaller than that of F_1 for all initial conditions. F_2 is said to dominate F_1 *strictly*.

Alternatively, for the dominance ordering of Example 2.10, if $K(F_2) <^* K(F_1)$ then [7]

$$J(x_0, F_2) \leqslant J(x_0, F_1) \quad \forall\, x_0, \qquad J(\bar{x}_0, F_2) < J(\bar{x}_0, F_1) \quad \text{for some } \bar{x}_0.$$

Hence, if F_2 either dominates or strictly dominates F_1, then the cost will never increase for any initial condition, and will actually decrease for at least one initial condition.

Some relations have special properties; this will be illustrated by the relation '$a \sim b$ iff $a - b$ is divisible by 3'. This relation $\sim$ on $\mathbb{R}$ is usually denoted by writing $a \sim b$ as $a \equiv b \pmod 3$, which means that the difference $a - b$ is an integral multiple of 3. More will be said on such relations in Section 2.3. The properties of importance are as follows.

(i) A relation $\sim$ on a set X is said to be *reflexive* if

$$a \sim a \quad \forall\, a \in X.$$

Example 2.11. We have $a \equiv a \pmod 3$ for all $a \in \{\ldots, -9, -6, -3, 0, 3, 6, 9, \ldots\}$, because $a - a = 0$, which is an integral multiple of 3 $(0 = 0 \times 3)$.

(ii) A relation $\sim$ on a set X is said to be *symmetric* if

$$a \sim b \Rightarrow b \sim a \quad \forall\, a, b \leftarrow X. \text{ [8]}$$

Example 2.12. The property that $a \equiv b \pmod 3$ for all $a, b \in \{\ldots, -9, -6, -3, 0, 3, 6, 9, \ldots\}$ implies that $a - b = 3k$ for some $k \in \mathbb{Z}$. Thus, $b - a = -3k$

$= 3(-k)$, so that the difference $b-a$ is also divisible by 3. Therefore $b \equiv a \pmod 3$ and $\equiv$ is symmetric.

(iii) A relation $\sim$ on a set X is said to be *transitive* if

$$a \sim b \text{ and } b \sim c \quad \Rightarrow \quad a \sim c \qquad \forall\, a,b,c \in X.$$

Example 2.13. Since $a \equiv b \pmod 3$ and $b \equiv c \pmod 3$ for all $a,b,c \in \{\ldots, -9, -6, -3, 0, 3, 6, 9, \ldots\}$, we have that $a-b = 3k_1$ and $b-c = 3k_2$, where $k_1, k_2 \in \mathbb{Z}$. It follows that $a-c = (a-b)+(b-c) = 3(k_1+k_2)$. Since $k_1+k_2 \in \mathbb{Z}$, we have $a \equiv c \pmod 3$, and so $\equiv$ is transitive.

A relation $\sim$ on a set X which is reflexive, symmetric, and transitive is called an *equivalence relation.* Thus, the relation $\equiv$ in Examples 2.11, 2.12, and 2.13 is an equivalence relation. A relation which is not an equivalence relation is given in the next example.

Example 2.14

A partial order ($\preccurlyeq$) on a set A is not an equivalence relation because it is not a symmetric relation. If $a \preccurlyeq b$ for $a,b \in A$, then it is not true that $b \preccurlyeq a$, unless $a = b$. A partial order is said to be an *antisymmetric* relation since $a \preccurlyeq b$ and $b \preccurlyeq a$ together imply $a = b$.

Table 2.1 shows a number of different relations (on the set of integers $\mathbb{Z}$) for which some of the properties of an equivalence relation are true and some are false. A tick (√) means that the property is satisfied whereas a cross (×) means

Table 2.1 Relations on set $\mathbb{Z}$

Property $a \sim b$ iff	reflexive	symmetric	transitive
(i) $a=b$	√	√	√
(ii) $a \leqslant b$	√	×	√
(iii) ab a perfect square	√	√	×
(iv) $ab \neq 0$	×	√	√
(v) $a-b$ odd	×	√	×
(vi) $a<b$	×	×	√
(vii) $a-b \in \{0, 1\}$	√	×	×
(viii) $a-b=1$	×	×	×

that it is not. For example, in (iv), the relation is reflexive if $a^2 \neq 0$ for all $a \in \mathbb{Z}$. But this is not true for $a = 0$, and so the relation is not reflexive. Again, in (vii), the transitive property is satisfied if, for all a, b, $c \in \mathbb{Z}$, the conditions $a - b \in \{0, 1\}$ and $b - c \in \{0, 1\}$ together imply $a - c \in \{0, 1\}$. But, when $a - b = 1$ and $b - c = 1$, then $a - c = (a - b) + (b - c) = 1 + 1 = 2 \notin \{0, 1\}$ so the relation is not transitive. The reader should check the other results in Table 2.1.

If $\sim$ is an equivalence relation on a set X and if $a \in X$, then the elements $b \in X$ such that $b \sim a$ form a subset of X denoted by $[a]$. This subset $[a]$ is called an *equivalence class*. Thus,

$$[a] = \{b \in X : b \sim a\}.$$

Since $\sim$ is an equivalence relation, we have $a \sim a$, so that $a \in [a]$. If $b \in [a]$ then $[b] = [a]$. [9]

A given equivalence relation $\sim$ on a set X divides X into a number of equivalence classes. These classes are such that

(i) every element of X is in one and only one of the classes,
(ii) if a and b are in the same class then $a \sim b$,
(iii) if a and b are in different classes then $a \sim b$

is false ($a \not\sim b$).

Any division of a set X into non-empty subsets $X_1, \ldots, X_n$ such that [10]

$$\bigcup_{i=1}^{n} X_i = X$$

and such that

$$i \neq j \quad \Rightarrow \quad X_i \cap X_j = \varnothing$$

for all $i, j = 1, \ldots, n$ (i.e. the X_i are all disjoint sets) is called a *partition* of X. From what has been said above, an equivalence relation $\sim$ on a set X leads to a partition of X in the form of the equivalence classes generated by $\sim$. Conversely, a partition of X always defines an equivalence relation on X. Also, whenever a set A is a proper subset of a set X, so that $A \subset X$, then A forms a partition of X with two disjoint sets A and A'.

Example 2.15

Equivalence classes of some importance in dynamical system theory arise when one considers the set of all continuous trajectories defined by $t \mapsto x(t)$ $(t \in I_\varepsilon)$, where $I_\varepsilon = \{t \in \mathbb{R} : |t| < \varepsilon\}$, such that $x(0) = a$ (Fig. 2.8). An equivalence relation $\sim$ arises when attention is focused on the set of all trajectories that have the same value for their gradients at $t = 0$. Thus, $x(t) \sim y(t)$ iff $\dot{x}(0) = \dot{y}(0)$. Of course, a different set of trajectories is obtained

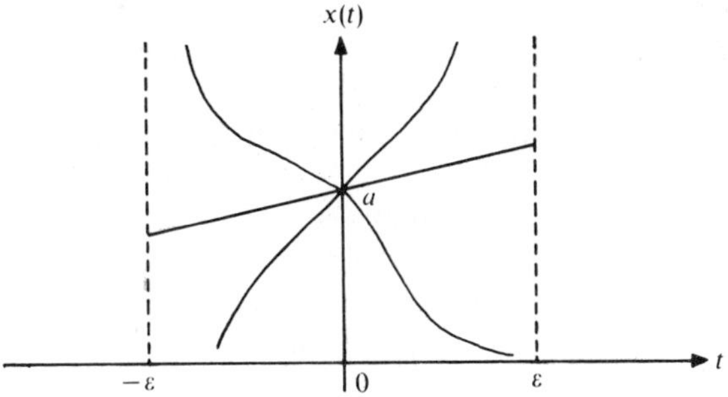

Fig. 2.8 Equivalence classes of trajectories

for each particular value of $\dot{x}(0)$ and these sets are the equivalence classes $[x(\bullet)]$ corresponding to each value $\dot{x}(0)$. That is, to each value $\dot{x}(0)$ there corresponds an equivalence class $[x(\bullet)]$ and vice versa.

2.3 Congruence modulo n (with $n \in \mathbb{N}$)

There are special sets of integers in which the difference between any two elements of a set is an integral multiple of some fixed $n \in \mathbb{N}$.

Example 2.16

Suppose that $n = 3$ and consider the set of all integers of the form $3k$, where $k \in \mathbb{Z}$. This is the set $\{\ldots, -9, -6, -3, 0, 3, 6, 9, \ldots\}$ used in Section 2.2 to introduce equivalence relations.

An equivalence relation $\sim$ on $\mathbb{Z}$ can be defined by

$$a \sim b \text{ iff } a - b \text{ is divisible by } 3.$$

Then the equivalence class [0] is $\{\ldots, -9, -6, -3, 0, 3, 6, 9, \ldots\}$. Again, all integers $3k+1$ form the equivalence class $[1] = \{\ldots, -8, -5, -2, 1, 4, 7, \ldots\}$. Integers $3k+2$ form the equivalence class $[2] = \{\ldots, -7, -4, -1, 2, 5, 8, \ldots]$. The numbers 0, 1, 2 in the equivalence classes [0], [1], [2] are simply the remainders, or residues, when the elements of the classes are divided by 3. [11] The set $\{0, 1, 2\}$ of residues consists of a set of representative elements, one from each class [0], [1], [2], and is denoted by $\mathbb{Z}_3$. The equivalence classes [0], [1], [2] are called *congruence* (or *residue*) *classes, modulo 3.* These classes form a complete set of equivalence classes with respect to $\sim$, i.e. no two elements of the set $\mathbb{Z}_3$ are equivalent, and every integer is equivalent to just one member of that set. Each element of $\mathbb{Z}_3$ can be identified with the class it represents: 1 with [1], etc. and the complete set $\{[0], [1], [2]\}$ of equivalence classes is sometimes also denoted by $\mathbb{Z}_3$ (to see why this is permissible refer to Section 3.3).

The equivalence on $\mathbb{Z}$, with equivalence classes $\mathbb{Z}_3 = \{[0], [1], [2]\}$, is called a congruence modulo 3 with the notation $a \equiv b \pmod 3$. If a and b are two integers, then the following statements are equivalent:

$a \sim b$,

a and b are congruent modulo 3,

a and b lie in the same class in $\mathbb{Z}_3$,

$a - b$ is a multiple of 3,

a and b have the same remainders on division by 3,

$a \equiv b \pmod 3$.

The same follows for congruence modulo n, with 3 replaced by any integer $n > 1$. In general there are n equivalence classes associated with the integer n so that $\mathbb{Z}_n = \{[0] ,..., [n-1]\}$. Again, the remainders obtained by dividing all integers by n form the set $\mathbb{Z}_n = \{0 ,..., n-1\}$. A congruence class $[r]$, modulo n, is called a *coset* of r, modulo n. Cosets also exist for the algebraic systems discussed in Chapters 4, 5, and 6.

Example 2.17

Dividing any integer by 5 yields a remainder which is 0, 1, 2, 3, or 4 and $\{0, 1, 2, 3, 4\} = \mathbb{Z}_5$. The cosets of these remainders are [0] ,..., [4].

Example 2.18

$32 \equiv 8 \pmod{24}$ means 32 is congruent to 8 modulo 24, i.e. $32-8$ is a multiple of 24. In fact, $32-8 = 1 \times 24$ (or $32-24 = 8$). Such an example would arise using a 24-hour clock if, at 17.00 on Wednesday, 6 May, the time was required fifteen hours later. The calculation would involve $17+15 = 32$ and $32-24 = 8$ so that the required time would be 08.00 on Thursday, 7 May.

Congruence modulo 7 for the days of a given month has as classes the seven days of the week so if the 8th day of the month is a Wednesday then the 12th day is not a Wednesday because $8 \not\equiv 12 \pmod 7$.

It is worth mentioning at this point that the following notation is also used. [12]

$$n\mathbb{Z} = \{a \in \mathbb{Z} : n|a,\ n \in \mathbb{N}\} = \{0, \pm n, \pm 2n, \pm 3n, ...\}.$$

Example 2.19

$3\mathbb{Z} = \{a \in \mathbb{Z} : 3|a\} = \{0, \pm 3, \pm 6, \pm 9, ...\}$ and similarly $2\mathbb{N} = \{m \in \mathbb{N} : 2|m\} = \{2, 4, 6, ...\}$ and so on. Again, the coset of r modulo n is given by

$$[r] = \{nq + r : \forall q \in \mathbb{Z}\} = n\mathbb{Z} + r.$$

Notice here there is a peculiarity of notation which is potentially confusing; $n\mathbb{Z} + r$ is the 'addition' of a set to a number, which is meaningless in the usual sense of addition, so strictly it is *defined* as

$$n\mathbb{Z} + r \triangleq \{nq + r : \forall\ q \in \mathbb{Z}\}.$$

Exercises

1. Show that the set $\mathrm{So}(n, \mathbb{R})$ is a proper subset of $\mathrm{G}\ell(n, \mathbb{R})$.
2. Consider the relation $\sim$ on $\mathbb{R}^{n \times n}$, where $\sim$ means 'is similar to'. Thus, for $\boldsymbol{P}, \boldsymbol{Q} \in \mathbb{R}^{n \times n}$, the statement $\boldsymbol{P} \sim \boldsymbol{Q}$ means there exists a nonsingular matrix $\boldsymbol{T} \in \mathbb{R}^{n \times n}$ such that $\boldsymbol{T}^{-1}\boldsymbol{P}\boldsymbol{T} = \boldsymbol{Q}$. Show that this relation on $\mathbb{R}^{n \times n}$ is an equivalence relation.
3. If $S = \{(a, b) : a, b \in \mathbb{Z}, b \neq 0\}$, show that

$$(a, b) \sim (c, d) \quad \Leftrightarrow \quad ad = bc$$

defines an equivalence relation on S.

4. Show that a partial order $\leqslant$ can be defined on the set $\mathbb{R}^n$ of all ordered n-tuples by letting $\boldsymbol{x} \leqslant \boldsymbol{y}$ if and only if $x_i \leqslant y_i$ $(i = 1, ..., n)$ where $\boldsymbol{x} = (x_1, ..., x_n)$, $\boldsymbol{y} = (y_1, ..., y_n) \in \mathbb{R}^n$, and $x_i \leqslant y_i$ has the usual meaning in $\mathbb{R}$.
5. A relation $\sim$ is defined on $\mathbb{R}^2$ by

$$(a, b) \sim (c, d) \quad \Leftrightarrow \quad a + d = b + c.$$

Describe the geometrical interpretation of the partition of $\mathbb{R}^2$ resulting from the equivalence classes of this relation.

6. Show that the cosets [0] and [1], modulo 2, associated with $\mathbb{Z}_2 = \{0, 1\}$ are $2\mathbb{Z}$ and $2\mathbb{Z} + 1$ respectively.

Notes for Chapter 2

1. The set of all real numbers that lie between the two real numbers a and b and which include the end numbers a and b is called a *closed* interval on $\mathbb{R}$ and is denoted by $[a, b]$. On the other hand, the same set of real numbers but with a and b excluded is called an *open* interval on $\mathbb{R}$ and is denoted by (a, b). From this, the half-open intervals $[a, b)$ and $(a, b]$ have obvious interpretations although neither is open or closed. Sometimes the open interval (a, b) is written $]a, b[$. $\mathbb{R}$ can be represented by $(-\infty, \infty)$.

2. The symbol $\Leftrightarrow$ stands for 'if and only if' (see Note 3). Two other associated signs are $\Rightarrow$, meaning 'implies', and $\Leftarrow$, meaning 'is implied by'. Thus, $\alpha \Rightarrow \beta$ means 'α implies β', i.e. if α is a true statement then statement β is also true. Similarly, $\alpha \Leftarrow \beta$ means 'α is implied by β' or 'β implies α'. These two symbols $\Rightarrow$ and $\Leftarrow$ correspond to statements of a necessary condition and a sufficient condition (see Note 3). Finally, $\alpha \Leftrightarrow \beta$ means 'α implies β and β implies α' and the symbol $\Leftrightarrow$ is equivalent to the statement 'if and only if' discussed in Note 3 below. Yet another symbol $\rightleftarrows$ has been used for many years to represent British Rail—nothing to do with mathematics!
3. The phrase 'if and only if' (often written 'iff' or symbolically as $\Leftrightarrow$) is important in mathematics. A simple example will suffice to explain the concepts involved in the phrase. For $x,y \in \mathbb{R}$, we have

$$0 < x < y \quad \Rightarrow \quad 0 < x^2 < y^2.$$

In other words, $x^2 < y^2$ if $0 < x < y$. The statement $0 < x < y$ is sufficient to ensure that x^2 is less than y^2, and $0 < x < y$ is said to be a sufficient condition for $x^2 < y^2$ to be true. However, it is not necessary for x and y to satisfy $0 < x < y$ in order that $x^2 < y^2$ be true. For example, the values $x = -\frac{1}{2}$ and $y = -1$ are such that $x^2 < y^2$ $(\frac{1}{4} < 1)$, but these values of x and y do not satisfy $0 < x < y$, since both values are negative. Thus, although $0 < x < y$ is a sufficient condition for $x^2 < y^2$, it is not a necessary one. Put another way, although it is true to say that $x^2 < y^2$ if $0 < x < y$, it is not true to say that $x^2 < y^2$ *only if* $0 < x < y$.

Now consider a condition that is necessary but not sufficient. For $x,y \in \mathbb{R}$, the condition $0 < x < y$ holds only if $x^2 < y^2$. That is to say, $x^2 < y^2$ is necessary for the inequalities $0 < x < y$ to be true, and $x^2 < y^2$ is called a necessary condition for $0 < x < y$. But $x^2 < y^2$ is not a sufficient condition for $0 < x < y$, as the example $x = -\frac{1}{2}$ with $y = -1$ again illustrates, since in this case $y < x < 0$. Thus, $x^2 < y^2$ is a necessary condition for $0 < x < y$ to be true, but it is not a sufficient condition. In other words, although it is true to say that $0 < x < y$ only if $x^2 < y^2$, it is not true to say that $0 < x < y$ if $x^2 < y^2$ (i.e. not true for all $x,y \in \mathbb{R}$).

Finally, for $x,y \in \mathbb{R}$, we have

$$|x| < |y| \quad \text{if} \quad x^2 < y^2 \quad (\text{i.e. } x^2 < y^2 \Rightarrow |x| < |y|),$$

$$|x| < |y| \quad \text{only if} \quad x^2 < y^2 \quad (\text{i.e. } x^2 < y^2 \Leftarrow |x| < |y|).$$

In this case, $x^2 < y^2$ is both a sufficient condition and a necessary condition for $|x| < |y|$, written $x^2 < y^2 \Leftrightarrow |x| < |y|$.
4. The reader should be warned that both S and E are replaced by R in some of the older literature (e.g. Hermes and Haynes 1963).
5. Given a binary relation A, some authors write xAy to denote $(x, y) \in A$.
6. The solution of the linear vector differential equation

$$\dot{\boldsymbol{x}}(t) = \boldsymbol{A}(t)\boldsymbol{x}(t), \ \boldsymbol{x}(0) = \boldsymbol{x}_0,$$

where $\boldsymbol{x}(t) \in \mathbb{R}^n$ and $\boldsymbol{A}(t) \in \mathbb{R}^{n \times n}$, is given by $\boldsymbol{x}(t) = \boldsymbol{\Phi}(t, 0)\boldsymbol{x}_0$, where $\boldsymbol{\Phi}(t, 0) \in \mathbb{R}^{n \times n}$ is a time-varying matrix known as the *state transition matrix* and depends on $\boldsymbol{A}(t)$. The matrix $\boldsymbol{\Phi}(t, 0)$ is often written as e^{At} when $\boldsymbol{A}(t) = \boldsymbol{A}$ is a constant matrix. This matrix is discussed in Section 4.3 (see also Example 1.4). Since the transition matrix for the system $\dot{\boldsymbol{x}} = (\boldsymbol{A} + \boldsymbol{BFC})\boldsymbol{x}$ is dependent upon the unknown matrix $\boldsymbol{F}$, it has been written as $\boldsymbol{X}(\boldsymbol{F}, t)$.

7. The sign $\forall$ means ' for all ', so that ' $\forall\ x \in X$ ' is read as ' for all x belonging to X '.
8. In fact, if $\sim$ is symmetric, then $a \sim b \Leftrightarrow b \sim a$ because if $a \sim b \Rightarrow b \sim a$ for all $a,b \in X$, then $b \sim a \Rightarrow a \sim b$.
9. To prove this statement, note that $b \in [a] \Rightarrow b \sim a$, and that $c \sim b$ for any $c \in [b]$. Now, $c \sim b$ and $b \sim a$ together imply that $c \sim a$, by the transitive property of the equivalence relation $\sim$. But $c \sim a \Rightarrow c \in [a]$, so it has been shown that any c belonging to $[b]$ also belongs to $[a]$. Therefore $[b] \subseteq [a]$.

Since $b \sim a \Rightarrow a \sim b$, the above argument can be repeated with the interchange of a and b to give $[a] \subseteq [b]$. The two results $[b] \subseteq [a]$ and $[a] \subseteq [b]$ yield the result $[a] = [b]$ (by (2.1)).
10. A set X containing all the elements that occur in the sets $X_1 ,..., X_n$ can be written as $X = X_1 \cup \cdots \cup X_n$, or more concisely as

$$X = \bigcup_{i=1}^{n} X_i,$$

which is the union of all the sets $X_1 ,..., X_n$.

A similar interpretation applies to

$$X = \bigcap_{i=1}^{n} X_i,$$

which is the intersection of all the sets $X_1 ,..., X_n$ and contains only those elements common to $X_1 ,..., X_n$.
11. Remainders are normally taken as non-negative integers. For example, when considering the division of the integer -34 by 3, it is usual to select $-34 = 3(-12) + 2$ with remainder 2 rather than $-34 = 3(-11) - 1$ with remainder -1.
12. The notation $n|a$ means ' n divides a '. If $n,a \in A$ for some set A, then $n|a$ if $a = nb$ for some $b \in A$.

3

MAPPINGS

The idea of a function defined by the equation $y = f(x)$, where x and y are real numbers, is very well known in elementary mathematics. This equation assigns a unique number y respectively to each number x, the assignment being through the function f. Then f is said to map the real number x onto the real number y. This idea may be extended to mappings [1] which map elements of one general set A into elements of another general set B. Functions or mappings may themselves be elements of a set. An example is the set $C^0(\mathbb{R})$ of all continuous functions on $\mathbb{R}$ mentioned in Section 1.2. In this case, a function f is thought of as an element of $C^0(\mathbb{R})$ (i.e. $f \in C^0(\mathbb{R})$) in the same way as a real number x is an element of the set $\mathbb{R}$. A very large number of concepts in mathematics, some with strange-sounding names such as homomorphisms and diffeomorphisms, are examples of mappings.

3.1 Introduction

Given the equation $y = x^3$, where $x, y \in \mathbb{R}$, the corresponding graph is shown in Fig. 3.1. To obtain any point on this graph, a value of x is chosen, say x_1, and the value of y is calculated, y_1 say, from the fact that $y_1 = x_1^3$. This process of cubing real numbers can be thought of as a simple input–output system f (Fig. 3.2) in which a chosen value of x is the input and the process or rule f is the cubing of the input to give the output x^3. The specification of f is given either by the equation $f(x) = x^3$ which means that f, operating on x, gives x^3, or by using a barred arrow notation $x \mapsto x^3$, which means that x is transformed to x^3 under an operation f. When f assigns to each real number $x \in \mathbb{R}$ a (unique) real number $f(x) \in \mathbb{R}$, then f is written as $f: \mathbb{R} \to \mathbb{R}$ and is called a function. The element $f(x)$ is called the *image* of x under f and may be thought of as the value of the function f at x (Fig. 3.3). It is often convenient to refer to ‘ the function $f: \mathbb{R} \to \mathbb{R}$ defined by the equation $f(x) = x^3$ ’ simply as ‘ the function $f(x) = x^3$ ’. There is no harm in this, provided that it is well understood from whence x is taken and that really f is the function and $f(x)$ the image of x under f. Notice that $f(x)$ must be unique for f to be a function, i.e. there cannot be two different values $f_1(x)$ and $f_2(x)$ corresponding to the same x. For example, the solution of the equation $y^2 = x$ does not define a function because, if there were such a function f, then $y = f(x) = \pm\sqrt{x}$ and so $f(x_1)$ has two possible values $\sqrt{x_1}$ and $-\sqrt{x_1}$ which are different unless

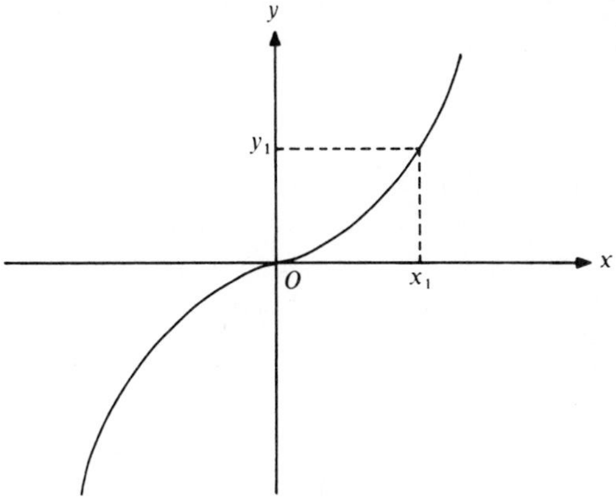

Fig. 3.1 Graph of $y = x^3$

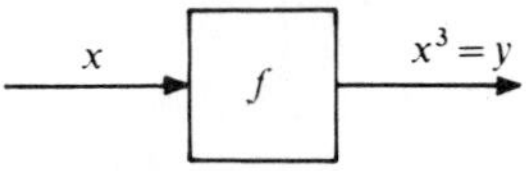

Fig. 3.2 Input/output system f

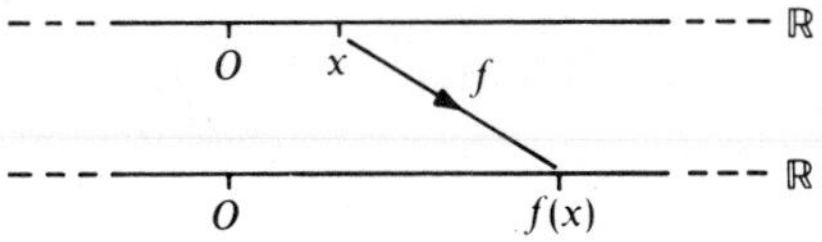

Fig. 3.3 Function $f: \mathbb{R} \to \mathbb{R}$

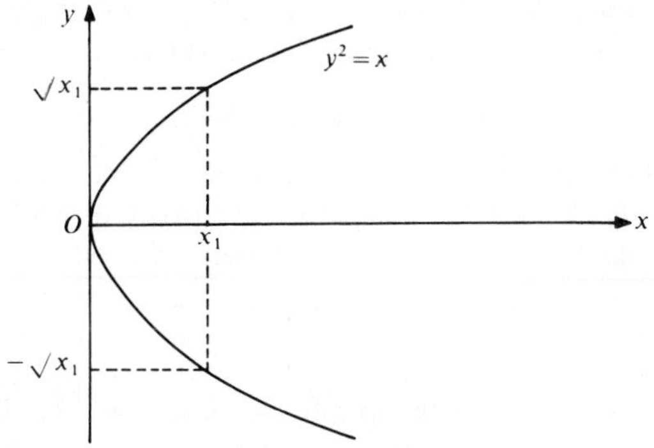

Fig. 3.4 Graph of $y^2 = x$

$x_1 = 0$ (Fig. 3.4). [2] However, it is permissible to have an image $f(x)$ corresponding to more than one x, and an example of this is now given.

Example 3.1

The function $f: \mathbb{R} \to \mathbb{R}$ defined by $f(x) = x^2$ has the graph shown in Fig. 3.5. To each $f(x_1) > 0$ there correspond two values of x, namely x_1 and $-x_1$. Furthermore, for any $x \in \mathbb{R}$, the image $f(x)$ is never negative, so that all images belong to $\mathbb{R}^+$. [3] Thus, if the function $f: \mathbb{R} \to \mathbb{R}$ is defined by $x \mapsto x^2$, then one could also write $f: \mathbb{R} \to \mathbb{R}^+$.

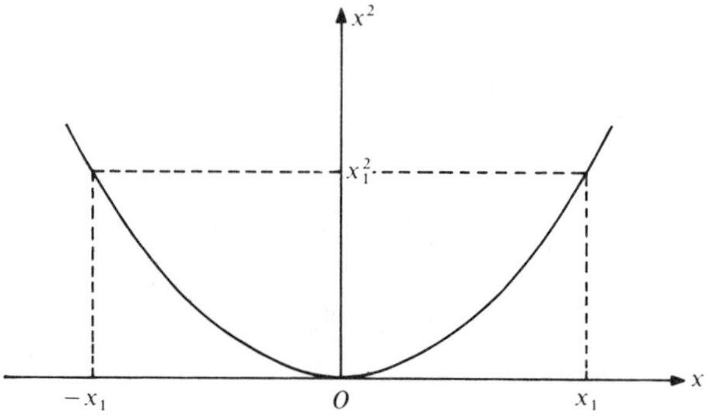

Fig. 3.5 Graph of $f(x) = x^2$

Example 3.1 immediately leads to a further definition. The *image set* (or *range*) of a function $f: \mathbb{R} \to \mathbb{R}$ is, for all $x \in \mathbb{R}$, the set of all images $f(x)$ and is written as im f (sometimes as $\Re(f)$ or $f(\mathbb{R})$). Thus,

$$\text{im}\, f \triangleq \{f(x) : x \in \mathbb{R}\}. \text{ [4]}$$

When a function f is written as $f: \mathbb{R} \to \mathbb{R}$, it means that for every $x \in \mathbb{R}$ there exists an image $f(x) \in \mathbb{R}$. Of course, there are many functions that are not defined for *all* real numbers. The next example gives one such function.

Example 3.2

The equation $f(x) = 1/x$ has the graph shown in Fig. 3.6. Here $f(x)$ is not defined at $x = 0$. [5] The set of real numbers on which f is defined is $\mathbb{R} \setminus \{0\}$, and this property of f may be written as $f: \mathbb{R} \setminus \{0\} \to \mathbb{R}$.

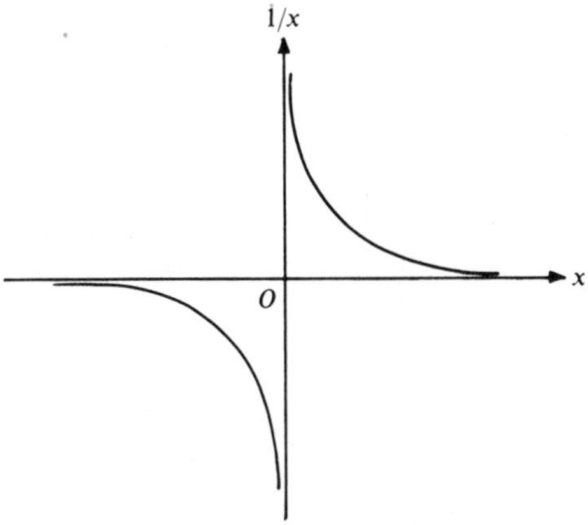

Fig. 3.6 Graph of $f(x) = 1/x$

In order to include such functions in the general theory, it is convenient to write X as the set of all $x \in \mathbb{R}$ for which $f(x)$ is defined. Then $f: X \to \mathbb{R}$ will always imply that the function f is defined at all points $x \in X$, and X is called the *domain* of f. All images $f(x)$ of $x \in X$ are such that $f(x) \in \mathbb{R}$ and $\mathbb{R}$ is a possible *codomain* of f. As seen in Example 3.1, in which $\mathfrak{R}(f) = \mathbb{R}^+ \subset \mathbb{R}$, the range of a function may be a proper subset of a codomain. The range of a function f is the image set of its domain.

The graph of $y = x^3$ shown in Fig. 3.1 is that of the function $f: \mathbb{R} \to \mathbb{R}$ defined by $f(x) = x^3$. Every operation of f relates an element $f(x)$ in the range of f with an element x in the domain of f. These two elements form an ordered pair $(x, f(x))$ and the set of all such ordered pairs, for all elements $x \in \mathbb{R}$, actually defines the function f.[6] Note that the set of ordered pairs $\{(x, x^3): x \in \mathbb{R}\}$ is a proper subset of the cartesian product $\mathbb{R} \times \mathbb{R}$, i.e. of $\mathbb{R}^2$. This leads to the usual idea of expressing an equation $y = f(x)$ $(x, y \in \mathbb{R})$ in terms of its graph, which may be defined by the relation (as defined in Section 2.2):

$$\text{graph}\, f \triangleq \{(x, y) \in \mathbb{R}^2 : y = f(x)\}.$$

The concept of a function f of one variable x, discussed above, can be generalized in a fairly obvious way to a function of several variables. For example, a function $f: X \to \mathbb{R}$, with $X \subseteq \mathbb{R}^2$, assigns to each element $\boldsymbol{x} = (x_1, x_2) \in X$ a unique real number $f(x_1, x_2)$, the image of (x_1, x_2) under f. This function f can be specified either by $(x_1, x_2) \mapsto f(x_1, x_2)$ or, in terms of the vector $\boldsymbol{x} = (x_1, x_2)$, by the equivalent form $\boldsymbol{x} \mapsto f(\boldsymbol{x})$. The function $f: X \to \mathbb{R}$ may be defined as a set of ordered triples $(x_1, x_2, f(x_1, x_2))$, each

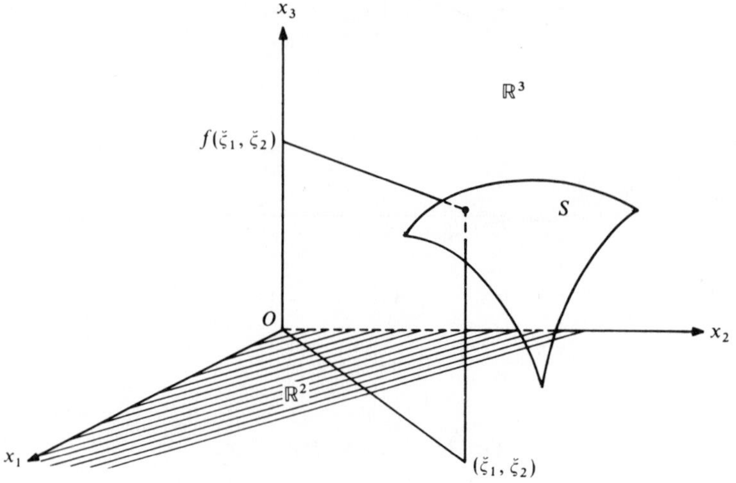

Fig. 3.7 Graph of $f: X \to \mathbb{R}$, with $X \subseteq \mathbb{R}^2$

triple belonging to $\mathbb{R}^3$. This definition again leads to the idea of a graph, in this case a surface S in 3-dimensional space, representing the function f (Fig. 3.7), and

$$\text{graph } f = \{(x_1, x_2, x_3) \in \mathbb{R}^3 : x_3 = f(x_1, x_2)\}.$$

Example 3.3

$f: \mathbb{R}^2 \to \mathbb{R}$ given by $f(x_1, x_2) = x_1^2 + x_2^2$. Notice that the domain of f is the whole of $\mathbb{R}^2$; so $X = \mathbb{R}^2$ in this case. The graph of f is the bowl-shaped

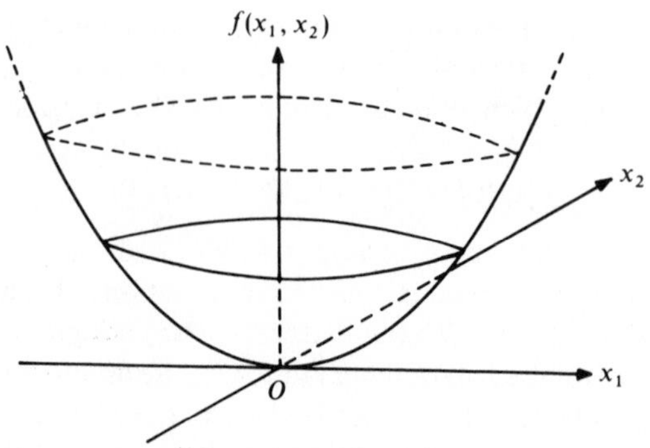

Fig. 3.8 Graph of $f(x_1, x_2) = x_1^2 + x_2^2$

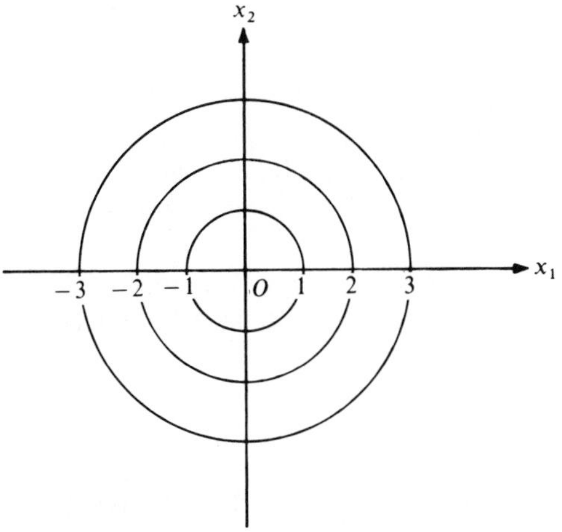

Fig. 3.9 Contours of $f(x_1, x_2) = x_1^2 + x_2^2$ in $\mathbb{R}^2$

surface shown in Fig. 3.8. Such a graph is often represented by contour lines in the (x_1, x_2)-plane as shown in Fig. 3.9.

Example 3.4

The modulus function $f: \mathbb{R}^2 \to \mathbb{R}$ defined by $(a, b) \mapsto |a - b|$. This function f represents the distance between two real numbers a and b.

In general, a function $f: X \to \mathbb{R}$, with $X \subset \mathbb{R}^n$, assigns to each element $\boldsymbol{x} = (x_1, \ldots, x_n) \in \mathbb{R}^n$ a unique real number $f(x_1, \ldots, x_n)$ and is defined by $\boldsymbol{x} \mapsto f(\boldsymbol{x})$ $(\boldsymbol{x} \in X)$, with $f(\boldsymbol{x}) \in \mathbb{R}$.

When the result of an operation f is not a real number, then f is often called a *mapping*. The mapping $\boldsymbol{f}: \mathbb{R}^n \to \mathbb{R}^n$ in (1.27) takes a vector $\boldsymbol{x} \in \mathbb{R}^n$ to a unique vector $\boldsymbol{f}(\boldsymbol{x}) \in \mathbb{R}^n$, which is then equated to the vector $\dot{\boldsymbol{x}}$. Some authors use the term 'vector-valued function' for a mapping that takes an element $\boldsymbol{x}$ of a subset $X \subseteq \mathbb{R}^n$ to an element of $\mathbb{R}^m$, where $m \geqslant 2$.

Example 3.5

$\boldsymbol{f}: \mathbb{R}^n \to \mathbb{R}^m$ defined by $\boldsymbol{f}(\boldsymbol{x}) = \boldsymbol{A}\boldsymbol{x}$, with $\boldsymbol{A} \in \mathbb{R}^{m \times n}$. This mapping maps a vector $\boldsymbol{x} = (x_1, \ldots, x_n) \in \mathbb{R}^n$ to a vector $\boldsymbol{f}(\boldsymbol{x}) = \big(f_1(\boldsymbol{x}), \ldots, f_m(\boldsymbol{x})\big) \in \mathbb{R}^m$. Here

$$f_i(\boldsymbol{x}) = \sum_{j=1}^{n} a_{ij} x_j$$

where a_{ij} $(j = 1, \ldots, n)$ are the elements of the ith row of matrix $\boldsymbol{A}$. The matrix

A is a representation of the mapping f, but is sometimes thought of as the mapping itself.

Single letters such as f, g ,..., F, G ,..., ϕ, ψ ,..., are normally used to denote mappings. In the subject known as functional analysis, the parentheses in $f(x)$ are often omitted and the operation of f acting on x is written as fx (Kreyszig 1978).

Many different types of mappings appear in this book, so it will be useful to consider mappings from a general domain X to a general codomain Y.

3.2 General mappings

A mapping f from a non-empty set X to a non-empty set Y, written as $f\colon X \to Y$ (or occasionally as $X \xrightarrow{f} Y$), assigns to each element $x \in X$ a unique element $f(x) \in Y$, called the image of x under f. The set X is the domain of f, by which is meant that $f(x)$ is defined for every element x in X. The set Y is a codomain of f and, if Y is $\mathbb{R}$ or a subset of $\mathbb{R}$, then the mapping f is again called a function. The image set (or range) of f is the set $\operatorname{im} f$ (sometimes written as $\mathfrak{R}(f)$, $f(X)$, or $X(f)$), defined as

$$\operatorname{im} f \triangleq \{y \in Y\colon y = f(x) \text{ for some } x \in X\}$$

and $\operatorname{im} f \subseteq Y$. As before, the notation $x \mapsto f(x)$ is often used to show the effect of f on an element $x \in X$.

The transfer function $s \mapsto g(s)$ of Example 1.5 represents a mapping $f\colon U \to Y$ from an input space U to an output space Y, since (1.9) is equivalent to

$$\bar{y}(s) = g(s)\bar{u}(s).$$

Then the mapping $f\colon U \to Y$ can be defined by $\bar{u}(s) \mapsto g(s)\bar{u}(s)$.

Two mappings f and g are defined to be equal, written $f = g$, if they have the same domain, the same codomain and, for all x in the common domain, $f(x) = g(x)$.

Example 3.6

$f\colon \mathbb{R} \to \mathbb{R}^+$ defined by $f(x) = |x|$ and $g\colon \mathbb{R} \to \mathbb{R}^+$ defined by $g(x) = \sqrt{(x^2)}$ are two functions which are equal, i.e. $f = g$.

Example 3.7

Let $f\colon \mathbb{R} \to \mathbb{R}^+$ be defined by $f(x) = x^2$ and $g\colon \mathbb{R}^+ \to \mathbb{R}^+$ be defined by $g(x) = x^2$. These two functions f and g are not equal, because their domains $\mathbb{R}$ and $\mathbb{R}^+$ are different.

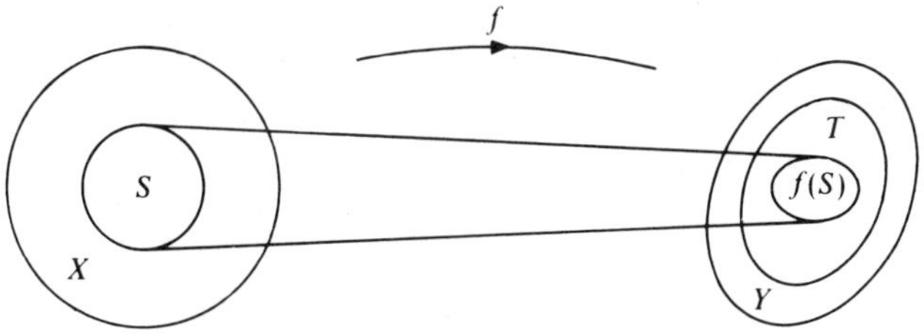

Fig. 3.10 Restriction of f to $S \subset X$

In Example 3.7, the function g is called the *restriction* of f to $\mathbb{R}^+$ and is written $f \upharpoonright \mathbb{R}^+$. This idea of a restriction of a function is useful for general mappings. Given a mapping $f: X \to Y$, consider the operation of f on a proper subset S of X such that $f(S) \subseteq T$, with $T \subseteq Y$ (Fig. 3.10). Then a mapping $g: S \to T$ such that $g(x) = f(x)$ for all $x \in S$ is the restriction of f to S as domain and T as codomain. In the case where the codomains are the same, i.e. $T = Y$, then the restriction of f to S is written $f \upharpoonright S$. Similarly, a mapping $f: X \to Y$ is called an *extension* of a mapping $g: S \to T$ if g is the restriction of f to S as domain and T as codomain.

The interpretation of a function in terms of its graph can be generalized to an arbitrary mapping $f: X \to Y$. The graph of such a map may be defined as

$$\text{graph}\, f \triangleq \{(x, y): x \in X, y = f(x)\}$$

and is a binary relation (Section 2.2) from X to Y.

As already illustrated, the image set $\text{im}\, f$ of a function $f: X \to \mathbb{R}$, with $X \subseteq \mathbb{R}$, may be the whole codomain $\mathbb{R}$ (as in $f(x) = x^3$) or a proper subset of $\mathbb{R}$ (as in $f(x) = x^2$). These and other important differences between functions also occur between general mappings and these considerations lead to several special types of mappings.

3.3 Special mappings

A mapping $f: X \to Y$ is said to be an 'into' mapping if $\text{im}\, f$ is a proper subset of the set Y (Fig. 3.11). The importance of an 'into' mapping is that not all

Fig. 3.11 An 'into' mapping

points of set Y correspond to points of X under the mapping f. Here the codomain Y must be part of the definition of the 'into' mapping and the range of f is a proper subset of codomain Y.

Example 3.8

$f:\mathbb{Z} \to \mathbb{Z}$ defined by $n \mapsto n^2$ is an 'into' mapping. Integers such as 2, 3, 5, 6, ... are not elements of the image set of f. The set im f contains only integers that can be expressed as n^2 for $n \in \mathbb{Z}$, namely 0, 1, 4, 9, 16, ...

A mapping $f:X \to Y$ is said to be an '*onto*' mapping (or *surjective*) if $\text{im}\, f = Y$ (Fig. 3.12). Here every point of Y corresponds to at least one point of set X under the mapping f. Again, with an 'onto' mapping f, the codomain Y must be part of the definition of the mapping.

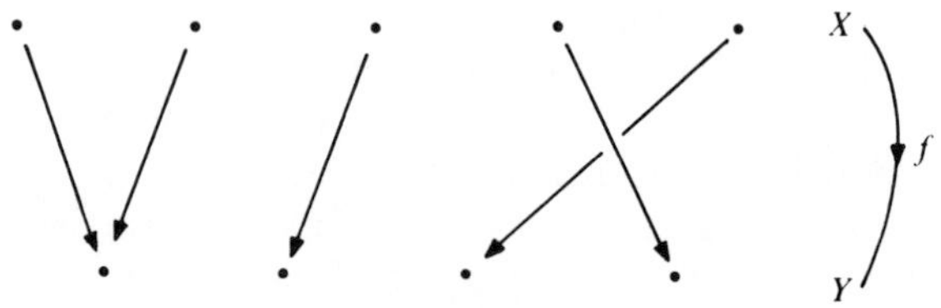

Fig. 3.12 An 'onto' mapping (a surjection)

Example 3.9

$f:\mathbb{R} \to \mathbb{R}^+$ defined by $x \mapsto x^2$ is an 'onto' mapping because $\text{im}\, f = \mathbb{R}^+$.

A mapping $f:X \to Y$ is said to be '*one-to-one* (1–1) (or *injective*) if, for $x_1, x_2 \in X$,

$$x_1 \neq x_2 \quad \Rightarrow \quad f(x_1) \neq f(x_2)$$

(Fig. 3.13). Here, no point of Y corresponds to more than one point of X. It

Fig. 3.13 A '1–1 mapping' (an injection)

excludes the behaviour, exhibited in Fig. 3.11 and Fig. 3.12, in which two different elements of X are mapped onto the same element in Y.

Example 3.10

$f:\mathbb{Z} \to \mathbb{Z}$ defined by $n \mapsto 2n$ is an injective mapping. For $n_1, n_2 \in \mathbb{Z}$, we have that $n_1 \neq n_2 \Rightarrow 2n_1 \neq 2n_2$, so that f is 1–1. The image set of f includes only the even integers and so is a proper subset of $\mathbb{Z}$.

A mapping $f:X \to Y$ is said to be '*one-to-one and onto*' (or *bijective*) if f is both injective and surjective (Fig. 3.14). This is the most attractive mapping available. To each element of X there corresponds one and only one element of Y (all mappings are single-valued) and to each element of Y corresponds one and only one element of X.

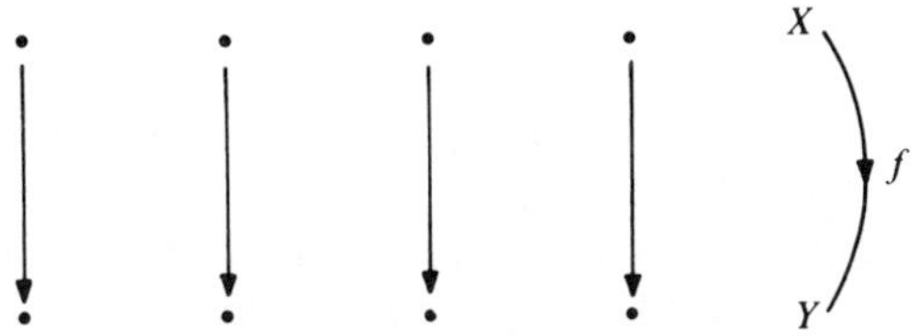

Fig. 3.14 A '1–1 and onto' mapping (a bijection)

Example 3.11

$f:\mathbb{Z} \to \mathbb{Z}$ defined by $n \mapsto -n$ is a bijection, because (a) $n_1 \neq n_2$ implies that $-n_1 \neq -n_2$ and so f is injective and (b) any $-n \in \mathbb{Z}$ has a corresponding $n \in \mathbb{Z}$ and so f is also surjective.

Example 3.12

$f:\mathbb{Z} \to \mathbb{Z}$ given by $n \mapsto 2n$ is an injective mapping, but is not bijective because it is not surjective (Example 3.10). However, the function $g:\mathbb{N} \to E$, given by $g(n) = 2n$, where E is the set of all even positive integers, is surjective as well as being injective; g is therefore bijective. Note that $g \neq f$ because g and f have different domains.

Example 3.13

$f:\mathbb{R} \to \mathbb{R}$ given by $x \mapsto x$ is known as the identity function from $\mathbb{R}$ to $\mathbb{R}$ and is denoted by $\mathrm{id}_{\mathbb{R}}$. This identity function is clearly bijective; since it is defined by $f(x) = x$, the graph of f is as shown in Fig. 3.15.

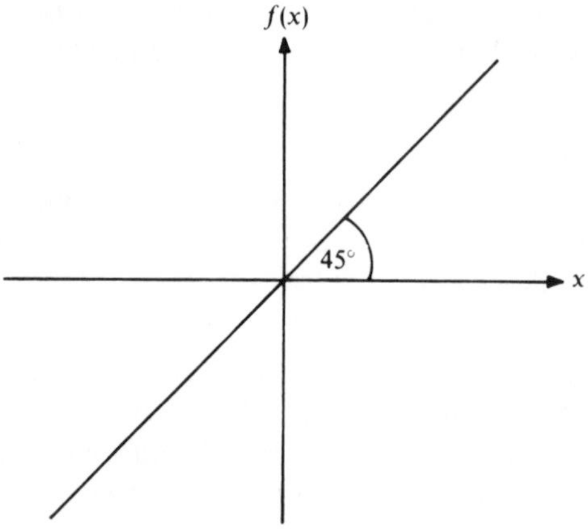

Fig. 3.15 Identity function

In general, the identity mapping $\mathrm{id}_X : X \to X$ from a non-empty set X to itself is defined for all $x \in X$ as $\mathrm{id}_X(x) = x$. Very often the letter I is used in place of id_X in the case of operator-style notation: $Ix = x$. Such a case arises when $\boldsymbol{x} \in \mathbb{R}^n$ and I is represented by the $n \times n$ unit matrix $\mathbf{I}_n$.

The idea of a bijective mapping is used to determine whether two given sets A and B are equivalent (cf. Section 2.2). Now, two sets A and B are said to be equivalent, written $A \sim B$, if there exists a bijective mapping $f : A \to B$. Since a bijective mapping is 1–1 and onto, this means that a one-to-one correspondence can be established between the elements of the equivalent sets A and B. A set is finite if it is either the empty set $\varnothing$ or equivalent to the set $\{1, \ldots, n\}$ for some $n \in \mathbb{N}$; all other sets are infinite. A set equivalent to $\mathbb{N}$ is called *countably infinite* (or *denumerable*). A set is *countable* if it is either finite or countably infinite. Alternatively, a set is countable if its elements can be arranged as the terms of a sequence (with possibly repeated elements), as in Exercise 8.

Example 3.14

Let E be the set of all even positive integers so that

$$E = \{n : n \in \mathbb{N}, n \text{ even}\}.$$

The function $g : \mathbb{N} \to E$ defined by $n \mapsto 2n$ is bijective by Example 3.12, and so E is a countable set.

If (1.27) is expressed in the form $\dot{\boldsymbol{x}}(t) = \boldsymbol{f}(\boldsymbol{x}(t))$, it is seen that the right-hand side of this equation involves two mappings, $\boldsymbol{x} : \mathbb{R} \to \mathbb{R}^n$ defined by

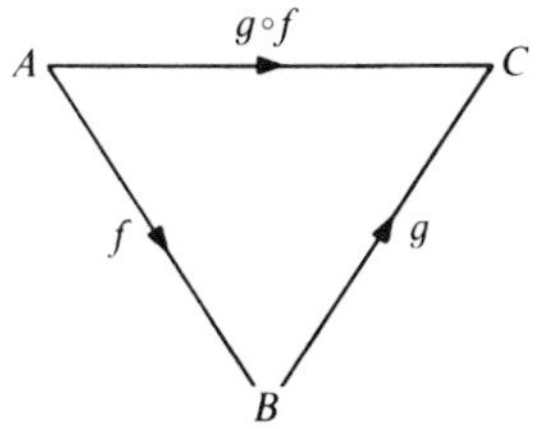

Fig. 3.16 Commutative diagram

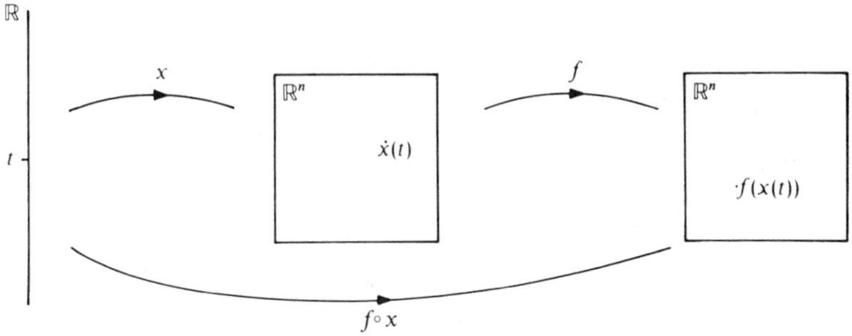

Fig. 3.17 Composite mapping

$t \mapsto \boldsymbol{x}(t)$ and $\boldsymbol{f}: \mathbb{R}^n \to \mathbb{R}^n$-given by $\boldsymbol{x} \mapsto \boldsymbol{f}(\boldsymbol{x})$. This is an example of a composite mapping which is the operation of two (or more) mappings; first $f: A \to B$ defined by $x \mapsto f(x)$ followed by $g: B \to C$ defined by $y \mapsto g(y)$. This composite mapping is sometimes written as gf with operator-style notation, i.e. $(gf)x = gfx \triangleq gy$, where $y = fx$, but more often as $g \circ f$ with parenthetic notation, i.e. $(g \circ f)(x) \triangleq g(f(x))$. Notice that the application of the two mappings is in the reverse order to that given in the notation $g \circ f$. In other words, f is applied first and then g. Clearly the composite mapping $(g \circ f): A \to C$ is defined by $x \mapsto g(f(x))$. It can be represented by the diagram shown in Fig. 3.16, called a *commutative diagram* because the direct path from A to C represents the same mapping as the indirect path from A to C via B. Figure 3.17 illustrates the composite mapping $(\boldsymbol{f} \circ \boldsymbol{x}): \mathbb{R} \to \mathbb{R}^n$ which was defined above as $t \mapsto \boldsymbol{f}(\boldsymbol{x}(t))$.

Example 3.15

$\sin x^2$ is the value of the composite function $g \circ f$ in which $f: \mathbb{R} \to \mathbb{R}^+$ is the squaring function defined by $x \mapsto x^2$ and $g: \mathbb{R}^+ \to [-1, 1]$ is the sine function defined by $y \mapsto \sin y$.

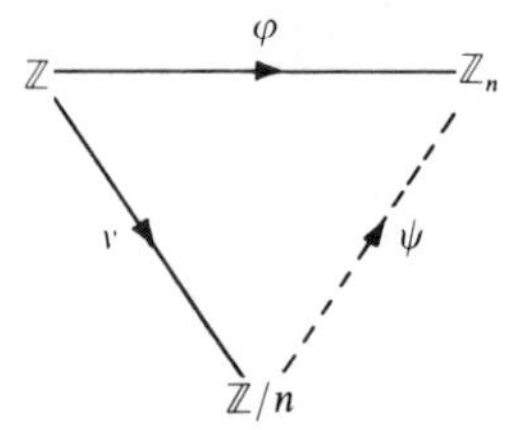

Fig. 3.18 $\psi \circ \nu = \varphi$

There are certain technicalities concerning the image of f and the domain of g in order for the composite map $g \circ f$ to be well defined. However, these are of little consequence from a practical point of view. [7]

The composition of mappings is an *associative* operation. Suppose that we are given the mappings $f: A \to B$, $g: B \to C$, and $h: C \to D$. Then, provided that all necessary composite mappings exist, the associative property of mapping composition is that $f \circ (g \circ h) = (f \circ g) \circ h$.

Consider the set $\{[0], \ldots, [n-1]\}$ of cosets of integers modulo n associated with $\mathbb{Z}_n = \{0, \ldots, n-1\}$ as discussed in Section 2.3. Instead of denoting the set of cosets also by $\mathbb{Z}_n$, suppose we denote it by $\mathbb{Z}/n$. A mapping $\varphi: \mathbb{Z} \to \mathbb{Z}_n$ can be defined by $k \mapsto \varphi(k)$, where $\varphi(k)$ is the remainder on dividing k by n. For example, if $n = 6$ and $k = 29$, then $29 = 4 \times 6 + 5$ and $5 \in \mathbb{Z}_6$. The mapping φ is surjective. A mapping $\nu: \mathbb{Z} \to \mathbb{Z}/n$ can also be defined by $k \mapsto [r]$, where $[r] = n\mathbb{Z} + r \triangleq \{k : k = nq + r, q \in \mathbb{Z}\}$. With $n = 6$, the integers 29, 35, 41, ... all map to [5] under ν. There is then a 1–1 and onto mapping between the remainders, modulo n, in $\mathbb{Z}_n$ and the cosets, modulo n, in $\mathbb{Z}/n$. Thus, there exists a further bijective mapping $\psi: \mathbb{Z}/n \to \mathbb{Z}_n$, defined by $[r] \to r$ such that $\psi \circ \nu = \varphi$.

This is illustrated by the commutative diagram shown in Fig. 3.18.

3.4 Inverse mappings

A given dynamical input–output system is said to be *invertible* if the output uniquely determines the input (Silverman 1969, Hirschorn 1979). The practical problem associated with invertibility of systems is to determine the input of a system when only the output is known. Problems of this type are important in many areas of application such as decoding (Hill 1986). The transfer-function matrix $\boldsymbol{G}(s)$ in (1.14) of a system Σ represents a mapping $f: U \to Y$, defined by (1.13), from an input space U to an output space Y, the range of f. If one exists, an inverse system Σ_1 is essentially represented by a further mapping $\boldsymbol{g}: Y \to U$ which maps the output of Σ to the input of Σ (Fig. 3.19). [8] With the input $\bar{\boldsymbol{u}}$ of Σ unknown, the output $\bar{\boldsymbol{y}}$ is observed and fed

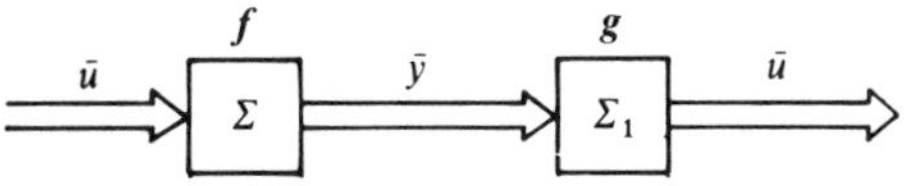

Fig. 3.19 System inverse

as input into Σ_1. Thus, a system Σ_1 is to be found such that its output is guaranteed to be the known input $\bar{\boldsymbol{u}}$ of Σ. The problem is somewhat more complicated than described here, but the essential fact is that the mapping $\boldsymbol{g}$ above clearly reverses the action of mapping $\boldsymbol{f}$ and is called the *inverse mapping* of $\boldsymbol{f}$. Having seen how inverse mappings can arise in practice, we are ready to look at the underlying mathematical theory.

Consider the function $f:\mathbb{R} \to \mathbb{R}$ defined by the equation $y = mx + c$, with $m,c \in \mathbb{R}$. Solving this equation for x yields $x = (y - c)/m$, and this new equation also defines a function $g:\mathbb{R} \to \mathbb{R}$ by $g(y) = (y - c)/m$, provided that $m \neq 0$. When this function g exists, it is called the *inverse function* of f. It is usual to denote this inverse function by f^{-1} rather than by some other letter like g.

An inverse function f^{-1} does not exist for the function $f:\mathbb{R} \to \mathbb{R}^+$ defined by $y = x^2$ since solving for x gives $x = \pm\sqrt{y}$ and so there are always two image points $\sqrt{y}$ and $-\sqrt{y}$ for every $y > 0$. This violates the single-valued property of a function (see Section 3.1). However, the two squaring functions $f:\mathbb{R}^+ \to \mathbb{R}^+$ and $g:\mathbb{R}^- \to \mathbb{R}^+$, each defined by $y = x^2$, do yield inverse functions, namely $f^{-1}:\mathbb{R}^+ \to \mathbb{R}^+$ defined by $x = \sqrt{y}$ and $g^{-1}:\mathbb{R}^+ \to \mathbb{R}^-$ defined by $x = -\sqrt{y}$ respectively (Fig. 3.20).

From the above examples, it is clear that not all functions give rise to inverse functions. Conditions are required under which such inverses exist. Given a function $f:A \to \mathbb{R}$ defined by $y = f(x)$, where $A \subseteq \mathbb{R}$, when is it possible to define a function $f^{-1}:B \to A$, given by $x = f^{-1}(y)$, in which $B \subseteq \mathbb{R}$? To answer this question, consider again the graph of the function $f:\mathbb{R} \to \mathbb{R}^+$ defined by $y = f(x) = x^2$ (Fig. 3.21(a)). The inverse of graph 3.21(a) is obtained by rotating the x and y axes and the graph through 90 degrees in an anticlockwise direction about the origin O, and then reflecting the graph and the y axis in the x axis.[9] The resulting graph in Fig. 3.21(b) does not represent a function because, for each $y > 0$, there exist two different values of x (i.e. the image of any $y > 0$ is not unique). The clue to resolving the difficulty lies in noting that, for any $c > 0$, the line $y = c$ cuts the graph in Fig. 3.21(b) in more than one point (actually two). It is this which leads to the nonuniqueness of the inverse x for any $y > 0$. If the domain of f is restricted to $(-\infty, 0]$ or $[0, \infty)$, then the restrictions $f\restriction\mathbb{R}^+$ and $f\restriction\mathbb{R}^-$, shown as f and g respectively in Fig. 3.20(a, b), both have inverses (Fig. 3.20(c, d)).

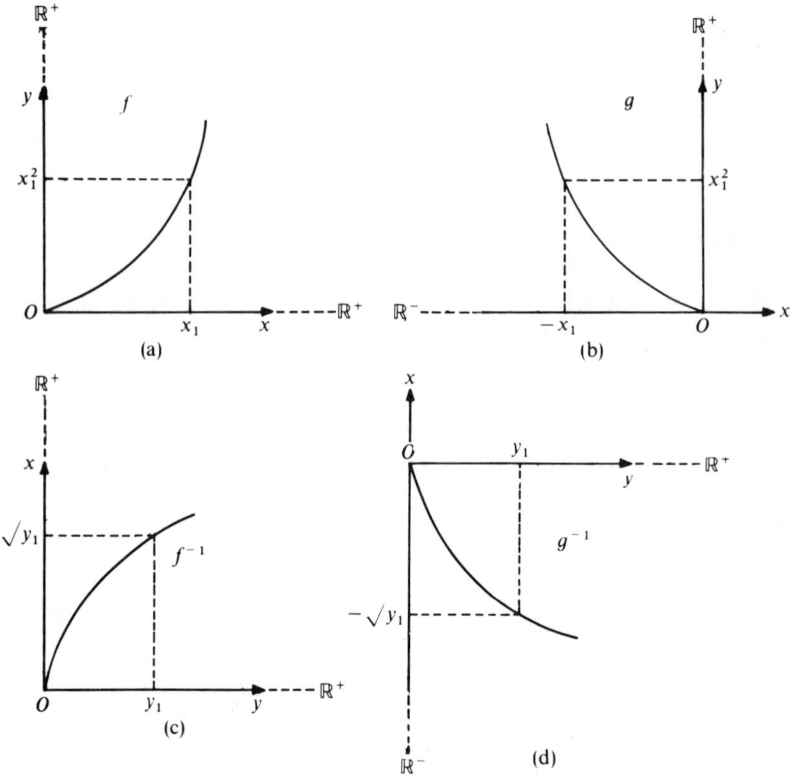

Fig. 3.20 (a) $y = x^2$: $x_1 \mapsto x_1^2$, (b) $y = x^2$: $-x_1 \mapsto x_1^2$, (c) $x = \sqrt{y}$: $y_1 \mapsto \sqrt{y_1}$, (d) $x = -\sqrt{y}$: $y_1 \mapsto -\sqrt{y_1}$,

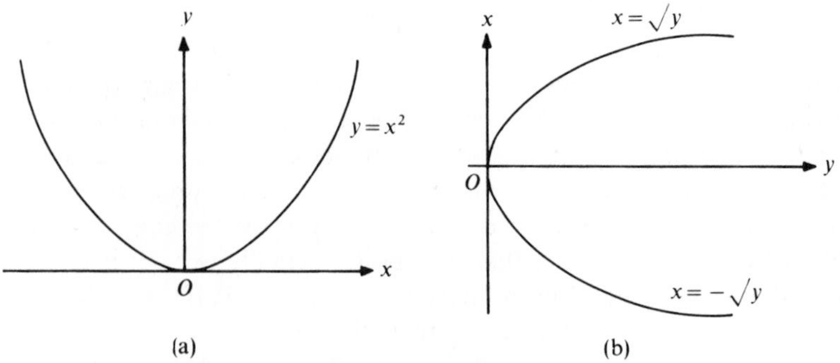

Fig. 3.21 (a) $x \mapsto x^2$, (b) $y \mapsto \{\sqrt{y}, -\sqrt{y}\}$

Note that the function $f: A \to \mathbb{R}$ need not be surjective in order to have an inverse but it must be 1–1. Thus, injectivity is sufficient to define the inverse, i.e. when the domain of f^{-1} is the range of f. For example, the identity function $f:[0,1] \to \mathbb{R}$ defined by $x \mapsto x$ is injective and has inverse $f^{-1}:[0,1] \to [0,1]$. What is important in finding the inverse of an injective function is the *range* of f rather than its *codomain*. In particular, a bijective function always has an inverse.

Certain functions that have special properties automatically have inverses. For example, if a function f is continuous (no breaks in its graph) and monotonic (either increasing or decreasing) [10] in an interval $[a, b]$, then the inverse function f^{-1} exists and is itself continuous and monotonic in the interval $[f(a), f(b)]$. Note that the condition of continuity is just as important as monotonicity. For example, the graph representing $f(x) = \tan x$ (Fig. 3.22) is monotonic increasing everywhere except at points where the graph exhibits infinite discontinuities (e.g. at $-\frac{1}{2}\pi$ and $\frac{1}{2}\pi$, Fig. 3.22) but the inverse function f^{-1} does not exist.

The above discussion has been restricted to functions of one variable. However, the basic ideas of inverse functions can be carried over to general mappings. As above, when a mapping $f: X \to Y$, defined by $x \mapsto f(x)$, is bijective, then there always exists an inverse mapping $f^{-1}: Y \to X$, given by $x = f^{-1}(y)$. This mapping f^{-1} undoes what the mapping f did, and the composite mappings $f^{-1} \circ f$ and $f \circ f^{-1}$ are the identity maps id_X and id_Y respectively.

Again, with a mapping $f: X \to Y$ which is injective but not bijective, it is still possible to define the inverse mapping $f^{-1}: T \to X$, with $T \subset \Re(f)$, such

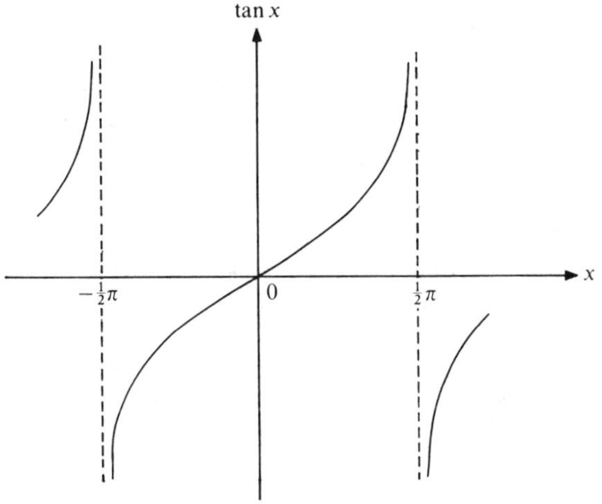

Fig. 3.22 Graph of $f(x) = \tan x$

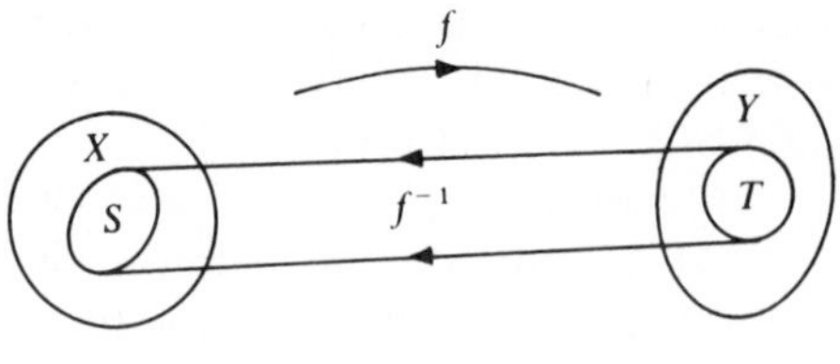

Fig. 3.23 Inverse image of T under f

that $t \in T$ is mapped onto $x \in S \subseteq X$ for which $f(x) = t$ (Fig. 3.23). The set S is the inverse image $f^{-1}(T)$, given by

$$S = f^{-1}(T) = \{x \in X : f(x) \in T\}.$$

Example 3.16

$f:\mathbb{R} \to \mathbb{R}$, defined by $y = f(x) = x^3$, is a bijective function. The inverse function $f^{-1}:\mathbb{R} \to \mathbb{R}$ exists and is defined by $f^{-1}(y) = y^{\frac{1}{3}}$, where $y^{\frac{1}{3}}$ is the unique real root of the equation $x^3 = y$.

Inverses of composite mappings can also be defined under certain conditions. In fact,

$$(g \circ f)^{-1} = f^{-1} \circ g^{-1}, \tag{3.1}$$

provided that the mappings f and g are bijective and that both composite mappings in (3.1) are defined. Note that, if both f and g are bijective, then so is $g \circ f$ (when it exists).

Consider a linear (see Section 6.1) mapping $\boldsymbol{f}:\mathbb{R}^n \to \mathbb{R}^n$ defined by the equation $\boldsymbol{x} \mapsto \boldsymbol{y} = \boldsymbol{A}\boldsymbol{x}$, in which the $n \times n$ matrix $\boldsymbol{A}$ is nonsingular ($\det \boldsymbol{A} \neq 0$). For a given vector $\boldsymbol{y} \in \mathbb{R}^n$, the usual problem arising in practice is to solve the set of n equations, given by $\boldsymbol{A}\boldsymbol{x} = \boldsymbol{y}$, for $\boldsymbol{x}$ in terms of the known vector $\boldsymbol{y}$. The unique solution is $\boldsymbol{x} = \boldsymbol{A}^{-1}\boldsymbol{y}$, where the inverse matrix $\boldsymbol{A}^{-1}$ exists because $\boldsymbol{A}$ is nonsingular. The equation $\boldsymbol{x} = \boldsymbol{A}^{-1}\boldsymbol{y}$ defines the inverse mapping $\boldsymbol{f}^{-1}:\mathbb{R}^n \to \mathbb{R}^n$ by $\boldsymbol{y} \mapsto \boldsymbol{A}^{-1}\boldsymbol{y}$.

Things are not always as straightforward. An obvious difficulty arises when the matrix $\boldsymbol{A}$ is singular ($\det \boldsymbol{A} = 0$). In this case, suppose the vector $\boldsymbol{y}$ is the null vector, so that the equation to be solved is $\boldsymbol{A}\boldsymbol{x} = \boldsymbol{0}$. Since $\boldsymbol{A}$ is a singular matrix, there are an infinite number of solutions for $\boldsymbol{x}$. The set $\{\boldsymbol{x}:\boldsymbol{A}\boldsymbol{x} = \boldsymbol{0}\}$ of all these solutions is called the *nullspace* or *kernel* of the mapping $\boldsymbol{f}$ (or $\boldsymbol{A}$), written $\ker \boldsymbol{f}$. The mapping $\boldsymbol{g}:\ker \boldsymbol{f} \to \{\boldsymbol{0}\}$ is defined by $\boldsymbol{x} \mapsto \boldsymbol{0}$. The inverse mapping $\boldsymbol{g}^{-1}$ does not exist, since there are an infinite number of vectors $\boldsymbol{x} \in \ker \boldsymbol{f}$ corresponding to the null vector $\boldsymbol{0}$.

Likewise, if $\boldsymbol{x}_0$ is one solution of the equation $\boldsymbol{A}\boldsymbol{x} = \boldsymbol{y}$ for general $\boldsymbol{y}$, and K is the kernel of $\boldsymbol{f}$, with $\boldsymbol{f}$ defined by $\boldsymbol{x} \mapsto \boldsymbol{A}\boldsymbol{x}$, then $K + \boldsymbol{x}_0$ is the set of all solutions of the equation $\boldsymbol{A}\boldsymbol{x} = \boldsymbol{y}$. To prove this, note that $\boldsymbol{k} + \boldsymbol{x}_0 \in K + \boldsymbol{x}_0$ and

$$\boldsymbol{A}(\boldsymbol{k} + \boldsymbol{x}_0) = \boldsymbol{A}\boldsymbol{k} + \boldsymbol{A}\boldsymbol{x}_0 = \boldsymbol{0} + \boldsymbol{y} = \boldsymbol{y}$$

for any $\boldsymbol{k} \in K$, so that any element of $K + \boldsymbol{x}_0$ is a solution of $\boldsymbol{A}\boldsymbol{x} = \boldsymbol{y}$. Furthermore, if $\boldsymbol{A}\boldsymbol{x} = \boldsymbol{y}$, then

$$\boldsymbol{A}(\boldsymbol{x} - \boldsymbol{x}_0) = \boldsymbol{A}\boldsymbol{x} - \boldsymbol{A}\boldsymbol{x}_0 = \boldsymbol{y} - \boldsymbol{y} = \boldsymbol{0},$$

so that $\boldsymbol{x} - \boldsymbol{x}_0 \in K$, whence $\boldsymbol{x} \in K + \boldsymbol{x}_0$. Thus, every solution of $\boldsymbol{A}\boldsymbol{x} = \boldsymbol{y}$ belongs to $K + \boldsymbol{x}_0$.[11]

Example 3.17

$f: \mathbb{R}^3 \to \mathbb{R}^3$ is defined by $\boldsymbol{x} \mapsto \boldsymbol{A}\boldsymbol{x}$, where

$$\boldsymbol{A} = \begin{bmatrix} 1 & 0 & 1 \\ 5 & 4 & 9 \\ 2 & 4 & 6 \end{bmatrix}.$$

The rank of $\boldsymbol{A}$ is 2, and so $\det \boldsymbol{A} = 0$. It is not difficult to show that any solution of the equation $\boldsymbol{A}\boldsymbol{x} = \boldsymbol{0}$ is given by $(\alpha, \alpha, -\alpha)$, with α arbitrary, i.e.

$$\ker f = \{\boldsymbol{x} \in \mathbb{R}^3 : \boldsymbol{x} = (\alpha, \alpha, -\alpha), \alpha \in \mathbb{R}\}.$$

Clearly $\ker f$ is the straight line in $\mathbb{R}^3$ which passes through the origin (0, 0, 0) and the point (1, 1, − 1). It is a subspace of the space $\mathbb{R}^3$ (see Section 6.1 for definition of a subspace).

Exercises

1. Write down the largest domain of the function f defined by $f(x) = 3x^2/(x^2 - 3x + 2)$ for $x \in \mathbb{R}$.
2. A map $f: \mathbb{R} \to \mathbb{R}$ is defined by $f(t) = t^2 - 1$. Find the image of 3 under f, the image set under f of $[-1, 2]$ and the set im f. Is f surjective?
3. $f: \mathbb{R} \to \mathbb{R}$ is defined by $x \mapsto \sin x$. Explain why f is neither surjective nor injective. How can f be made a surjective mapping? How can a bijective restriction be obtained from f?
4. A map $\phi: \mathbb{R}^{n \times n} \to \mathbb{R}$ is defined by $\phi(\boldsymbol{A}) = \det \boldsymbol{A}$. Show that ϕ is surjective but not injective, and describe the inverse set $\phi^{-1}(\mathbb{R} \backslash \{\boldsymbol{0}\})$.
5. If $f: X \to Y$ is a mapping from X to Y, prove that

$$x \sim y \quad \Leftrightarrow \quad f(x) = f(y)$$

is an equivalence relation on X.

6. Given a mapping $f:X \to Y$, prove that f is injective iff

$$f(A) \cap f(X \backslash A) = \emptyset \quad \forall\, A \subseteq X.$$

7. Show that the function α mapping the set of all equivalence classes of Example 2.15 to $\mathbb{R}$, defined by $[x] \mapsto \dot{x}(0)$, is a bijection.

8. Show that the set S of rational numbers in the interval $[0\,,1]$ is countable.

9. A mapping $f:\mathbb{R} \to \mathbb{R}$ is defined by $t \mapsto t-1$, and another mapping $g:\mathbb{R} \to \mathbb{R}$ is defined by $t \mapsto \sin t$. Find formulae specifying the composite maps $g \circ f$ and $f \circ g$.

10. Show that the function $f:\mathbb{R} \to \mathbb{R}$ given by $f(x) = \sin x$ does not have an inverse function f^{-1}, but that the inverse function g^{-1} does exist for the function $g:[-\frac{1}{2}\pi\,,\frac{1}{2}\pi] \to \mathbb{R}$ defined by $x \mapsto \sin x$.

Notes for Chapter 3

1. The words 'function' and 'mapping' are synonymous; other equivalent terms often used are 'operator' and 'transformation'. In many texts, the term 'function' is restricted to the case where $f(x) \in \mathbb{R}$.
2. Notice that, by excluding from the graph in Fig. 3.4 that portion lying below the x axis, a function can be defined because there then corresponds only one value $\sqrt{x}$ for each nonnegative x (see also Section 3.4).
3. $\mathbb{R}^+$ is defined as the semi-infinite interval $[0\,,\infty)$. Notice the open nature of this interval at the infinite end in order to avoid dealings with infinity itself (see Note 5 below). $\mathbb{R}^-$ is similarly defined as the semi-infinite interval $(-\infty\,,0]$.
4. Remember (Note 1.11) sign $\triangleq$ indicates a definition, so that im f is defined as the set of all images $\{f(x):x \in \mathbb{R}\}$.
5. The 'division' of any nonzero real number by zero can be thought of as having a meaning 'infinity' in a limited context, and there are topics in mathematics where 'points at infinity' is a very useful concept. The only undefined operations are then $0/0$, ∞/∞, and $\infty \times 0$. However, in this book $1/x$ is defined for any nonzero real number as small or as large as desired, but it is not defined for the value $x = 0$. A statement such as '$1/x = \infty$ when $x = 0$' will not be used, although '$1/x$ tends to infinity (written as $1/x \to \infty$) as x tends to zero (written $x \to 0$)' is acceptable. Many of the concepts introduced in this book are solely to keep infinity (∞) at arm's length.
6. By defining a function f in this way, codomains do not form part of the definition of a function, and f can be written as $f:X \to \Re(f)$. Nevertheless, this definition will not always be used in this book, because the idea of a codomain for a function f containing a proper subset $\Re(f)$ is useful in describing certain special types of function (see Section 3.3). In particular, if the properties 'into' and 'onto' are defined as properties of the mapping, then the codomain must be part of the definition of the mapping. The codomain is then also part of the definition of equality between functions (see Section 3.2).
7. Scientists and engineers adopt the convention that the domain of a function is always as large as possible and then never mention the word 'domain' again. For

example,

$$\begin{aligned} f(x) &= 1/(x-1) && \text{domain } \mathbb{R}\backslash\{1\} \\ g(x) &= \sqrt{x} && \text{domain } \mathbb{R}^+ \\ (f\circ g)(x) &= 1/(\sqrt{x}-1) && \text{domain } \mathbb{R}^+\backslash\{1\} \\ (g\circ f)(x) &= \sqrt{(1/(x-1))} && \text{domain } \{x : x > 1\}. \end{aligned}$$

8. The use of thick arrows for inputs and outputs in a figure indicates a multi-input/multi-output system.
9. Alternatively, reflect the graph in the line $y = x$.
10. A function $f: X \to \mathbb{R}$, with $X \subseteq \mathbb{R}$, defined by $x \mapsto f(x)$ is called a monotonic increasing function if, throughout the domain X, any increase in x always produces an increase in $f(x)$. Similarly, a monotonic decreasing function is one for which any increase in x always produces a decrease in $f(x)$.
11. Note that the kernel of $\boldsymbol{f}$ ($K = \ker \boldsymbol{f}$) is a space; in fact a subspace of $\mathbb{R}^n$ (subspace is defined in Section 6.1). However, $K + \boldsymbol{x}_0$ is not a space but a hyperplane.

4

SEMIGROUPS AND GROUPS

The examples of Chapter 1 illustrate the importance of certain sets of mathematical objects such as the set $\mathbb{R}$ of all real numbers (Fig. 1.4) and the set of all polynomials with coefficients in $\mathbb{R}$ (Example 1.3). In the present chapter, some structure will be put on such sets in the form of a single operation (e.g. addition or multiplication of the elements). The rules governing such an operation will differ according to the nature of the elements of the sets. For example, multiplication on $\mathbb{R}$ is commutative (i.e. $ab = ba$ for all $a,b \in \mathbb{R}$) whereas on $\mathbb{R}^{n \times n}$ $(n > 1)$ it is not (there exist $\boldsymbol{A},\boldsymbol{B} \in \mathbb{R}^{n \times n}$ such that $\boldsymbol{AB} \neq \boldsymbol{BA}$). It is also important to realize that the names of operations such as 'multiplication' can describe quite different operations (e.g. ordinary multiplication ab in $\mathbb{R}$, the vector product $\boldsymbol{u} \times \boldsymbol{v}$ in $\mathbb{R}^3$, or the composition $g \circ f$ of two functions f and g).

4.1 Binary operations

Consider a set S which may be finite or infinite. An operation (in general denoted by $\circ$) which takes any two elements $a,b \in S$ and transforms them to another element $c = a \circ b$, which is not necessarily in S, is called a *binary operation* because it operates on two elements of S. Such an operation is said to be *closed* if, for all ordered pairs $(a, b) \in S \times S$, the element $a \circ b$ also belongs to S. Thus, a closed binary operation is a mapping $\phi: S \times S \to S$ defined by $\phi(a, b) = a \circ b$. A non-empty set S on which is defined a binary operation is called an *algebraic system* and denoted by $(S, \circ)$. Often the set S itself is referred to as an algebraic system, provided that it is clear which binary operation is being used.

Example 4.1

Multiplication on $\mathbb{R}$ is a closed binary operation: if $a,b \in \mathbb{R}$ then $ab \in \mathbb{R}$.[1] As mentioned above, the commutative law of multiplication is valid in this case. Multiplication on $\mathbb{R}^{n \times n}$ is also a closed binary operation, since $\boldsymbol{A},\boldsymbol{B} \in \mathbb{R}^{n \times n} \Rightarrow \boldsymbol{AB} \in \mathbb{R}^{n \times n}$. However, the commutative law is not valid on $\mathbb{R}^{n \times n}$ for $n > 1$.

It is sometimes useful to express a binary operation on elements of a finite set S in tabular form as illustrated in the next example.

Example 4.2

$S = \{\pm \boldsymbol{I}, \pm \boldsymbol{J}, \pm \boldsymbol{K}, \pm \boldsymbol{L}\}$ where

$$\boldsymbol{I} = \begin{bmatrix} 1 & 0 \\ 0 & 1 \end{bmatrix}, \quad \boldsymbol{J} = \begin{bmatrix} \mathrm{j} & 0 \\ 0 & -\mathrm{j} \end{bmatrix}, \quad \boldsymbol{K} = \begin{bmatrix} 0 & 1 \\ -1 & 0 \end{bmatrix}, \quad \boldsymbol{L} = \begin{bmatrix} 0 & \mathrm{j} \\ \mathrm{j} & 0 \end{bmatrix},$$

and $\mathrm{j} = \sqrt{(-1)}$. If multiplication of matrices on $\mathbb{C}^{2 \times 2}$ (Note 1.14) is chosen as the binary operation on S, then the products of pairs of elements from S can be presented in a multiplication table—see Table 4.1—to be interpreted in the following way. Any element in the far left-hand column (say $\boldsymbol{K}$) postmultiplied by any element in the top row (say $\boldsymbol{J}$) yields a product element ($\boldsymbol{KJ}$), the 'value' of which ($-\boldsymbol{L}$) can be found in the ($\boldsymbol{K}, \boldsymbol{J}$) position in the body of the table, so that $\boldsymbol{KJ} = -\boldsymbol{L}$. The commutative law is not valid, because multiplication is not commutative on $\mathbb{C}^{2 \times 2}$. So, for example, $\boldsymbol{KJ} = -\boldsymbol{L}$ but $\boldsymbol{JK} = \boldsymbol{L}$.

Table 4.1 Multiplication table for Example 4.2

$\bullet$	I	$-I$	J	$-J$	K	$-K$	L	$-L$
I	I	$-I$	J	$-J$	K	$-K$	L	$-L$
$-I$	$-I$	I	$-J$	J	$-K$	K	$-L$	L
J	J	$-J$	$-I$	I	L	$-L$	$-K$	K
$-J$	$-J$	J	I	$-I$	$-L$	L	K	$-K$
K	K	$-K$	$-L$	L	$-I$	I	$-J$	J
$-K$	$-K$	K	L	$-L$	I	$-I$	J	$-J$
L	L	$-L$	K	$-K$	$-J$	J	$-I$	I
$-L$	$-L$	L	$-K$	K	J	$-J$	I	$-I$

There are many different kinds of algebraic systems, but one of the simplest (i.e. the binary operation has very few properties) is a semigroup.

4.2 Semigroups

A semigroup $(S, \circ)$ is a non-empty set S with an associative binary operation $\circ$ such that

(i) S is closed under $\circ$, i.e.—$a \circ b$ is defined, and $a \circ b \in S$, for all $a, b \in S$.
(ii) $\circ$ is associative, i.e. $a \circ (b \circ c) = (a \circ b) \circ c$ for all $a, b, c \in S$.

Example 4.3

The set $\mathbb{N}$ with ordinary addition of integers is a semigroup $(\mathbb{N}, +)$ (more precisely an additive semigroup because the binary operation is addition) since

$$a + b \in \mathbb{N} \quad \text{and} \quad a + (b + c) = (a + b) + c \quad \forall\, a,b,c \in \mathbb{N}.$$

Example 4.4

The set $G\ell(2, \mathbb{R})$ of real nonsingular 2×2 matrices under matrix addition is not a semigroup because $G\ell(2, \mathbb{R})$ is not closed under addition. For example,

$$\begin{bmatrix} 1 & 1 \\ 2 & 1 \end{bmatrix} \text{ and } \begin{bmatrix} 1 & 1 \\ -2 & -1 \end{bmatrix}$$

are both real nonsingular 2×2 matrices and so belong to $G\ell(2, \mathbb{R})$. However,

$$\begin{bmatrix} 1 & 1 \\ 2 & 1 \end{bmatrix} + \begin{bmatrix} 1 & 1 \\ -2 & -1 \end{bmatrix} = \begin{bmatrix} 2 & 2 \\ 0 & 0 \end{bmatrix}$$

is a singular matrix and so is not an element of $G\ell(2, \mathbb{R})$.

If a semigroup $(S, \circ)$ contains an element e such that

$$a \circ e = a = e \circ a$$

for all $a \in S$, then e is called an *identity*. Note that, taking $a = e$, it follows that $e \circ e = e$. A semigroup containing such an element is called a *semigroup with identity* or a *monoid*. It turns out that, if e is an identity element in a semigroup $(S, \circ)$, then e is unique and can legitimately be called *the* identity of $(S, \circ)$. [2]

Example 4.5

$(\mathbb{N}, \cdot)$ is a (multiplicative) semigroup in which the binary operation $\cdot$ is the ordinary multiplication of real numbers (the reader should check this). $(\mathbb{N}, \cdot)$ has the identity 1 because

$$n \cdot 1 = n = 1 \cdot n$$

for all $n \in \mathbb{N}$. However, it is important to note that not all semigroups have identity elements. For example, the additive semigroup $(\mathbb{N}, +)$ is not a monoid although $(\mathbb{N} \cup \{0\}, +)$ is, with identity 0.

Semigroups play a major role in the theory of machines or finite automata (Kalman, Falb, and Arbib 1969) and the following short discussion will illustrate how an input space for a dynamical system may be formulated as a semigroup with identity. Suppose that the two infinite sequences

$$\boldsymbol{u} = (\dots , 0, 0, u_1 ,\dots, u_p, 0, 0, \dots),$$

$$\boldsymbol{v} = (\dots , 0, 0, v_1 ,\dots, v_q, 0, 0, \dots),$$

in which all but a finite number of terms are zero, are inputs to a dynamical system and $u_i, v_j \in \mathbb{R}$ (see Example 1.3). Infinite sequences are chosen here because, in some applications, the time domain is taken as $(-\infty, \infty)$ and also they are analogous to the infinite sequences used to represent polynomials in Section 5.2. However, in many instances, as in Example 4.6 below, finite sequences are sufficient.

Let the input space Ω be the set of all such infinite sequences. A binary operation on Ω can then be defined by

$$\boldsymbol{u} \circ \boldsymbol{v} \triangleq (\dots , 0, 0, u_1,\dots, u_p, v_1 ,\dots, v_q, 0, 0, \dots) \in \Omega. \tag{4.1}$$

This operation, whereby input sequences are strung together as in (4.1), is known as *concatenation*. It is an associative operation since, for any $\boldsymbol{u}, \boldsymbol{v}, \boldsymbol{w} \in \Omega$, it does not matter whether we concatenate $\boldsymbol{u}$ and $\boldsymbol{v}$ first to give the input $\boldsymbol{u} \circ \boldsymbol{v}$, and then combine with $\boldsymbol{w}$ to give $(\boldsymbol{u} \circ \boldsymbol{v}) \circ \boldsymbol{w}$, or concatenate $\boldsymbol{v}$ and $\boldsymbol{w}$ first and then form $\boldsymbol{u} \circ (\boldsymbol{v} \circ \boldsymbol{w})$. The resulting input sequence in either case is the same, i.e.

$$(\boldsymbol{u} \circ \boldsymbol{v}) \circ \boldsymbol{w} = \boldsymbol{u} \circ (\boldsymbol{v} \circ \boldsymbol{w}).$$

Note that concatenation is not a commutative operation because, in general, the sequence

$$\boldsymbol{v} \circ \boldsymbol{u} = (\dots , 0, 0, v_1 ,\dots, v_q, u_1 ,\dots, u_p, 0, 0, \dots)$$

is different from $u \circ v$ given in (4.1).

The input space Ω, together with concatenation, is a semigroup. It is actually a monoid, because the infinite sequence $\mathbf{0} = (\dots , 0, 0 ,\dots, 0, 0, \dots)$ containing only zeros is the identity element for concatenation on Ω. That is,

$$\boldsymbol{u} \circ \mathbf{0} = \boldsymbol{u} = \mathbf{0} \circ \boldsymbol{u} \quad \forall\ \boldsymbol{u} \in \Omega.$$

An example from dynamical system theory can now be given which brings together the concepts of an equivalence relation, a mapping, and concatenation. The transfer-function matrix $\boldsymbol{G}(z)$ in (1.15) is a representation of an input/output map $\boldsymbol{f}\colon \Omega \to \Gamma$ from an input space Ω to an output space Γ. Suppose the input space is defined as the set of all multi-input sequences as given in Example 1.3. Using $\circ$ to denote concatenation of these finite sequences in the same way as for the infinite sequences described above, define an equivalence relation on Ω such that, for all $\boldsymbol{u}, \boldsymbol{v} \in \Omega$, we have

$$\boldsymbol{u} \sim \boldsymbol{v} \quad \text{iff} \quad \boldsymbol{f}(\boldsymbol{u} \circ \boldsymbol{w}) = \boldsymbol{f}(\boldsymbol{v} \circ \boldsymbol{w}) \ \forall \ \boldsymbol{w} \in \Omega.$$

That is to say, if $\boldsymbol{u}$ and $\boldsymbol{v}$ are each concatenated with the same arbitrary input $\boldsymbol{w}$ and the resulting outputs are the same whatever the value of $\boldsymbol{w}$, then $\boldsymbol{u}$ is equivalent to $\boldsymbol{v}$. This is the *Nerode equivalence relation induced by* $\boldsymbol{f}$ and it is important in automata theory (Kalman et al 1969). If a system is to have a well-defined input–output map then the inputs must be applied to the system in the zero state. [3] Then two inputs are in the same Nerode equivalence class iff, starting from the zero state before the inputs are applied, they give the same output for $t > 0$.

The *Myhill equivalence relation* (Kalman et al 1969) is rather more special than the Nerode equivalence relation, in that two inputs $\boldsymbol{u}$ and $\boldsymbol{v}$ are equivalent iff they give the same final state whatever the initial state, so that

$$\boldsymbol{u} \sim \boldsymbol{v} \quad \text{iff} \quad \boldsymbol{f}(\boldsymbol{w}_1 \circ \boldsymbol{u} \circ \boldsymbol{w}_2) = \boldsymbol{f}(\boldsymbol{w}_1 \circ \boldsymbol{v} \circ \boldsymbol{w}_2) \ \forall \ \boldsymbol{w}_1, \boldsymbol{w}_2 \in \Omega.$$

The next example illustrates the Myhill equivalence relation.

Example 4.6

Consider the following third-order discrete system with two inputs and one output.

$$x_1(t+1) = x_2(t), \qquad x_1(-2) = 1,$$

$$x_2(t+1) = -x_1(t) - 2x_2(t) + u_1(t), \qquad x_2(-2) = 0,$$

$$x_3(t+1) = -2x_3(t) + u_2(t), \qquad x_3(-2) = 0,$$

$$y(t) = h(\boldsymbol{x}(t)) = x_1(t) + x_3(t).$$

Select an input sequence $\boldsymbol{u} = (\boldsymbol{u}(-2), \boldsymbol{u}(-1), \boldsymbol{u}(0))$, in which

$$\boldsymbol{u}(-2) = (1, 0), \qquad \boldsymbol{u}(-1) = (2, 0), \qquad \boldsymbol{u}(0) = (1, 0).$$

Then it is easy to calculate the output sequence $\boldsymbol{y} = (y(-1), y(0), y(1), \ldots)$ as $y(-1) = 0$, $y(0) = 0$, $y(1) = 2$,..., with state $\boldsymbol{x}(1) = (2, -3, 0)$.

Now select a different input $\boldsymbol{v} = (\boldsymbol{v}(-2), \boldsymbol{v}(-1), \boldsymbol{v}(0))$, in which

$$\boldsymbol{v}(-2) = (0, 0), \qquad \boldsymbol{v}(-1) = (0, 1), \qquad \boldsymbol{v}(0) = (0, 2).$$

From the same initial state $\boldsymbol{x}(-2) = (1, 0, 0)$, the output sequence is identical to the one obtained with input $\boldsymbol{u}$, and the state at $t = 1$ is again $\boldsymbol{x}(1) = (2, -3, 0)$. In fact, it is a simple matter to show that, for any initial state $x_1(-2) = \alpha$, $x_2(-2) = \beta$, $x_3(-2) = \gamma$, the two given inputs $\boldsymbol{u}$ and $\boldsymbol{v}$ will give final state

$$x_1(1) = 2\alpha + 3\beta, \qquad x_2(1) = -3\alpha - 4\beta, \qquad x_3(1) = -8\gamma.$$

Hence

$$f(w_1 \circ u \circ w_2) = f(w_1 \circ v \circ w_2) \quad \forall\ w_1, w_2 \in \Omega,$$

and $u \sim w$, where $\sim$ is the Myhill equivalence relation. In polynomial form, the inputs u and v are represented by

$$(z^2 + 2z + 1, 0) \quad \text{and} \quad (0, z + 2)$$

respectively. With g^{T} as the transfer function (1.15), $g^{\mathrm{T}}u = g^{\mathrm{T}}v = 1$, which counts as zero in Kalman's formulation (Rosenbrock 1970b). As regards output for $t > 0$, neither u nor v has any effect, since the output is due entirely to the non-zero initial state $x(-2)$.[4]

Suppose $(S, \circ)$ is a semigroup with identity e and that $a \in S$. If S contains an element b such that

$$a \circ b = e = b \circ a,$$

then b is called an *inverse* of a and is usually denoted by a^{-1}. When an element of S has an inverse, then that inverse is unique, so it is correct to talk of *the* inverse a^{-1} of a in S. In other words, no element $s \in S$ can have more than one inverse s^{-1}.[5]

Not all elements of a semigroup with identity have inverses, although the identity e always has itself as inverse since $e \circ e = e$. For example, in the monoid $(\mathbb{Z}, \cdot)$ only the identity 1 and the element -1 have inverses (1 and -1 respectively).

Let $(S, \circ)$ be a semigroup, and let T be a subset of S. Clearly $\circ$ is a binary operation on T as well as on S. The subset T is called a *subsemigroup* of S iff $(T, \circ)$ is a semigroup. A useful test for a subsemigroup is that a non-empty subset T of S is a subsemigroup of S iff T is closed under the operation $\circ$.

Example 4.7

$S = \mathbb{R}^{2 \times 2}$, and T is the set of all real matrices of the form

$$\begin{bmatrix} t & 0 \\ 0 & 0 \end{bmatrix} \quad (t \in \mathbb{R}).$$

Clearly $T \subset \mathbb{R}^{2 \times 2}$ and $T \neq \varnothing$. Also, $(\mathbb{R}^{2 \times 2}, \cdot)$ is a multiplicative semigroup (check this). For any two elements of T, say

$$\boldsymbol{T}_1 = \begin{bmatrix} t_1 & 0 \\ 0 & 0 \end{bmatrix}, \qquad \boldsymbol{T}_2 = \begin{bmatrix} t_2 & 0 \\ 0 & 0 \end{bmatrix},$$

we have

$$\boldsymbol{T}_1 \boldsymbol{T}_2 = \begin{bmatrix} t_1 t_2 & 0 \\ 0 & 0 \end{bmatrix} \in T.$$

Thus, T is closed under the operation of matrix multiplication and so T is a subsemigroup of $\mathbb{R}^{2\times 2}$.

It is important to note that, if a semigroup with identity has a subsemigroup which itself has an identity, then these two identities are not necessarily the same.

Example 4.8

The semigroup $(\mathbb{R}^{2\times 2}, \cdot)$ has identity $\mathbf{I}_2$ and gives rise to the subsemigroup T as defined in Example 4.7. The identity in T is easily seen to be the matrix

$$\boldsymbol{J} = \begin{bmatrix} 1 & 0 \\ 0 & 0 \end{bmatrix}$$

since for all $A \in T$, $\boldsymbol{J}A = A = A\boldsymbol{J}$. Thus, both the semigroup and the subsemigroup have an identity, $\mathbf{I}_2$ and $\boldsymbol{J}$ respectively, but $\boldsymbol{J} \neq \mathbf{I}_2$.

As was pointed out in Section 3.3, the composition of mappings from some space X to itself has the associative property. This property and the existence of the identity map $\mathrm{id}_X : X \to X$ ensures that the set of all mappings $f : X \to X$ is a semigroup with identity. An important special case is the set of all bijective functions $f : X \to X$ which, together with the operation of composition, becomes an algebraic system known as a *group*.

4.3 Groups

Suppose that $(G, \circ)$ is a semigroup with identity e and that, for each element $a \in G$, there exists an element $a^{-1} \in G$ such that

$$a \circ a^{-1} = e = a^{-1} \circ a. \tag{4.2}$$

Then a^{-1} is the unique [5] inverse of a, and the algebraic system $(G, \circ)$ is called a *group*. Thus, a group is a non-empty set G with a binary operation denoted by $\circ$ such that

(i) G is closed under $\circ$,
(ii) $\circ$ is associative in G,
(iii) the identity element [6] $e \in G$ satisfies

$$a \circ e = a = e \circ a \quad \forall\ a \in G.$$

(iv) for each $a \in G$, there exists a unique inverse a^{-1} satisfying (4.2).

A multiplicative group $(G, \cdot)$ in which $G = \{e\}$ (i.e. G is of order 1, containing only the identity element e) is called a *trivial* group. The equation $e \cdot e = e$ completely describes such a group.

Example 4.9

$(\mathbb{Z}, +)$ is a group with identity $e = 0$ (the zero element) and, for each $a \in \mathbb{Z}$, the inverse a^{-1} equals $-a$ (the negative of a). Then

$$a + 0 = a = 0 + a \quad \text{and} \quad a + (-a) = 0 = (-a) + a$$

for all $a \in \mathbb{Z}$. This group $(\mathbb{Z}, +)$ is called an *additive* group because of the $+$ operation, and is sometimes denoted by $\mathbb{Z}^+$ in the literature. However, there is a danger of confusion with this notation because $\mathbb{R}^+$ is a standard symbol for the set of non-negative real numbers $[0, \infty)$.

An *abelian* group is a group $(G, \circ)$ in which the binary operation $\circ$ is commutative (i.e. $a \circ b = b \circ a$ for all $a,b \in G$). The binary operation in an abelian group is often written as addition $+$ (an additive abelian group) with identity 0 and inverse $-a$ of element a. Then, for all $a,b,c \in G$, we have

$$\begin{aligned}
a + b &\in G && \text{(closed under } +\text{)},\\
(a + b) + c &= a + (b + c) && \text{(associative law)},\\
a + b &= b + a && \text{(commutative law)},\\
a + 0 &= a = 0 + a && \text{(identity 0)},\\
a + (-a) &= 0 = (-a) + a && \text{(inverse } (-a)\text{)}.
\end{aligned}$$

Example 4.10

The space $\mathbb{R}^2$ is an abelian group with (componentwise, or pointwise) addition as the group operation: (i) for all $\boldsymbol{x} = (x_1, x_2)$ and $\boldsymbol{y} = (y_1, y_2)$ in $\mathbb{R}^2$, we have

$$\boldsymbol{x} + \boldsymbol{y} = \begin{bmatrix} x_1 \\ x_2 \end{bmatrix} + \begin{bmatrix} y_1 \\ y_2 \end{bmatrix} \triangleq \begin{bmatrix} x_1 + y_1 \\ x_2 + y_2 \end{bmatrix} \in \mathbb{R}^2.$$

Thus, $\mathbb{R}^2$ is closed under addition.

(ii)

$$\begin{aligned}
\boldsymbol{x} + (\boldsymbol{y} + \boldsymbol{z}) &= \begin{bmatrix} x_1 \\ x_2 \end{bmatrix} + \begin{bmatrix} y_1 + z_1 \\ y_2 + z_2 \end{bmatrix} = \begin{bmatrix} x_1 + (y_1 + z_1) \\ x_2 + (y_2 + z_2) \end{bmatrix} \\
&= \begin{bmatrix} x_1 + y_1 + z_1 \\ x_2 + y_2 + z_2 \end{bmatrix} \quad \forall\, \boldsymbol{x},\boldsymbol{y},\boldsymbol{z} \in \mathbb{R}^2,
\end{aligned}$$

the last step being possible because addition on $\mathbb{R}$ is associative. Similarly,

$$(\boldsymbol{x}+\boldsymbol{y})+\boldsymbol{z}=\begin{bmatrix} x_1+y_1 \\ x_2+y_2 \end{bmatrix}+\begin{bmatrix} z_1 \\ z_2 \end{bmatrix}=\begin{bmatrix} (x_1+y_1)+z_1 \\ (x_2+y_2)+z_2 \end{bmatrix}$$

$$=\begin{bmatrix} x_1+y_1+z_1 \\ x_2+y_2+z_2 \end{bmatrix}.$$

Hence $\boldsymbol{x}+(\boldsymbol{y}+\boldsymbol{z})=(\boldsymbol{x}+\boldsymbol{y})+\boldsymbol{z}$ for all $\boldsymbol{x},\boldsymbol{y},\boldsymbol{z}\in\mathbb{R}^2$, and so addition is associative on $\mathbb{R}^2$.

(iii) $\boldsymbol{x}+\boldsymbol{y}=(x_1+y_1,\, x_2+y_2)=(y_1+x_1,\, y_2+x_2)$, because addition is commutative on $\mathbb{R}$. But $(y_1+x_1,\, y_2+x_2)=\boldsymbol{y}+\boldsymbol{x}$, and so addition is also commutative on $\mathbb{R}^2$.

(iv) The identity element $\mathbf{0}$ is equal to (0, 0) since

$$\boldsymbol{x}+\mathbf{0}=\begin{bmatrix} x_1+0 \\ x_2+0 \end{bmatrix}=\begin{bmatrix} x_1 \\ x_2 \end{bmatrix}=\begin{bmatrix} 0+x_1 \\ 0+x_2 \end{bmatrix}=\mathbf{0}+\boldsymbol{x} \quad \forall\, \boldsymbol{x}\in\mathbb{R}^2.$$

(v) For each $\boldsymbol{x}\in\mathbb{R}^2$, there exists a unique inverse $\boldsymbol{x}^{-1}\in\mathbb{R}^2$ such that, if $\boldsymbol{x}=(x_1,\, x_2)$, then $\boldsymbol{x}^{-1}=(-x_1,\, -x_2)$, because

$$\boldsymbol{x}+\boldsymbol{x}^{-1}=\begin{bmatrix} x_1 \\ x_2 \end{bmatrix}+\begin{bmatrix} -x_1 \\ -x_2 \end{bmatrix}=\begin{bmatrix} x_1+(-x_1) \\ x_2+(-x_2) \end{bmatrix}=\begin{bmatrix} x_1-x_1 \\ x_2-x_2 \end{bmatrix}=\mathbf{0}.$$

Also, $\boldsymbol{x}^{-1}+\boldsymbol{x}=\mathbf{0}$ by commutativity.

This example extends to the case $\mathbb{R}^n$, for any $n\in\mathbb{N}$, in a straightforward manner.

Some examples illustrating how groups arise in dynamical system theory will now be discussed. Consider a system governed by the scalar linear differential equation

$$\dot{x}(t)=x(t), \qquad x(0)=x_0, \tag{4.3}$$

so that $x(t)\in\mathbb{R}$ and $t\in[0\,,\,\infty)$. The solution of this equation can be written as

$$x(t)=\mathrm{e}^t x_0. \tag{4.4}$$

The real number e^t may be interpreted as an operator acting on the initial point x_0 and which moves the system state x from x_0 to $x(t)$ at time t. Thus (Fig. 4.1), if $x(t_1)=x_1$ at some time $t_1>0$, then (4.4) gives

$$x_1=\mathrm{e}^{t_1}x_0. \tag{4.5}$$

Suppose now x_1 is taken as a new initial value of x and time τ is measured from t_1, so that $t=t_1+\tau$. With this change of time variable from t to τ, the differential equation in (4.3) together with the new initial condition can be

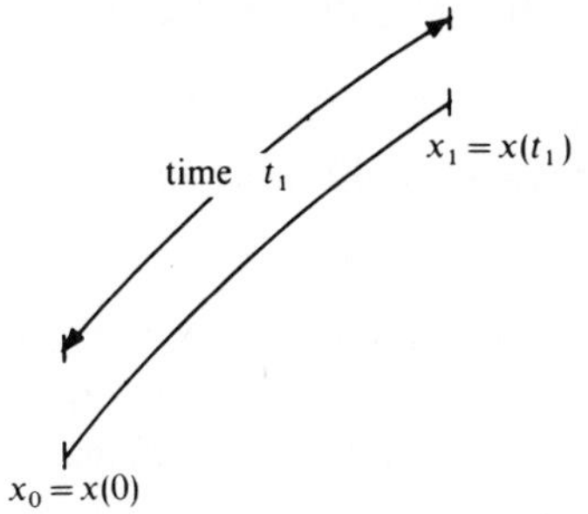

Fig. 4.1 Operator e^{t_1}

written as

$$\frac{dz(\tau)}{d\tau} = z(\tau), \qquad z(0) = x_1. \tag{4.6}$$

The solution of (4.6) has the same form as (4.4) and is given by

$$z(\tau) = e^{\tau} x_1. \tag{4.7}$$

At a time $\tau = t_2$, we have

$$z(t_2) = e^{t_2} x_1 \tag{4.8}$$

and, since $z(t_2) = x(t_1 + t_2) = x_2$ say, (4.8) can be written as (Fig. 4.2)

$$x_2 = e^{t_2} x_1.$$

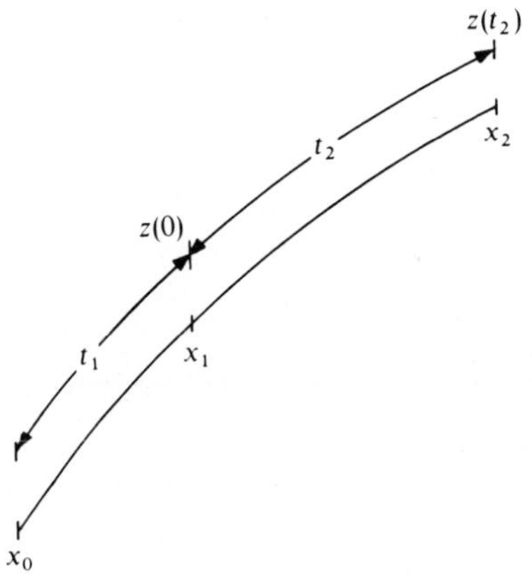

Fig. 4.2 Operators e^{t_1} and e^{t_2}

If we substitute for x_1 from (4.5), this last equation becomes

$$x_2 = e^{t_2}(e^{t_1}x_0)$$

and, by the associative law of multiplication on $\mathbb{R}$, this is the same as

$$x_2 = x(t_1 + t_2) = (e^{t_2}e^{t_1})x_0. \tag{4.9}$$

Equation (4.9) simply states the obvious fact that the result of moving the state x from x_0 to x_2 is the same as moving first from x_0 to x_1 and then from x_1 to x_2. The operator $e^{t_2}e^{t_1}$, thought of as a single operator, moves x_0 to x_2. On the other hand, this operator may be thought of as the composition of the two separate operators e^{t_1} and e^{t_2}, with e^{t_1} operating first on x_0 to give x_1 and then e^{t_2} operating on x_1 to give x_2. Notice that, in the composite operator $e^{t_2}e^{t_1}$, the operations are written in reverse order because e^{t_1} is the first operator to be used and so is written adjacent to x_0 (cf. composite functions in Chapter 3).

Now consider a third time interval of duration t_3, during which time the state x moves from x_2 to x_3. As before, the result of this movement can be written in the form (Fig. 4.3)

$$x_3 = x(t_1 + t_2 + t_3) = e^{t_3}x_2 = e^{t_3}(e^{t_2}e^{t_1})x_0 \tag{4.10}$$

by (4.9). The composite operator $e^{t_3}(e^{t_2}e^{t_1})$ is essentially the multiplication of the three real numbers e^{t_3}, e^{t_2}, e^{t_1}; by the associative law of multiplication on $\mathbb{R}$, this operator may therefore be written in the alternative form $(e^{t_3}e^{t_2})e^{t_1}$, so that

$$x_3 = (e^{t_3}e^{t_2})e^{t_1}x_0. \tag{4.11}$$

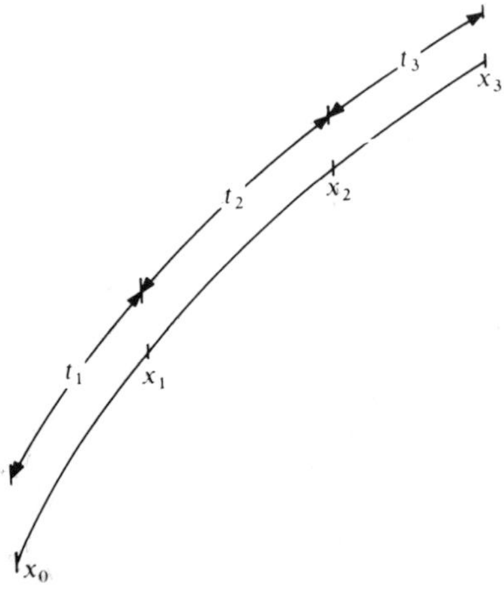

Fig. 4.3 Operators e^{t_1}, e^{t_2}, and e^{t_3}

Equations (4.10) and (4.11) show that, under the binary operation of composition, the set of operators $\{e^t : t \in \mathbb{R}\}$ satisfy the associative law. These equations are also saying that, in order to reach x_3 from x_0, one can either move first from x_0 to x_2 and then to x_3 (4.10), or first move from x_0 to x_1 and then from x_1 to x_3 (4.11). All this may seem terribly obvious, yet similar ideas have proved very useful in nonlinear control theory (Krener 1977).

Suppose time could run backwards and that the initial condition for the differential equation in (4.3) was $x(t_1) = x_1$ instead of $x(0) = x_0$. What would be the value of x if the system was to run backwards in time from t_1 to zero? Well, the solution of the differential equation under these circumstances is

$$x(0) = e^{-t_1} x_1, \tag{4.12}$$

an equation which could have been obtained from (4.5) immediately. Thus, the operator e^{-t_1} takes x_1 back to the point x_0 and is called the operator inverse to e^{t_1}. Substituting in (4.12) for x_1 from (4.5) gives

$$x(0) = e^{-t_1}(e^{t_1} x_0) = (e^{-t_1} e^{t_1}) x_0 = e^0 x_0 = x_0.$$

The operator e^0 $(= 1)$ leaves a point x_0 at x_0, and is the identity operator.

From the above results concerning the set G of operators $\{e^t : t \in \mathbb{R}\}$ associated with the differential equation (4.3), the following facts may be observed:

(i) For all $e^{t_1}, e^{t_2} \in G$, the composite operator $e^{t_2} e^{t_1}$ belongs to G, i.e. the set G is closed under the binary operation of composition.

(ii) $$e^{t_3}(e^{t_2} e^{t_1}) = (e^{t_3} e^{t_2}) e^{t_1} \quad \forall\, e^{t_1}, e^{t_2}, e^{t_3} \in G,$$

i.e. composition is associative in G.

(iii) For all $e^t \in G$, the identity operator e^0 (i.e. the real number 1) satisfies

$$e^t e^0 = e^t = e^0 e^t.$$

(iv) For all $e^t \in G$, there exists an inverse $e^{-t} \in G$ such that

$$e^t e^{-t} = e^0 = e^{-t} e^t.$$

Thus, $G = \{e^t : t \in \mathbb{R}\}$ is a group; furthermore it is an abelian group because, for all $e^{t_1}, e^{t_2} \in G$, we have

$$e^{t_2} e^{t_1} = e^{(t_2 + t_1)} = e^{(t_1 + t_2)} = e^{t_1} e^{t_2}.$$

It is easy to show that the scalar linear differential equation $\dot{x}(t) = ax(t)$, with $a \in \mathbb{R}$ and $x(0) = x_0$, has a solution $x(t) = e^{at} x_0$ and that the set of operators $\{e^{at} : t \in \mathbb{R}\}$ is also an abelian group. This is left as an exercise for the reader (Exercise 6). The result of Exercise 6 can be generalized to a set of n linear differential equations

$$\dot{\boldsymbol{x}} = \boldsymbol{A}\boldsymbol{x}, \qquad \boldsymbol{x}(0) = \boldsymbol{x}_0, \tag{4.13}$$

where $x(t) \in \mathbb{R}^n$, $A \in \mathbb{R}^{n \times n}$. Consider first a fundamental set of independent solutions $z_1(\cdot), \dots, z_n(\cdot)$ which satisfy

$$\dot{z}_i(t) = Az_i(t) \quad \text{with} \quad z_i(0) = e_i \quad (i = 1, \dots, n),$$

where e_i is the ith column of the $n \times n$ identity matrix $\mathbf{I}_n$. Then a matrix $\Phi(t, 0)$ can be formed as a function of time t and initial time (here chosen as $t = 0$), with independent columns which are the vectors $z_i(t)$, so that

$$\Phi(t, 0) = [z_1(t), \dots, z_n(t)],$$

where

$$\Phi(0, 0) = [z_1(0), \dots, z_n(0)] = [e_1, \dots, e_n] = \mathbf{I}_n. \tag{4.14}$$

The time-varying matrix $\Phi(t, 0)$ is called the *transition matrix*. Since

$$[\dot{z}_1, \dots, \dot{z}_n] = [Az_1, \dots, Az_n] = A[z_1, \dots, z_n],$$

the transition matrix satisfies the differential equation

$$\dot{\Phi}(t, 0) = A\Phi(t, 0).$$

In other words, $\Phi(t, 0)$ satisfies a matrix differential equation which has the same form as the original equation (4.13) but with initial condition $\Phi(0, 0) = \mathbf{I}_n$ from (4.14). To obtain the solution of (4.13), simply note that

$$\frac{\mathrm{d}}{\mathrm{d}t}(\Phi x_0) = \dot{\Phi} x_0 = A(\Phi x_0)$$

and $\Phi(0, 0)x_0 = \mathbf{I}_n x_0 = x_0$. Thus, the vector $\Phi(t, 0)x_0$ satisfies the differential equation and initial condition (4.13) and hence is the solution, i.e.

$$x(t) = \Phi(t, 0)x_0. \tag{4.15}$$

Now this last equation bears a striking resemblance to the solution $x(t) = \mathrm{e}^{at}x_0$ of Exercise 6, except that $\Phi(t, 0)$ replaces the exponential e^{at}. In fact, the transition matrix $\Phi(t, 0)$ is often denoted by e^{At}, which is a generalization of the scalar exponential e^{at} (cf. Example 1.4).

The set of fundamental solutions $z_i(t)$ $(i = 1, \dots, n)$ are independent for each t, and so the transition matrix $\Phi(t, 0)$ is always nonsingular. It follows from (4.15) that

$$\Phi^{-1}(t, 0)x(t) = x_0$$

and, if $G = \{\Phi(t, 0): t \in \mathbb{R}\}$ is considered as a set of operators analogous to the scalar case, then every operator $\Phi \in G$ has a (unique) inverse Φ^{-1}, the inverse matrix of Φ. The composition of two operators $\Phi_1, \Phi_2 \in G$ is the usual multiplication of two $n \times n$ matrices, and so G is closed under this binary operation. The identity operator is $\mathbf{I}_n$, and the upshot of all this is that the set $G = \{\Phi(t, 0): t \in \mathbb{R}\}$ is also a group. However, unlike the scalar case, it is not an abelian group since matrix multiplication is not commutative.

The above ideas can be generalized still further to the vector differential equation

$$\dot{\boldsymbol{x}} = \boldsymbol{f}(\boldsymbol{x}), \qquad \boldsymbol{x}(t_0) = \boldsymbol{x}_0 \tag{4.16}$$

where $\boldsymbol{x}(t) \in \mathbb{R}^n$ for $t \in [t_0, \infty)$. Guided by the linear case discussed above, suppose that the solution of (4.16) is written in the form

$$\boldsymbol{x}(t) = \boldsymbol{\phi}_t \boldsymbol{x}_0, \tag{4.17}$$

in which the operator $\boldsymbol{\phi}_t$ moves the state from $\boldsymbol{x}_0$ to $\boldsymbol{x}(t)$ in time $t - t_0$. In differential geometry, the exponential notation is often retained, and the operator $\boldsymbol{\phi}_t$ is written as e^{tf} (Brockett 1976). It is easy to verify that the set of operators $\{\boldsymbol{\phi}_t : t \in \mathbb{R}\}$ form a group. The binary operation will again be the composition of operators. To see this more clearly, write (4.17) in the form

$$\boldsymbol{x}(t) = \boldsymbol{\phi}(t, \boldsymbol{x}_0).$$

Then (Fig. 4.2) $\boldsymbol{\phi}(t_1, \boldsymbol{x}_0) = \boldsymbol{x}_1$ and

$$\boldsymbol{\phi}_{t_2}\boldsymbol{\phi}_{t_1}\boldsymbol{x}_0 = \boldsymbol{\phi}(t_2, \boldsymbol{\phi}(t_1, \boldsymbol{x}_0)).$$

Also (Krener 1977, Bell 1984)

$$\begin{aligned}
\boldsymbol{\phi}_{t_2}\boldsymbol{\phi}_{t_1}\boldsymbol{x}_0 &= \boldsymbol{\phi}_{t_2 + t_1}\boldsymbol{x}_0,\\
\boldsymbol{\phi}_0\boldsymbol{x}_0 &= \boldsymbol{x}_0,\\
\boldsymbol{\phi}_{-t_1}\boldsymbol{\phi}_{t_1}\boldsymbol{x}_0 &= \boldsymbol{x}_0,\\
\boldsymbol{\phi}_{t_3}(\boldsymbol{\phi}_{t_2}\boldsymbol{\phi}_{t_1}) &= (\boldsymbol{\phi}_{t_3}\boldsymbol{\phi}_{t_2})\boldsymbol{\phi}_{t_1}.
\end{aligned}$$

Another example where a group appears is the rotation of rectangular cartesian axes OX and OY in $\mathbb{R}^2$. Suppose new axes OX' and OY' are obtained by simply rotating the axes OX and OY through an angle α about the fixed

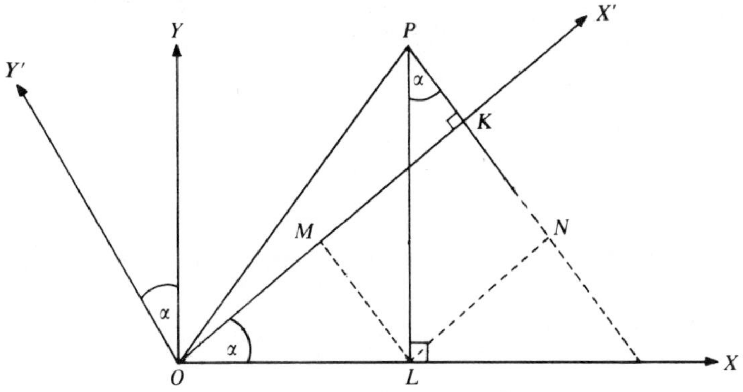

Fig. 4.4 Rotation of axes

point O in a counterclockwise direction (Fig. 4.4). Let (x, y) be the coordinates of a point P with respect to axes XOY and (x', y') with respect to axes $X'OY'$. Then from Fig. 4.4

$$OL = x, \qquad PL = y, \qquad OK = x', \qquad PK = y',$$

$$OM = x\cos\alpha, \qquad LN = y\sin\alpha, \qquad PN = y\cos\alpha, \qquad LM = x\sin\alpha.$$

It follows that

$$x' = x\cos\alpha + y\sin\alpha,$$

$$y' = -x\sin\alpha + y\cos\alpha,$$

or, in matrix notation,

$$\begin{bmatrix} x' \\ y' \end{bmatrix} = \begin{bmatrix} \cos\alpha & \sin\alpha \\ -\sin\alpha & \cos\alpha \end{bmatrix} \begin{bmatrix} x \\ y \end{bmatrix}. \tag{4.18}$$

The matrix

$$\boldsymbol{A} = \begin{bmatrix} \cos\alpha & \sin\alpha \\ -\sin\alpha & \cos\alpha \end{bmatrix}$$

is orthogonal, since $\boldsymbol{A}\boldsymbol{A}^{\mathrm{T}} = \boldsymbol{I} = \boldsymbol{A}^{\mathrm{T}}\boldsymbol{A}$ and $\det \boldsymbol{A} = 1$. In (4.18), the matrix $\boldsymbol{A}$ can be thought of as representing a rotation r_α of axes through the angle α, with $-\pi < \alpha \leqslant \pi$. Such a rotation r_α followed by a similar one r_β through an additional angle β may be denoted by the product rotation $r_\beta r_\alpha$. This operation of r_α followed by r_β is clearly equivalent to a single rotation $r_{\alpha+\beta}$ through the angle $\alpha + \beta$. Some care is needed here because, if the values α and β are such that $\alpha + \beta$ lies outside of the interval $(-\pi, \pi]$, then the rotation $r_{\alpha+\beta}$ must be interpreted as $r_{\alpha+\beta-2\pi}$ or $r_{\alpha+\beta+2\pi}$ depending on α and β (e.g. $\alpha = \pi$, $\beta = \frac{1}{2}\pi$, $\alpha + \beta = \frac{3}{2}\pi \notin (-\pi, \pi]$ so $r_\beta r_\alpha = r_{-\frac{1}{2}\pi}$ (Fig. 4.5)). The set $G = \{r_\alpha : -\pi < \alpha \leqslant \pi\}$ of all such rotations form a group because, for all $r_\alpha, r_\beta \in G$, we have $r_\beta r_\alpha \in G$; the identity rotation is r_0 (rotation through zero angle) and the inverse rotation to r_α is $r_{-\alpha}$ when $\alpha \neq \pi$ and $r_\pi^{-1} = r_\pi$. Note

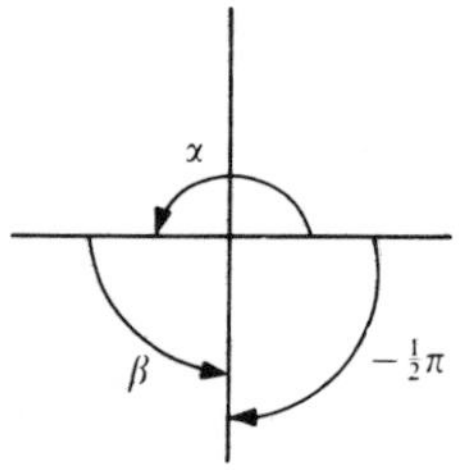

Fig. 4.5 Rotation $\alpha + \beta \notin (-\pi, \pi]$

that the inverse $r_{-\alpha}$ to r_α is represented by the inverse of the matrix A in (4.18), i.e.

$$A^{-1} = \begin{bmatrix} \cos\alpha & -\sin\alpha \\ \sin\alpha & \cos\alpha \end{bmatrix},$$

such that

$$\begin{bmatrix} x \\ y \end{bmatrix} = A^{-1}\begin{bmatrix} x' \\ y' \end{bmatrix}.$$

Now consider reflections in a line OL in $\mathbb{R}^2$ (Fig. 4.6). Suppose OL makes an angle α with the OX axis, and that a point P' with coordinates (x', y') with respect to axes XOY is the reflection of the point P with coordinates (x, y) in the line OL. From the constructions shown in Fig. 4.6,

$$ON' = ON = x, \qquad N'P' = NP = y,$$

where N' is the reflection of N in OL. Since $L\hat{O}N = \alpha$, it follows that $L\hat{O}N' = \alpha$. Then

$$x' = x\cos 2\alpha + y\sin 2\alpha,$$

$$y' = x\sin 2\alpha - y\cos 2\alpha.$$

In matrix notation, this reflection is represented by the equation

$$\begin{bmatrix} x' \\ y' \end{bmatrix} = \begin{bmatrix} \cos 2\alpha & \sin 2\alpha \\ \sin 2\alpha & -\cos 2\alpha \end{bmatrix}\begin{bmatrix} x \\ y \end{bmatrix}.$$

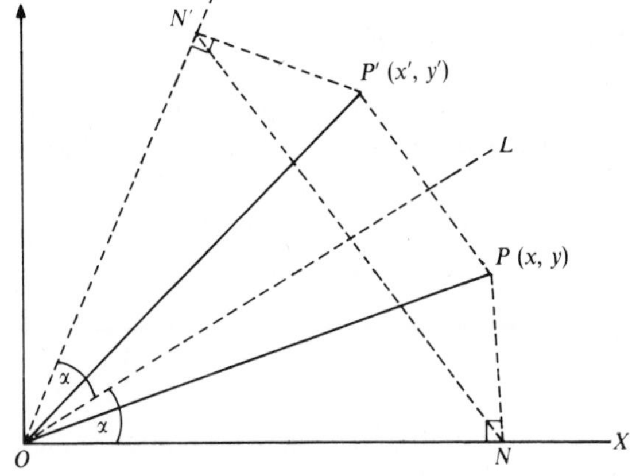

Fig. 4.6 Reflection in line OL

The matrix

$$B = \begin{bmatrix} \cos 2\alpha & \sin 2\alpha \\ \sin 2\alpha & -\cos 2\alpha \end{bmatrix}$$

is orthogonal but, unlike the case of rotations, $\det B = -1$.

Now the set of all real orthogonal 2×2 matrices $\mathrm{So}(2, \mathbb{R})$ (or $\mathrm{So}(2)$), i.e.

$$\mathrm{So}(2, \mathbb{R}) = \{X \in \mathrm{G}\ell\,(2, \mathbb{R}) : X^{\mathsf{T}} X = I\},$$

is a group (the special orthogonal group) with respect to matrix multiplication. From the last two examples, it is seen that $\mathrm{So}(2)$ consists of two distinct parts: namely G_1, comprising all orthogonal matrices A with $\det A = 1$, and G_2, which is the set of all orthogonal matrices B with $\det B = -1$.[7] Clearly, multiplication of matrices is defined on both G_1 and G_2 since they are both subsets of $\mathrm{So}(2)$. It is not difficult to show that G_1 is a group with identity $\mathbf{I}_2$ ($\det \mathbf{I}_2 = 1$). However, G_2 is not a group, since the identity element $\mathbf{I}_2$ does not belong to G_2 ($\det \mathbf{I}_2 \neq -1$) and so G_2 does not contain an identity element. When a group G contains a subset G_1 that itself is a group with respect to the binary operation in G, then G_1 is called a *subgroup* of G. In the above discussion, G_2 is a subset of $\mathrm{So}(2)$ but not a subgroup. The subgroup $G_1 = \{X \in \mathrm{G}\ell\,(2, \mathbb{R}) : X^{\mathsf{T}} X = \mathbf{I}_2, \det X = 1\}$ is usually denoted by $\mathrm{S}\ell\,(2)$ (or $\mathrm{S}\ell\,(2, \mathbb{R})$): the special linear group.

For every group G, the group G itself and the set consisting of only the identity element e are both subgroups of G. A subgroup of G other than G and $\{e\}$ is called a *proper* subgroup (see Exercise 9 for another example of a subgroup). Since every subgroup is a group in its own right, a subgroup is automatically closed under its group operation.

4.4 Isomorphisms and homomorphisms

The set of real numbers $\mathbb{R}$ with the binary operation of addition $(+)$ is a group with identity 0 and inverse $-a$ for all $a \in \mathbb{R}$. Addition of real numbers is associative. In the discussion earlier of the group of operators $\{\phi_t : t \in \mathbb{R}\}$, it was shown that $\phi_{t_2}\phi_{t_1} = \phi_{t_2 + t_1}$ for all $t_1, t_2 \in \mathbb{R}$. For example,

$$\phi_3 \phi_4 = \phi_7.$$

This equation is clearly akin to

$$3 + 4 = 7.$$

In the first equation, the result of the binary operation is the composition of the operators ϕ_4 and ϕ_3, which is equivalent to the operator ϕ_7. In the second equation, the binary operation is addition, and the sum of the two real numbers 3 and 4 is equal to the real number 7. Clearly, associated with any equation of the form $\phi_{t_2}\phi_{t_1} = \phi_{t_2+t_1} = \phi_{t_3}$ (say), involving operators from the group $\{\phi_t : t \in \mathbb{R}\}$, is the equation $t_2 + t_1 = t_3$ involving real numbers from the group $(\mathbb{R}, +)$. Because of this equivalence between the two groups, the group $\{\phi_t : t \in \mathbb{R}\}$ is said to be *isomorphic* (meaning 'of equal form') to the additive group of real numbers $(\mathbb{R}, +)$. Any two algebraic systems that are structurally identical are said to be isomorphic. Thus, provided that it is known how elements of a group G combine together under the binary operation of G, then everything is known about all groups isomorphic to G.

In general, if G_1 and G_2 are groups, then G_1 is said to be isomorphic to G_2 (written $G_1 \cong G_2$) iff there exists [8] a bijective mapping $\theta: G_1 \to G_2$ such that (Fig. 4.7)

$$\theta(x \circ y) = \theta(x) \circ \theta(y) \quad \forall\ x, y \in G_1. \tag{4.19}$$

The bijective mapping θ in (4.19) is called a *(group) isomorphism* from G_1 to G_2. The interpretation of (4.19) is that, as x and y in G_1 correspond to $\theta(x)$ and $\theta(y)$ respectively in G_2, so the element $x \circ y$ in G_1 corresponds to the element $\theta(x) \circ \theta(y)$ in G_2. Notice that the operation $\circ$ on the left-hand side of (4.19) denotes the operation in G_1 whereas, on the right-hand side, it denotes the binary operation in G_2; these two operations are in general quite different.

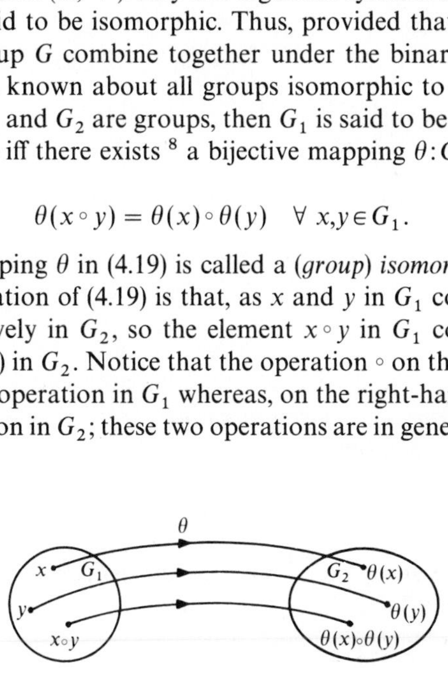

Fig. 4.7 A group isomorphism (θ bijective)

Example 4.11

$G_1 = (\mathbb{R}, +)$, the additive group of real numbers. $G_2 = (\mathbb{R}^+ \backslash \{0\}, \cdot)$, the multiplicative group of positive numbers. The exponential function $\theta:(\mathbb{R}, +) \to (\mathbb{R}^+ \backslash \{0\}, \cdot)$ defined by $\theta(r) = \mathrm{e}^r$ is an isomorphism from G_1 to G_2 because, for all $x,y \in \mathbb{R}$, we have

$$\mathrm{e}^{x+y} = \mathrm{e}^x \cdot \mathrm{e}^y,$$

i.e. $\theta(x + y) = \theta(x) \cdot \theta(y)$, so (4.19) is satisfied. Furthermore, θ is bijective since, for all $s \in \mathbb{R}^+ \backslash \{0\}$, the unique inverse of θ is defined by $\theta^{-1}(s) = \ln s$ (see Section 3.4). Of course, the inverse function

$$\theta^{-1}:(\mathbb{R}^+ \backslash \{0\}, \cdot) \to (\mathbb{R}, +)$$

is an isomorphism and illustrates the fact that multiplication in $(\mathbb{R}^+\backslash\{0\},\cdot)$ can be carried out more easily using logarithms in $(\mathbb{R},+)$.

Equation (4.19) can also be represented by the commutative diagram shown in Fig. 4.8.

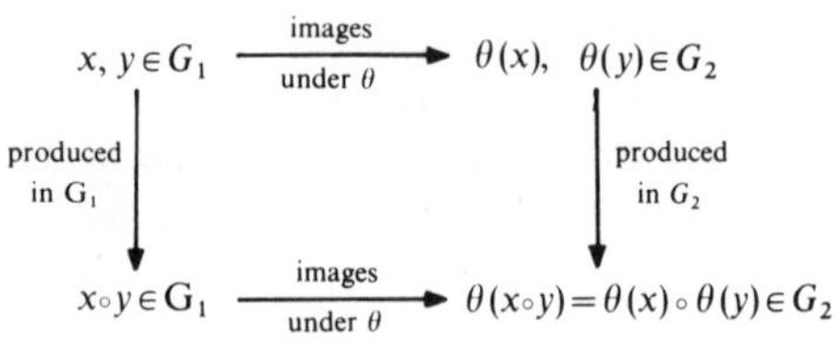

Fig. 4.8 Commutative diagram for a group isomorphism (θ bijective)

A mapping $\theta: G_1 \to G_2$ that satisfies (4.19) for all $x, y \in G_1$ (whether bijective or not) is called a *(group) homomorphism* from G_1 to G_2. So an isomorphism is a bijective homomorphism.

Example 4.12

Let $\theta: G_1 \to G_2$ be a group homomorphism, with e_1 and e_2 the identities in G_1 and G_2 respectively. Then $\theta(e_1) = e_2$, because

$$\begin{aligned}\theta(e_1) &= \theta(e_1 e_1) && \text{(dropping } \circ \text{ for convenience)}\\ &= \theta(e_1)\theta(e_1) && \text{(since } \theta \text{ is a homomorphism, (4.19)).}\end{aligned}$$

But, since e_2 is the identity in G_2, we have $\theta(e_1)e_2 = \theta(e_1)$. Therefore

$$\theta(e_1)e_2 = \theta(e_1)\theta(e_1),$$

and cancellation [9] gives

$$e_2 = \theta(e_1). \tag{4.20}$$

Again, $\theta(g^{-1}) = [\theta(g)]^{-1}$ for any $g \in G_1$, because

$$\begin{aligned}\theta(g)\theta(g^{-1}) &= \theta(gg^{-1}) && \text{(by (4.19) again),}\\ &= \theta(e_1)\\ &= e_2 && \text{(by the result proved above).}\end{aligned}$$

Thus [9] $\theta(g^{-1})$ is the inverse of $\theta(g)$ in G_2, i.e.

$$\theta(g^{-1}) = [\theta(g)]^{-1}. \tag{4.21}$$

These results are true for any homomorphism, not just group homomorphisms (see Chapter 5 for ring homomorphisms and Chapter 6 for module homomorphisms).

Example 4.13

Let G_1 be the additive group modulo 4 on $\mathbb{Z}_4 = \{0, 1, 2, 3\}$ and G_2 be the multiplicative group modulo 5 on $\mathbb{Z}_5 \backslash \{0\} = \{1, 2, 3, 4\}$. These two finite groups can be represented by Tables 4.2 and 4.3. A 1–1 and onto mapping $\theta: G_1 \to G_2$ is defined by $\theta(0) = 1$, $\theta(1) = 3$, $\theta(2) = 4$, $\theta(3) = 2$. In G_1, the result $2 + 3 = 1$ holds, and so $\theta(2 + 3) = \theta(1) = 3$. In G_2, we have

$$\theta(2) \cdot \theta(3) = 4 \cdot 2 = 3.$$

Table 4.2 Additive group $\mathbb{Z}_4$

G_1: +	0	1	2	3
0	0	1	2	3
1	1	2	3	0
2	2	3	0	1
3	3	0	1	2

Table 4.3 Multiplicative group $\mathbb{Z}_5 \backslash \{0\}$

G_2: •	1	3	4	2
1	1	3	4	2
3	3	4	2	1
4	4	2	1	3
2	2	1	3	4

Hence $\theta(2 + 3) = \theta(2) \cdot \theta(3)$. Similarly,

$$\theta(g_1 + g_2) = \theta(g_1) \cdot \theta(g_2) \quad \forall\, g_1, g_2 \in G_1,$$

and so $G_1 \cong G_2$. Furthermore, the results of Example 4.12 can be verified for this example. Here $e_1 = 0$, $e_2 = 1$, and $\theta(0) = 1$. Again, with $g = 3$, we get $g^{-1} = 1$ and $\theta(g^{-1}) = \theta(1) = 3$. But $\theta(g) = \theta(3) = 2$ and $[\theta(g)]^{-1} = (2)^{-1} = 3$. Hence

$$\theta(g^{-1}) = [\theta(g)]^{-1}.$$

The relationship of an isomorphism $\theta: G_1 \to G_2$ between two groups G_1 and G_2 is a symmetric one, i.e.

$$G_1 \cong G_2 \Rightarrow G_2 \cong G_1.$$

To see this, let $y_1 = \theta(x_1)$ and $y_2 = \theta(x_2)$. Then

$$\begin{aligned}\theta^{-1}(y_1 y_2) = \theta^{-1}(\theta(x_1)\theta(x_2)) &= \theta^{-1}\theta(x_1 x_2)\\ &= x_1 x_2 = \theta^{-1}(y_1)\theta^{-1}(y_2).\end{aligned}$$

Furthermore, $G_1 \cong G_1$ and

$$G_1 \cong G_2, \qquad G_2 \cong G_3 \quad \Rightarrow \quad G_1 \cong G_3.$$

Thus, isomorphism determines an equivalence relation for groups, and it is the equivalence classes arising from the relation of isomorphism that are of importance when studying the abstract properties of groups.

Example 4.14

Suppose $G_1 = G\ell(n, \mathbb{R})$ with matrix multiplication and $G_2 = \mathbb{R}\backslash\{0\}$ with multiplication in $\mathbb{R}$ for which the identity is 1. Then the mapping $\theta: G\ell(n, \mathbb{R}) \to \mathbb{R}\backslash\{0\}$ defined by $\theta(\boldsymbol{A}) = \det \boldsymbol{A}$ is a nonbijective group homomorphism. It is a homomorphism because, for all $\boldsymbol{A}, \boldsymbol{B} \in G\ell(n, \mathbb{R})$, we have $\det(\boldsymbol{AB}) = (\det \boldsymbol{A})(\det \boldsymbol{B})$. It is not an injective mapping (and therefore not bijective) since there are many nonsingular matrices having the same determinant. For example (Fig. 4.9),

$$\boldsymbol{A} = \begin{bmatrix} 4 & 3 \\ 3 & 4 \end{bmatrix}, \qquad \boldsymbol{B} = \begin{bmatrix} 10 & 1 \\ 3 & 1 \end{bmatrix}.$$

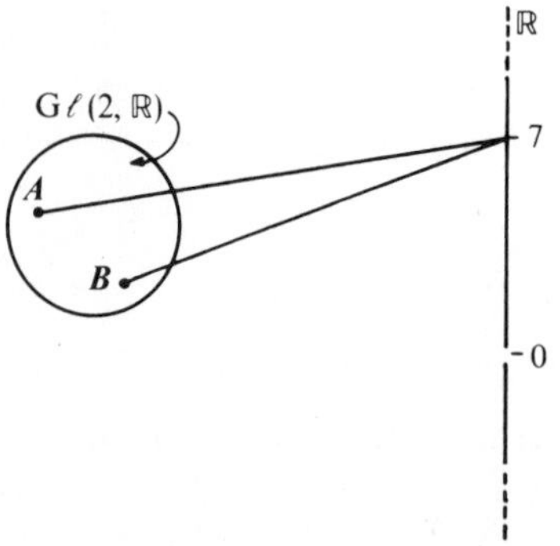

Fig. 4.9 A nonbijective group homomorphism

Thus, a homomorphism does not need to be either an onto map or a 1–1 map. A homomorphism of G_1 onto G_2 is called an *epimorphism.* A homomorphism which is 1–1 is called a *monomorphism.* A homomorphism of G into G is called an *endomorphism* and an isomorphism of G onto itself is called an *automorphism.*

The kernel of a group homomorphism $\theta: G_1 \to G_2$ is the subset of G_1 given by

$$\ker \theta \triangleq \{g \in G_1 : \theta(g) = e_2\},$$

where e_2 is the identity element of G_2. It can be shown that $\ker \theta$ is a subgroup of G_1 (Exercise 12).

Example 4.15

Let G_1, G_2 and $\theta: G_1 \to G_2$ be as given in Example 4.14. Then $\ker \theta$ is the subgroup of $G\ell(n, \mathbb{R})$ that contains all real $n \times n$ matrices having determinant 1, i.e.

$$\ker \theta = \{A \in G\ell(n, \mathbb{R}) : \det A = 1\}.$$

In most of what follows, G will denote a multiplicative group with identity 1. It is important to note that, for a subgroup H of G, the identity element in G and H are the same. [10]

A simple test to determine whether a subset H of a group G is a subgroup of G is the following (Whitelaw 1978). H is a subgroup iff

$$\text{(i) } H \neq \emptyset, \qquad \text{(ii) } x, y \in H \quad \Rightarrow \quad x^{-1} y \in H. \tag{4.22}$$

4.5 Cosets, normal subgroups, and quotient groups

Some subgroups of a group G give rise to special subsets of G known as *cosets.* In order to describe these cosets, multiplication of subsets of G must be defined. Suppose X and Y are non-empty subsets of a group G. Then the multiplication of X and Y is defined by

$$XY \triangleq \{xy : x \in X, y \in Y\}$$

and is a subset of G. [11] Two special cases of this are particularly important. If X is the singleton $\{x\}$, then

$$xY \triangleq \{xy : y \in Y\}.$$

Similarly, if Y is the singleton $\{y\}$, then

$$Xy = \{xy : x \in X\}.$$

Example 4.16

Let G be the multiplicative group $\{1, -1, \mathrm{j}, -\mathrm{j}\}$, where $\mathrm{j} = \sqrt{(-1)}$, and let $X = \{1, -1\}$ and $Y = \{\mathrm{j}, -\mathrm{j}\}$. Then

$$\begin{aligned} XY &= \{1\mathrm{j}, 1(-\mathrm{j}), (-1)\mathrm{j}, (-1)(-\mathrm{j})\} \\ &= \{\mathrm{j}, -\mathrm{j}, -\mathrm{j}, \mathrm{j}\} \\ &= \{\mathrm{j}, -\mathrm{j}\} = Y. \end{aligned}$$

On the other hand, if G and Y are the same as above, but $X = \{-1\}$, then

$$XY = (-1)Y = \{-\mathrm{j}, \mathrm{j}\} = Y.$$

Given that H is a proper subgroup of G then for any $g \in G$, the subset

$$gH = \{gh : h \in H\} \subseteq G$$

is called the *left coset* of H determined by g. Similarly, the *right coset* of H determined by g is the set

$$Hg = \{hg : h \in H\}.$$

Example 4.17

The group $(G, \circ)$ is given by $G = \{e, a, b, c, d, f\}$ and Table 4.4. It is easy to show that $H = \{e, c\}$ is a proper subgroup of G. The left cosets of H are (Fig. 4.10(a))

$$\begin{aligned} eH &= \{ee, ec\} = \{e, c\} = H = cH, \\ aH &= \{ae, ac\} = \{a, f\} = fH, \\ bH &= \{be, bc\} = \{b, d\} = dH. \end{aligned}$$

Table 4.4 Group G with subgroup $\{e, c\}$

$\circ$	e	a	b	c	d	f
e	e	a	b	c	d	f
a	a	b	e	f	c	d
b	b	e	a	d	f	c
c	c	d	f	e	a	b
d	d	f	c	b	e	a
f	f	c	d	a	b	e

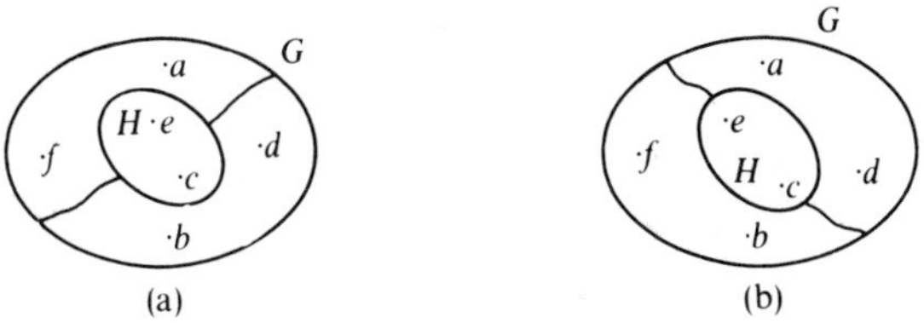

Fig. 4.10 Cosets of H: (a) left cosets, (b) right cosets

The right cosets are (Fig. 4.10(b))

$$He = \{ee, ce\} = \{e, c\} = H = Hc,$$

$$Ha = \{ea, ca\} = \{a, d\} = Hd,$$

$$Hb = \{eb, cb\} = \{b, f\} = Hf.$$

Notice that a coset can be referred to by different labels, e.g. $aH = fH$, even though $a \neq f$. The set of left cosets in Example 4.17 is a partition of G (Fig. 4.10(a)), and similarly for the set of right cosets (Fig. 4.10(b)). This is generally true, and so cosets may be considered as equivalence classes (Section 2.2).

Clearly, if G is an abelian group, then $gh = hg$ and so $gH = Hg$ for any $g \in G$ and $h \in H$. In this case, there is no need to distinguish between left and right cosets; both gH and Hg are termed the coset of H determined by g. The subgroup H is called a *normal subgroup of* G (denoted by $H \lhd G$) iff $gH = Hg$ for all $g \in G$.[12] A useful test on whether a given subgroup is normal is the following:

a subgroup H of a group G is a normal subgroup of G iff

$$ghg^{-1} \in H \quad \forall\, h \in H \quad \forall\, g \in G. \tag{4.23}$$

In the case when G is an abelian group,

$$ghg^{-1} = gg^{-1}h = 1h = h \in H \quad \forall\, h \in H \quad \forall\, g \in G$$

and so, by (4.23), every subgroup of an abelian group is normal.

Example 4.18

An abelian group $(G, \cdot)$ is given by $G = \{\boldsymbol{I}, -\boldsymbol{I}, \boldsymbol{A}, -\boldsymbol{A}\}$, where

$$\boldsymbol{I} = \begin{bmatrix} 1 & 0 \\ 0 & 1 \end{bmatrix}, \qquad \boldsymbol{A} = \begin{bmatrix} 1 & 0 \\ 0 & -1 \end{bmatrix},$$

with matrix multiplication as the group operation. The set $H = \{\boldsymbol{I}, \boldsymbol{A}\}$ is a subgroup.

The left cosets are

$$IH = \{I, A\} = H = AH, \qquad (-I)H = \{-I, -A\} = (-A)H,$$

and the right cosets are

$$HI = \{I, A\} = H = HA, \qquad H(-I) = \{-I, -A\} = H(-A).$$

So $IH = HI, (-I)H = H(-I), AH = HA, (-A)H = H(-A)$ and hence $H \lhd G$.

When $H \lhd G$, let $\mathscr{K} = \{gH : g \in G\}$ be the set of all cosets of H in G (left and right since $gH = Hg$). Suppose that K_1 and K_2 are any two cosets belonging to $\mathscr{K}$, so that $K_1 = g_1 H$ and $K_2 = g_2 H$ for some $g_1, g_2 \in G$. Under multiplication of subsets, consider the following products of cosets in which the associativity of subset multiplication in G is used (Exercise 13). [13]

$$K_1 K_2 = (g_1 H)(g_2 H) = g_1(Hg_2)H = g_1(g_2 H)H = (g_1 g_2)HH = (g_1 g_2)H,$$

the last step following from Exercise 14. Thus, $K_1 K_2 \in \mathscr{K}$, and $\mathscr{K}$ is closed under multiplication of subsets of G. Multiplication of cosets of H in G is thus governed by

$$(g_1 H)(g_2 H) = (g_1 g_2)H. \tag{4.24}$$

That is to say, given any two cosets of H, namely $g_1 H$ and $g_2 H$, determined by elements g_1 and g_2 of G respectively, the product of those two cosets is that coset of H determined by the element $g_1 g_2 \in G$. Multiplication of cosets in $\mathscr{K}$ is associative, by Exercise 13, since cosets are subsets of the group G. So, given any $K_3 \in \mathscr{K}$, we have that $K_3 = g_3 H$ for some $g_3 \in G$, and

$$\begin{aligned} K_1(K_2 K_3) &= (g_1 H)(g_2 H g_3 H) = (g_1 H g_2)(H g_3 H) = (g_1 g_2 H)(g_3 H) \\ &= (K_1 K_2)K_3. \end{aligned}$$

It has now been shown that $\mathscr{K}$ is a semigroup. In fact, it turns out that $\mathscr{K}$ is actually a group (Exercise 15), and this group $\mathscr{K}$ of cosets is called the *quotient group of G by H* and denoted by G/H. [14]

In Chapter 5 a similar discussion to the above will lead to the idea of a *quotient ring*. Part of the structure of a quotient ring involves an additive abelian group $(G, +)$ in which $(H, +)$ is a normal subgroup. [15] The corresponding quotient group $(G/H, +)$ is an additive abelian group, and each element of G/H is an additive coset of H in G. It will be helpful for the understanding of the work in Chapter 5 if the general results quoted above for a general quotient group are restated for the special case where the operation in G is addition. This is now done in the following paragraph.

For some $g \in G$, the corresponding coset $gH \in (G/H, +)$ is written as $g + H$. Just as gH was expressed earlier as the coset $\{gh : h \in H\}$, so

$$g + H = \{g + h : h \in H\}.$$

Table 4.5 Correspondence between multiplicative group G and additive group G

	Multiplicative group G	Additive group G
operation	xy	$x+y$
identity	1	0
inverse of x	x^{-1}	$-x$
set product	$XY=\{xy:x\in X, Y\in Y\}$	$X+Y=\{x+y:x\in X, y\in Y\}$
coset of subgroup H	xH, Hx	$x+H, H+x$
normal subgroup ($\forall\, x\in G$)	$xH=Hx$	$x+H=H+x$
quotient group G/H	$\{xH:x\in G\}$	$\{x+H:x\in G\}$
with identity	H	H
with inverse	$g^{-1}H$ of gH i.e. $(gH)^{-1}=g^{-1}H$	$(-g)+H$ of $g+H$ i.e. $-(g+H)=(-g)+H$

Equation (4.24) becomes

$$(g_1 + H) + (g_2 + H) = (g_1 + g_2) + H \quad \forall\, g_1, g_2 \in G.$$

The inverse of any element $g + H \in G/H$ is the negative of $g + H$ and is $(-g) + H$. The identity of G/H is, of course, the coset H as before (Exercise 15).

The correspondence between multiplication and addition in a group G is best illustrated in Table 4.5.

Example 4.19

Consider the additive group $\mathbb{Z}$ and the normal subgroup $3\mathbb{Z} = \{0, \pm 3, \pm 6, \pm 9, \dots\}$ of $\mathbb{Z}$ ($3\mathbb{Z} \lhd \mathbb{Z}$). The cosets of $3\mathbb{Z}$ are $0 + 3\mathbb{Z} = 3\mathbb{Z}$, $1 + 3\mathbb{Z} = \{\cdots, -8, -5, -2, 1, 4, 7, 10, \dots\}$, and $2 + 3\mathbb{Z} = \{\dots, -7, -4, -1, 2, 5, 8, 11, \dots\}$. Other cosets such as $3 + 3\mathbb{Z}$, $4 + 3\mathbb{Z}$, etc. are clearly alternative labels for the three cosets already obtained. Thus, the quotient group $\mathbb{Z}/3\mathbb{Z}$ is $\{0 + 3\mathbb{Z}, 1 + 3\mathbb{Z}, 2 + 3\mathbb{Z}\}$. Notice that the three cosets are just the congruence classes modulo 3, namely $[0], [1], [2]$ discussed in Section 2.3; this illustrates the remark made earlier that cosets are actually equivalence classes. It follows that, with a slight abuse of notation,

$$\mathbb{Z}/3\mathbb{Z} = \{[0], [1], [2]\} = (\mathbb{Z}_3, +).$$

Return now to the group homomorphism $\theta: G_1 \to G_2$ discussed earlier, with

$$H = \ker\theta = \{g \in G_1 : \theta(g) = e_2\} \subseteq G_1,$$

where e_2 is the identity in G_2. Then $\ker\theta$ is a normal subgroup of G_1. This is because, for all $g \in G_1$ and $h \in H$, we have

$$\theta(ghg^{-1}) = \theta(g)\theta(h)\theta(g^{-1}) = \theta(g)e_2[\theta(g)]^{-1} \text{ (by (4.21))}$$
$$= \theta(g)[\theta(g)]^{-1} = e_2.$$

So $ghg^{-1} \in H$ and hence, by (4.23), H is a normal subgroup of G_1, i.e. $\ker\theta \lhd G_1$. Thus the kernel of a group homomorphism is a normal subgroup. It follows that the quotient group $G_1/\ker\theta$ has identity $\ker\theta$.

Given any normal subgroup H of a group G (i.e. $H \lhd G$), define the map $\nu: G \to G/H$ by $\nu(g) = gH$ for all $g \in G$. Then ν is an epimorphism of G onto G/H with H as kernel (Exercise 16) and is called the *natural epimorphism from G onto $G/\ker\nu$*. Obviously $\operatorname{im}\nu = G/\ker\nu$, since ν is surjective.

Example 4.20

If $G = \mathbb{Z}$ and $H = 3\mathbb{Z}$, then the natural epimorphism $\nu: \mathbb{Z} \to \mathbb{Z}/3\mathbb{Z}$ is defined by $n \mapsto 3n$, and $\ker\nu = 3\mathbb{Z}$.

A generalization of the result concerning the natural epimorphism given above is that, if $\theta: G_1 \to G_2$ is a homomorphism between two groups G_1 and G_2, then $\operatorname{im}\theta = \theta(G_1) \cong G_1/\ker\theta$ (Exercise 17). This result is known as the *first isomorphism theorem.*

Example 4.21

Suppose that $G_1 = \mathrm{G}\ell(n, \mathbb{R})$ and $G_2 = \mathbb{R}\backslash\{0\}$. Consider the homomorphism $\theta: \mathrm{G}\ell(n, \mathbb{R}) \to \mathbb{R}\backslash\{0\}$ given by $X \mapsto \det X$. The kernel of this homomorphism is the normal subgroup

$$H = \{X \in \mathrm{G}\ell(n, \mathbb{R}): \det X = 1\} = \ker\theta,$$

and $\operatorname{im}\theta = \mathbb{R}\backslash\{0\}$. The first isomorphism theorem then states that

$$\mathbb{R}\backslash\{0\} \cong \mathrm{G}\ell(n, \mathbb{R})/\ker\theta.$$

There is a slight difficulty when defining operations on equivalence classes (or cosets) in order to form a group such as the quotient group G/H. This difficulty stems from the fact that equivalence classes and cosets can be referred to by different labels. For example, suppose the addition of equivalence classes is to be defined by

$$[a] + [b] \triangleq [a + b]. \tag{4.25}$$

If $p \in [a]$ and $q \in [b]$, so that $[p] = [a]$ and $[q] = [b]$, then

$$[a] + [b] = [p] + [q] = [p + q],$$

by (4.25). Is the equivalence class $[p+q]$ the same as the equivalence class $[a+b]$? In other words, does $p+q$ belong to $[a+b]$? If $[p+q] \neq [a+b]$, then the result of adding equivalence classes together will depend upon whether $[a]$ or $[p]$ is used to express one of the equivalence classes, and similarly with $[b]$ or $[q]$. This state of affairs would clearly be most unsatisfactory. However, because the equivalence classes are the cosets of a *normal* subgroup, different labelling of the classes yield the same results. In other words,

$$[p+q] = [a+b].$$ [16]

The reason for this is best illustrated by looking at a particular equivalence relation which occurs in linear dynamical system theory.

Example 4.22 (Kalman *et al.* 1969)

For the input–output homomorphism $f: \Omega \to \Gamma$ defined by $\boldsymbol{u} \mapsto f(\boldsymbol{u})$ from input sequence $\boldsymbol{u}$ to output sequence $f(\boldsymbol{u})$ discussed in Section 4.2, an equivalence relation $\sim$ can be defined for $\boldsymbol{u}, \boldsymbol{v} \in \Omega$ by

$$\boldsymbol{u} \sim \boldsymbol{v} \quad \text{iff} \quad f(\boldsymbol{u}) = f(\boldsymbol{v}) \tag{4.26}$$

or, what is the same thing, by

$$\boldsymbol{u} \sim \boldsymbol{v} \quad \text{iff} \quad \boldsymbol{u} - \boldsymbol{v} \in \ker f \subset \Omega. \tag{4.27}$$

Suppose $\boldsymbol{u}'$ and $\boldsymbol{v}'$ are arbitrary elements of $[\boldsymbol{u}]$ and $[\boldsymbol{v}]$ respectively. It is required to prove that $\boldsymbol{u}' + \boldsymbol{v}' \in [u+v]$. First, $\boldsymbol{u}, \boldsymbol{u}' \in [\boldsymbol{u}]$, so that $f(\boldsymbol{u}) = f(\boldsymbol{u}')$, i.e. $f(\boldsymbol{u}) - f(\boldsymbol{u}') = \mathbf{0}$. But f is a homomorphism, so that

$$f(\boldsymbol{u}) - f(\boldsymbol{u}') = f(\boldsymbol{u} - \boldsymbol{u}') \qquad \text{(cf. (4.19))}$$

and therefore

$$f(\boldsymbol{u} - \boldsymbol{u}') = \mathbf{0}.$$

Thus, $\boldsymbol{u} - \boldsymbol{u}' \in \ker f$. Similarly, $\boldsymbol{v} - \boldsymbol{v}' \in \ker f$. Since $\ker f$ is a subspace of Ω, we get

$$(\boldsymbol{u} - \boldsymbol{u}') + (\boldsymbol{v} - \boldsymbol{v}') \in \ker f,$$

i.e. $(\boldsymbol{u} + \boldsymbol{v}) - (\boldsymbol{u}' + \boldsymbol{v}') \in \ker f$. This implies that $\boldsymbol{u} + \boldsymbol{v} \sim \boldsymbol{u}' + \boldsymbol{v}'$, whence

$$\boldsymbol{u}' + \boldsymbol{v}' \in [\boldsymbol{u} + \boldsymbol{v}].$$

Exercises

1. Show that $(\mathbb{R}^{n \times n}, \cdot)$ is a multiplicative semigroup with identity $\mathbf{I}_n$.
2. Show that the set F of all mappings $f: X \to X$ from some non-empty set X into itself is a semigroup with identity under the operation of composition $(g, f) \mapsto g \circ f$.

3. Show that
   (i) $G\ell(n, \mathbb{R})$ with identity $\mathbf{I}_n$ and inverse A^{-1} for each $A \in G\ell(n, \mathbb{R})$ is a multiplicative group,[17]
   (ii) $\mathbb{R}^{m \times n}$ is an additive abelian group.

4. Consider the set G of all bijective mappings $g: X \to X$ under the binary operation of composition. Show that G is then a group with identity id_X and inverse g^{-1} for each $g \in G$.

5. Show that
   (i) $\mathbb{Q}$ is an abelian group with respect to addition on $\mathbb{R}$
   (ii) $\mathbb{Q} \backslash \{0\}$ is an abelian group with respect to multiplication on $\mathbb{R}$
   (iii) the multiplicative group $G\ell(n, \mathbb{R})$ is nonabelian for $n \geqslant 2$.

6. Prove that the scalar linear differential equation $\dot{x}(t) = ax(t)$, with $a \in \mathbb{R}$ and $x(0) = x_0$, has a solution $x(t) = \mathrm{e}^{at} x_0$, and that the set of operators $\{\mathrm{e}^{at} : t \in \mathbb{R}\}$ is an abelian group.

7. The set S is the same as that given in Example 4.2. Prove that S is a nonabelian group of order 8.

8. Show that the set $\mathrm{So}(n, \mathbb{R})$ of all real $n \times n$ orthogonal matrices forms a group with respect to matrix multiplication.

9. Consider the multiplicative group $(G, \cdot)$, where $G = \{1, -1, \mathrm{j}, -\mathrm{j}\}$ in which $\mathrm{j} = \sqrt{(-1)}$. Demonstrate that $(H, \cdot)$ is a subgroup of $(G, \cdot)$ when $H = \{1, -1\}$, and give the multiplication tables for $(H, \cdot)$ and $(G, \cdot)$. Show also that $(G, \cdot)/(H, \cdot) \cong (H, \cdot)$.

10. Show that the set of matrices

$$S = \left\{ \begin{bmatrix} a & -b \\ b & a \end{bmatrix} : a, b \in \mathbb{R} \right\},$$

with respect to addition of matrices, is isomorphic to the set of complex numbers $\mathbb{C}$, with respect to addition. Demonstrate that this result is also true when the operations on the two sets are multiplication on $\mathbb{R}^{2 \times 2}$ and multiplication on $\mathbb{C}$ respectively.

11. Prove that the mapping $\theta: \mathbb{R} \to G\ell(2, \mathbb{R})$ defined by

$$\theta(t) = \begin{bmatrix} \mathrm{e}^{at} & 0 \\ 0 & \mathrm{e}^{at} \end{bmatrix}, \qquad a \in \mathbb{R},$$

from the additive group of real numbers to the multiplicative group $G\ell(2, \mathbb{R})$, is a homomorphism.

12. Prove that the kernel of a group homomorphism $\theta: G_1 \to G_2$ is a subgroup of G_1.

13. For any group G prove that the power set $\mathscr{P}(G)$ is a semigroup with respect to multiplication of sets as defined in Section 4.5.

14. Given that H is a subgroup of a multiplicative group G, prove that $HH = H$.

15. $\mathscr{H} = \{gH : g \in G\}$ is the set of all cosets of H in a group G. Prove that $\mathscr{H}$ is a group.

16. Given that $H \lhd G$, prove that the map $\nu: G \to G/H$, defined by $\nu(g) = gH$ for all $g \in G$, is an epimorphism of G onto G/H with kernel H.

17. If $\theta: G_1 \to G_2$ is a group homomorphism, prove that $\mathrm{im}\,\theta$ is isomorphic to the quotient group $G_1/\ker\theta$.

18. Consider the homomorphism $\theta: \mathbb{R} \to \mathbb{C}\backslash\{0\}$ defined by $\theta(t) = \cos 2\pi t + \mathrm{j} \sin 2\pi t$ from the additive group $(\mathbb{R}, +)$ to the multiplicative group $\mathbb{C}\backslash\{0\}$. Prove that $\ker \theta = \mathbb{Z}$ and $\operatorname{im} \theta = G$, where G is the multiplicative group of all complex numbers on the unit circle in the argand diagram. Hence show that the additive quotient group $\mathbb{R}/\mathbb{Z}$ is isomorphic to the multiplicative group G.

Notes for Chapter 4

1. Often the symbol $\circ$ for a binary operation is omitted, and $a \circ b$ is written as ab. For example, composition of mappings is a binary operation and it has already been mentioned in Chapter 3 that a composite mapping involving two mappings f and g is generally written as $g \circ f$, but more often as gf when operator-style notation is used, i.e. when fx is used to denote $f(x)$.
2. Suppose that a semigroup $(S, \circ)$ has two identities $e,u \in S$. If it can be proved that $u = e$, then it follows that $(S, \circ)$ can have only one identity element. It is easy to prove that $u = e$ because, since e is an identity, $u \circ e = u$. Also, since u is an identity, $u \circ e = e$, and it follows that $u = e$. Thus, if a semigroup has an identity, then that identity is unique.
3. How to confirm that the system is in the zero state is an unexplained aspect of the Kalman theory.
4. In Kalman's formulation, a nonzero state before the input is applied has no meaning, because 'states' are defined as equivalence classes of inputs.
5. Suppose $\bar{a},a' \in S$ are both inverses of $a \in S$. To prove that $\bar{a} = a'$, let e be the identity element in $(S, \circ)$; then $a \circ a' = e$, since a' is an inverse of a. By associativity,

$$\bar{a} = \bar{a} \circ e = \bar{a} \circ (a \circ a') = (\bar{a} \circ a) \circ a'.$$

But $\bar{a} \circ a = e$, since $\bar{a}$ is an inverse of a. Hence $\bar{a} = e \circ a' = a'$. This proof is also valid when S is a group G (Section 4.3).
6. It is permissible to talk of *the* identity element e of a group G, because a group is a semigroup with unique identity element (see Note 2 above).
7. These are the only two possible values for $\det \boldsymbol{X}$ when $\boldsymbol{X}$ is an orthogonal matrix, because then $\boldsymbol{X}^{\mathrm{T}}\boldsymbol{X} = \boldsymbol{I}$. Taking determinants of this equation yields $\det \boldsymbol{X}^{\mathrm{T}}\boldsymbol{X} = \det \boldsymbol{I}$, i.e. $\det \boldsymbol{X}^{\mathrm{T}} \det \boldsymbol{X} = 1$. But $\det \boldsymbol{X}^{\mathrm{T}} = \det \boldsymbol{X}$, so that $(\det \boldsymbol{X})^2 = 1$, whence $\det \boldsymbol{X} = \pm 1$.
8. 'There exists' is often abbreviated to the sign $\exists$ in symbolic mathematical statements.
9. Arising from the properties of groups are a number of simple results which are easily proved (Whitelaw 1978). Among these are that, for any elements g, x, y belonging to a multiplicative group G, the cancellation laws are

$$gx = gy \quad \Rightarrow \quad x = y, \qquad xg = yg \quad \Rightarrow \quad x = y,$$

and the division laws are

$$xg = y \quad \Leftrightarrow \quad g = x^{-1}y, \qquad gx = y \quad \Leftrightarrow \quad g = yx^{-1}.$$

From the division laws it can be deduced that, if e is the identity in G, then $xy = e$ and $yx = e$ each imply that $y = x^{-1}$.

10. Suppose 1_H and 1_G are the identities of H and G respectively. Then, by definition, $1_H 1_H = 1_H$. But $1_H \in G$, so that $1_G 1_H = 1_H$. Thus, $1_H 1_H = 1_G 1_H$ and, by the cancellation law of Note 9, $1_H = 1_G$.
11. It is worth noting that $\varnothing Y = \varnothing$. Recall that XY is the set of all elements xy where $x \in X$ and $y \in Y$; hence, if X has no elements ($X = \varnothing$), then XY has no elements, and so $XY = \varnothing$, i.e. $\varnothing Y = \varnothing$. Thus, in the power set $\mathscr{P}(G)$ (see Example 2.9), the empty set acts as the zero element.
12. Note that $gH = Hg \Rightarrow gHg^{-1} = H$, where

$$gHg^{-1} = \{ghg^{-1} : h \in H\}.$$

For any two elements x and g belonging to a group G, the elements g and xgx^{-1} are called *conjugates* in G. Another name for a normal subgroup is a *self-conjugate subgroup*. The set gHg^{-1} is a conjugate of the set H.
13. Associativity in Exercise 13 generalizes naturally to non-empty subsets W, X, Y, Z of a group G and $(WX)(YZ)$ can be written as $WXYZ$ with the multiplications carried out in any order, e.g. as $(W(XY))Z$. When W is a singleton $\{g_1\} \subset G$, say, Y is a singleton $\{g_2\} \subset G$, and $X = Z = H$, then $(g_1 H)(g_2 H)$ can be written as $g_1(Hg_2)H$. Then $Hg_2 = g_2 H$, because $H \lhd G$.
14. Another name for a quotient group is *factor group*. The quotient group of G by H is sometimes called the quotient group G modulo H.
15. H must also be closed under multiplication by any element of G (i.e. $HG \subseteq H$ and $GH \subseteq H$) in order for a quotient ring to exist.
16. If a subgroup is not normal, one could still define an equivalence whose classes are the set of left cosets. However, on choosing any two of these left cosets, the group operation does not necessarily give another left coset. The smallest example of this is S_3, the symmetric group consisting of the six permutations of 1, 2, 3 (Patterson and Rutherford 1965).
17. A set containing singular matrices could not form a group under multiplication because $\boldsymbol{A}^{-1}$ does not exist when $\boldsymbol{A}$ is singular.
18. $\boldsymbol{A}\boldsymbol{A}^{-1} = \boldsymbol{I} \Rightarrow \det \boldsymbol{A}\boldsymbol{A}^{-1} = 1$, i.e. $\det \boldsymbol{A} \det \boldsymbol{A}^{-1} = 1$, whence $\det \boldsymbol{A}^{-1} = 1/\det \boldsymbol{A}$.

5

RINGS AND FIELDS

The algebraic systems discussed in Chapter 4 involve only one binary operation. But very often two operations can be associated with a set of mathematical objects. For example, real numbers in $\mathbb{R}$ can be added and multiplied together, as can real polynomials in $\mathbb{R}[z]$. By defining two binary operations on a general set, usually called addition and multiplication, such that the set becomes an abelian group with respect to addition and a semigroup with respect to multiplication, one is led to the concept of a *ring*. In a ring the two operations are linked by the concept of distributivity. A ring with the additional property that it is a multiplicative abelian group is called a *field*. There are a number of intermediate algebraic systems which must satisfy more properties than a general ring but less than a field (e.g. rings with identity, commutative rings, and integral domains). Many of the sets which arose in the examples of Chapter 1 turn out to be such algebraic systems and it is these systems which are now to be described in this chapter.

5.1 Rings

A ring is a non-empty set R with two binary operations of addition and multiplication which are connected by the distributive laws. These state that

$$a(b + c) = ab + ac, \qquad (b + c)a = ba + ca, \tag{5.1}$$

for all $a,b,c \in R$; furthermore, R is an abelian group with respect to addition and a semigroup with respect to multiplication. [1] Note that equations (5.1) are the only properties of a ring that relate the two binary operations of that ring.

Example 5.1

The set of integers $\mathbb{Z}$ give rise to the additive group $(\mathbb{Z}, +)$ by Example 4.9, and it is easy to prove that $\mathbb{Z}$ is a semigroup with respect to multiplication of integers. The distributive laws (5.1) are valid for all real numbers and so are true for integers. Thus, $(\mathbb{Z}, +, \cdot)$ is a ring—or simply $\mathbb{Z}$ is a ring, since the two operations on $\mathbb{Z}$ are clear. Note that the set of integers may be thought of either as a ring $\mathbb{Z}$ or as an additive (and abelian) group $(\mathbb{Z}, +)$: same set $\mathbb{Z}$ but different structures put on the set.

Example 5.2

The set of all real-valued functions $f\colon [0, 1] \to \mathbb{R}$ is a ring R when addition and multiplication of functions are defined (the so-called pointwise definitions) as follows:

$$(f+g)(x) \triangleq f(x)+g(x), \qquad (fg)(x) \triangleq f(x)g(x),$$

for all $x \in [0, 1]$ and $f, g \in R$. To prove this statement, all the properties of a ring must be checked—a laborious business. For example, what is the zero of the ring? what is the negative of $f(x)$? are the distributive laws satisfied? The reader should verify all the ring axioms for this example. Guidance can be obtained from MacLane and Birkoff (1979). As in an additive group, the identity 0 of addition in a ring R is called zero, and inverse of addition is called negative.

Some familiar results from arithmetic with real numbers carry over to rings in general. For example, suppose that a ring R has additive zero 0, i.e.

$$x+0=x \quad \forall\, x \in R.$$

Then [2], for all $x, y \in R$, we have

(i) $x0 = 0 = 0x$

(ii) $(-x)y = -(xy) = x(-y)$ (5.2)

(iii) $(-x)(-y) = xy.$

A ring which is a semigroup with identity (sometimes described as a *semigroup with a 1*) under multiplication is called a *ring with identity* (or a *ring with a 1*).

Example 5.3

$\mathbb{R}^{2\times 2}$ is a ring with identity, because under addition this set is an abelian group (Exercise 4.3(ii)) and under multiplication it is a semigroup with identity $\mathbf{I}_2$ (Exercise 4.1). The distributive laws (5.1) also hold under matrix addition and multiplication in $\mathbb{R}^{2\times 2}$.

The set of real-valued functions discussed in Example 5.2 is also a ring with identity because the identity function $\mathrm{id}_{[0,1]}\colon [0, 1] \to \mathbb{R}$ given by $\mathrm{id}_{[0,1]}(x) = x$ is a real-valued function.

Suppose that R is a ring with a 1 and an additive zero 0. If $0 = 1$, then $R = \{0\}$ because

$$x = 1x = 0x = 0 \quad \forall\, x \in R,$$

by (i) of (5.2). In this case R is called a *trivial* ring, similar to a trivial group mentioned in Section 4.3.

A commutative ring is a ring R with commutative multiplication so that $ab = ba$ for all $a, b \in R$. [3]

Example 5.4

The integers (modulo 3), i.e. $\mathbb{Z}_3 = \{0, 1, 2\}$, form a commutative ring. The addition and multiplication tables are given in Tables 5.1 and 5.2. From Table 5.1, $\mathbb{Z}_3$ is an abelian group with identity 0 and inverses

$$0^{-1} = 0, \qquad 1^{-1} = 2, \qquad 2^{-1} = 1.$$

Table 5.1 Addition table for $\mathbb{Z}_3$

+	0	1	2
0	0	1	2
1	1	2	0
2	2	0	1

Table 5.2 Multiplication table for $\mathbb{Z}_3$

•	0	1	2
0	0	0	0
1	0	1	2
2	0	2	1

From Table 5.2, $\mathbb{Z}_3$ is a semigroup with commutative multiplication, e.g. $1 \cdot 2 = 2$ and $2 \cdot 1 = 2$, etc. Proving associativity and distributivity directly is tedious. However, in this case they follow immediately, because these laws are true for $\mathbb{Z}$. Note that $\mathbb{Z}_3$ is also a commutative ring with identity 1.

A nonzero element a of a ring R is called a *left zero-divisor* if there exists a nonzero element b of R such that $ab = 0$. The element b is called a *right zero-divisor*. When R is commutative, the distinction between left and right zero-divisors disappears.

Example 5.5

In the noncommutative ring $\mathbb{R}^{2\times 2}$, we have

$$\boldsymbol{A} = \begin{bmatrix} 0 & 1 \\ 0 & 2 \end{bmatrix} \neq \mathbf{0}, \quad \boldsymbol{B} = \begin{bmatrix} -1 & 3 \\ 0 & 0 \end{bmatrix} \neq \mathbf{0},$$

but

$$\boldsymbol{AB} = \begin{bmatrix} 0 & 1 \\ 0 & 2 \end{bmatrix}\begin{bmatrix} -1 & 3 \\ 0 & 0 \end{bmatrix} = \begin{bmatrix} 0 & 0 \\ 0 & 0 \end{bmatrix} = \mathbf{0}.$$

Thus, $\boldsymbol{A}$ is a left zero-divisor and $\boldsymbol{B}$ a right zero-divisor in $\mathbb{R}^{2\times 2}$. Note that, although

$$\boldsymbol{BA} = \begin{bmatrix} -1 & 3 \\ 0 & 0 \end{bmatrix}\begin{bmatrix} 0 & 1 \\ 0 & 2 \end{bmatrix} = \begin{bmatrix} 0 & 5 \\ 0 & 0 \end{bmatrix} \neq \mathbf{0},$$

$\mathbf{A}$ is a right zero-divisor in $\mathbb{R}^{2\times 2}$, because

$$\begin{bmatrix} -2 & 1 \\ 2 & -1 \end{bmatrix}\begin{bmatrix} 0 & 1 \\ 0 & 2 \end{bmatrix} = \begin{bmatrix} 0 & 0 \\ 0 & 0 \end{bmatrix}.$$

An *integral domain* is a commutative ring with a nonzero identity 1 and with no zero-divisors.

Example 5.6

Consider the set $\mathbb{Z}_2 = \{0, 1\}$ with addition and multiplication given in Tables 5.3 and 5.4. The ring $\mathbb{Z}_2$ is an integral domain, because 0 is the zero

Table 5.3 Addition table for $\mathbb{Z}_2$

+	0	1
0	0	1
1	1	0

Table 5.4 Multiplication table for $\mathbb{Z}_2$

•	0	1
0	0	0
1	0	1

element, with 1 as the identity element, and each element is its own negative. The only nonzero element is 1 and $1 \cdot 1 = 1 \neq 0$, so there are no zero-divisors. The integral domain $\mathbb{Z}_2$ is very important in coding theory and computer theory (Green 1986).

Suppose that R is an integral domain and $x, y \in R$. Then it follows from the fact that there are no zero-divisors in R that

$$xy = 0 \quad \Rightarrow \quad x = 0 \text{ or } y = 0 \text{ (or both).}$$

Similarly, for any nonzero element $a \in R$,

$$ax = ay \quad \Rightarrow \quad x = y.$$

This is the cancellation law of multiplication [4] (cf. Note 4.9).

Subsets of a ring R can be added and multiplied together in the following way. If S and T are subsets of a ring R, then define

$$S + T \triangleq \{s + t: \quad s \in S, \quad t \in T\}$$

$$ST \triangleq \left\{ \sum_{i=1}^{n} s_i t_i : s_i \in S,\ t_i \in T,\ n \in \mathbb{N} \right\}.$$ [5]

Example 5.7

Take $R = \mathbb{Z}$, $S = \{0, 3\}$, $T = \{1, 2\}$. Then

$$\begin{aligned} S + T &= \{0 + 1,\ 0 + 2,\ 3 + 1,\ 3 + 2\} \\ &= \{1, 2, 4, 5\}, \\ ST &= \{0 \cdot 1,\ 0 \cdot 2,\ 3 \cdot 1,\ 3 \cdot 2,\ 0 \cdot 1 + 0 \cdot 2,\ 0 \cdot 2 + 3 \cdot 1,\ 3 \cdot 1 + 3 \cdot 1,\ 3 \cdot 1 + 3 \cdot 2, \ldots\} \\ &= \{0, 3, 6, 9, \ldots, 3n, \ldots\}. \end{aligned}$$

A *subring* of a ring R is a subset of R that is also a ring under the operations of R. Thus, a non-empty subset S of R is a subring of R iff it is closed under subtraction [6] and multiplication. That is, S is a subring of R iff

$$a - b \in S \quad \text{and} \quad ab \in S \quad \forall\ a, b \in S.$$

In other words, S is a subring of R iff

$$S - S = S \quad \text{and} \quad SS \subseteq S.$$

Note that $a + (-a) = 0 \in S$. (cf. H is a subgroup of group $(G, +)$ iff $H \neq \varnothing$ and $H - H \subseteq H$).

Example 5.8

$4\mathbb{Z} = \{a \in \mathbb{Z}: 4|a\} = \{0, \pm 4, \pm 8, \pm 12, \ldots\}$ is a subring of $\mathbb{Z}$.[7]

An important fact to remember is that, if a ring R with identity has a subring S, then the identity of S may not be the same as that of R.

Example 5.9

$R = \mathbb{R}^{2\times 2}$, $S = \left\{\begin{bmatrix} a & 0 \\ 0 & 0 \end{bmatrix}: \; a \in \mathbb{R}\right\}$. The identity in the ring R is $\begin{bmatrix} 1 & 0 \\ 0 & 1 \end{bmatrix}$, but the identity in the subring S is $\begin{bmatrix} 1 & 0 \\ 0 & 0 \end{bmatrix}$. This is essentially the same point as that made in Example 4.8.

The set of polynomials $\mathbb{R}[z]$ in the indeterminate z with real coefficients arose in the discussion of linear discrete systems in Example 1.2. The set $\mathbb{R}[z]$ is actually a ring and, because of its obvious importance in dynamical system theory, this ring will now be considered in more detail.

5.2 The polynomial ring $\mathbb{R}[z]$

An element $p \in \mathbb{R}[z]$ can be represented by

$$p(z) = p_0 + p_1 z + p_2 z^2 + \cdots + p_n z^n, \tag{5.3}$$

where $n \in \mathbb{N} \cup \{0\}$ and $p_i \in \mathbb{R}$ $(0 \leqslant i \leqslant n)$. Once the sequence of real numbers $(p_0, \ldots, p_n)$ is known, the polynomial p is uniquely determined in the above form. It is therefore reasonable to represent p by the infinite sequence $(p_0, \ldots, p_n, 0, 0, \ldots)$, denoted by (p_i), in which all but a finite number of entries are zero. Note that a polynomial p is here thought of as a single element in the space of all such real polynomials, rather than as the value $p(z)$ corresponding to some particular value of z (which, as the z in z-transforms, is an element of $\mathbb{C}$). This is why z is termed an indeterminate; it is the coefficients in (5.3) which determine the polynomial element p (see also Note 1.12).

The *degree of* p represented by (5.3), denoted by $\deg p$, is the integer n for which p_n is the last nonzero real number in the infinite sequence (p_i). A polynomial p of degree zero is just a real number p_0. If the degree of p is n and $p_n = 1$, then the polynomial p is said to be *monic*. Two polynomials $p = (p_i)$ and $q = (q_i)$ can be added together and multiplied according to the rules

$$p + q = (p_i + q_i), \qquad pq = (r_i), \tag{5.4}$$

where

$$r_i = p_0 q_i + p_1 q_{i-1} + \cdots + p_i q_0$$

$$= \sum_{k=0}^{i} p_k q_{i-k} = \sum_{k+j=i} p_k q_j.$$

Example 5.10

$p(z) = p_0 + p_1 z + \cdots + p_n z^n$ and $q(z) = q_0 + q_1 z + \cdots + q_m z^m$, with $m < n$. We have

$$p(z) + q(z) = (p_0 + q_0) + (p_1 + q_1)z + \cdots + (p_m + q_m)z^m$$

$$+ p_{m+1} z^{m+1} + \cdots + p_n z^n,$$

an element of $\mathbb{R}[z]$ which can be denoted by $p + q$.

$$p(z)q(z) = p_0 q_0 + (p_0 q_1 + p_1 q_0)z + (p_0 q_2 + p_1 q_1 + p_2 q_0)z^2 + \cdots + p_n q_m z^{n+m},$$

again an element of $\mathbb{R}[z]$ which can be denoted by pq.

$\mathbb{R}[z]$ is a commutative ring with identity. To see this, note that $\mathbb{R}[z]$ is an abelian group under addition. Also, it is easy to show that addition of polynomials is associative (Exercise 6), the zero polynomial $0 \in \mathbb{R}[z]$ is the infinite sequence (p_i) in which $p_i = 0$ for $i = 0, 1, \ldots,$ [8] and the inverse $-p$ of the polynomial p in (5.3) is $(-p_i)$ defined by

$$-p(z) = -p_0 - p_1 z - p_2 z^2 - \cdots - p_n z^n,$$

so that

$$p + (-p) = 0 = (-p) + p.$$

Multiplication in $\mathbb{R}[z]$ is commutative since, if $p = (p_i)$ and $q = (q_i)$ are in $\mathbb{R}[z]$, then

$$pq = (r_i),$$

where r_i is given by (5.4), and

$$qp = (s_i),$$

where

$$s_i = q_0 p_i + q_1 p_{i-1} + \cdots + q_i p_0$$

$$= p_i q_0 + p_{i-1} q_1 + \cdots + p_0 q_i \quad \text{(multiplication on } \mathbb{R} \text{ is commutative)}$$

$$= r_i \quad \text{(addition on } \mathbb{R} \text{ is commutative)}$$

Therefore $qp = pq$ for all $p, q \in \mathbb{R}[z]$, and so multiplication is commutative on $\mathbb{R}[z]$. Finally, multiplication on $\mathbb{R}[z]$ is associative (Exercise 7), and $\mathbb{R}[z]$ is a semigroup under multiplication with identity $e = (e_i) = (1, 0, 0, \ldots)$, so that $e(z) = 1 \in \mathbb{R}$. [9] Every $p \in \mathbb{R}[z]$ satisfies $pe = p = ep$. Thus, since the distributive law can be established by an argument analogous to that used in the solution to Exercise 7, $\mathbb{R}[z]$ is a commutative ring with a nonzero identity. It is also an integral domain, because it contains no zero-divisors, as will now be shown.

If two polynomials p and q are nonzero elements of $\mathbb{R}[z]$, then they can be represented by

$$p(z) = p_0 + p_1 z + \cdots + p_n z^n, \qquad q(z) = q_0 + q_1 z + \cdots + q_m z^m,$$

for some integers $m,n \geqslant 0$, where p_n and q_m are nonzero. Then the polynomial pq is not the zero polynomial 0, because the last term of pq is $p_n q_m z^{n+m}$ with $p_n q_m \neq 0$. Hence, if p and q are nonzero polynomials, then the product polynomial pq is also nonzero, i.e. $pq \neq 0$, and so $\mathbb{R}[z]$ has no zero-divisors. Thus $\mathbb{R}[z]$ is an integral domain.

Note that a polynomial of degree n does not have an inverse with respect to multiplication on $\mathbb{R}[z]$ unless $n = 0$. For example, if p is such that $p(z) = 1 - z \in \mathbb{R}[z]$, then

$$p(z) \cdot \frac{1}{1-z} = 1 \quad \text{but} \quad \frac{1}{1-z} \notin \mathbb{R}[z].$$

When a polynomial such as $z^3 + z^2 - z + 2 \in \mathbb{R}[z]$ of degree 3 is divided by another polynomial, say $z^2 - z - 6 \in \mathbb{R}[z]$ of degree 2, the result is

$$\frac{z^3 + z^2 - z + 2}{z^2 - z - 6} = z + 2 + \frac{7z + 4}{z^2 - z - 6}$$

or alternatively

$$z^3 + z^2 - z + 2 = (z + 2)(z^2 - z - 6) + 7z + 4.$$

In general, given $f,g \in \mathbb{R}[z]$ with $\deg g \leqslant \deg f$ and $g \neq 0$, there exist $q,r \in \mathbb{R}[z]$ such that

$$f = qg + r$$

and either $\deg r < \deg g$ or $r = 0$. This result is known as the *Euclidean division property*. The Euclidean function is a map $\phi\colon \mathbb{R}[z]\setminus\{0\} \to \mathbb{N} + \{0\}$ defined by $\phi(f) = \deg f$.

5.3 Homomorphisms, ideals, and quotient rings

A homomorphism of a ring R into a ring S is a map $\varphi\colon R \to S$ such that

$$\varphi(a + b) = \varphi(a) + \varphi(b), \qquad \varphi(ab) = \varphi(a)\varphi(b), \tag{5.5}$$

for all $a,b \in R$, i.e. a map that preserves the ring operations. Thus, a ring homomorphism is such that the image of a sum equals the sum of the images and the image of a product equals the product of the images (Fig. 5.1). In

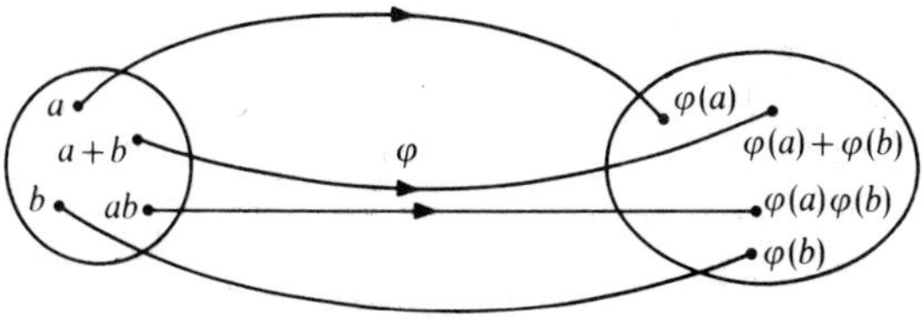

Fig. 5.1 A ring homomorphism

particular, if φ is a ring homomorphism, then it is automatically a group homomorphism from the group $(R, +)$ to the group $(S, +)$ and so, by (4.20) and (4.21),

$$\varphi(0_R) = 0_S \quad \text{and} \quad \varphi(-a) = -\varphi(a) \quad \forall\, a \in R,$$

where 0_R and 0_S are the zero elements of R and S respectively.

The image of a ring homomorphism $\varphi: R \to S$ is defined in the usual way as the set of all images of elements in R, i.e.

$$\text{im } \varphi = \varphi(R) \triangleq \{\varphi(r): r \in R\},$$

and is a subring of S.[10] To see this, let $x, y \in \text{im } \varphi$; then there exist $a, b \in R$ such that $\varphi(a) = x$ and $\varphi(b) = y$. Now (5.5) gives

$$x - y = \varphi(a) - \varphi(b) = \varphi(a - b) \in \text{im } \varphi,$$

$$xy = \varphi(a)\varphi(b) = \varphi(ab) \in \text{im } \varphi.$$

There are special types of ring homomorphisms, just as there are for group homomorphisms, and the same names are used for both (see Section 4.4). For example, a ring epimorphism is a ring homomorphism of R onto $S = \varphi(R)$. In particular, an isomorphism is a homomorphism which is 1–1 and onto (a bijective mapping). If $\varphi: R \to S$ is an isomorphism, then it has an inverse isomorphism $\varphi^{-1}: S \to R$. As in group theory, $R \cong S$ denotes that R is isomorphic to S.

As stated above, given a ring homomorphism $\varphi: R \to S$, then φ is also a group homomorphism from $(R, +)$ to $(S, +)$. Both these homomorphisms have a kernel defined by

$$\ker \varphi \triangleq \{r \in R: \varphi(r) = 0_S\}.$$ [11]

φ is a monomorphism (a 1–1 homomorphism) iff $\ker \varphi = \{0_R\}$. This is because φ is 1–1

$$\begin{array}{lll} \text{iff} & \varphi(a) = \varphi(b) & \Rightarrow \quad a = b \\ \text{iff} & \varphi(a) - \varphi(b) = 0 & \Rightarrow \quad a - b = 0_R \\ \text{iff} & \varphi(a - b) = 0_S & \Rightarrow \quad a - b = 0_R \\ \text{iff} & & \ker \varphi = \{0_R\}. \end{array}$$ [12]

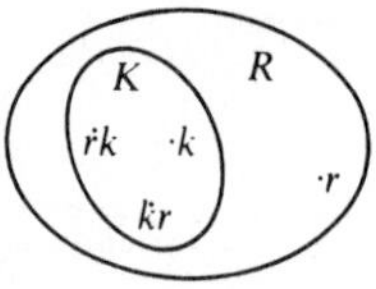

Fig. 5.2 An ideal K of a ring R

The kernel of φ is a subring of R. In fact it is more than that, as will be seen below.

An *ideal* of a ring R is a subring of R closed under multiplication by elements of R. Thus, a subring K of a ring R is an ideal of R (written $K \lhd R$, the same notation as for a normal subgroup H of a group G) iff, rk, kr, and $k - k'$ all belong to K, for all $k,k' \in K$ and all $r \in R$, i.e. $RK \subseteq K$ and $KR \subseteq K$ (Fig. 5.2).

Example 5.11

$K = 4\mathbb{Z}$ is an ideal of $\mathbb{Z}$. For any ring R, the sets $\{0\}$ and R itself are both ideals of R.

For a ring homomorphism $\varphi: R \to S$, the set $\ker \varphi$ is an ideal of R. To see this, let $K = \ker \varphi$. Since K is the kernel of the additive group homomorphism induced by φ, it is a subgroup of $(R, +)$, and it is only necessary to show that $KR \subseteq K$ and $RK \subseteq K$. If $r \in R$ and $a \in K$, then

$$\varphi(ra) = \varphi(r)\varphi(a) = 0 = \varphi(ar),$$

and so $ra, ar \in K$. Thus, K is an ideal of R (i.e. $K \lhd R$). The converse is also true, i.e. an ideal of a ring R is the kernel of one or more homomorphisms.

Suppose that $K \lhd R$, and let $R/K \triangleq \{(K + r): r \in R\}$ represent the set of additive cosets (see Section 4.5) with addition and multiplication defined as

$$\begin{aligned} (K + r) + (K + s) &\triangleq K + (r + s), \\ (K + r)(K + s) &\triangleq K + (rs). \end{aligned} \tag{5.6}$$

In Section 4.5, it was seen that a coset may possibly be expressed in more than one way (e.g. $He = Hc$ in Example 4.17). Any additive coset in (5.6) may also have many different labels: $K + x$ and $K + x'$, where $x,x' \in R$, may both represent the same coset just as $[a]$ and $[b]$ can both represent the same equivalence class (Section 2.2). It is necessary to show that the addition and multiplication of cosets as defined in (5.6) are valid irrespective of which labels are being used for the cosets.[13] For example, if the coset $K + x$ can be relabelled $K + x'$ and $K + y$ by $K + y'$ it has to be shown that $K + x'y'$ is the same coset as $K + xy$, which we do as follows. Since $K + x = K + x'$, then $x - x' \in K$, and similarly $K + y = K + y'$ implies that $y - y' \in K$. But K is an ideal of R so that $(x - x')y \in K$ and $x'(y - y') \in K$. Since K is closed under addition,

$$(x - x')y + x'(y - y') \in K,$$

i.e. $xy - x'y' \in K$. Hence

$$K + xy = K + x'y'.$$

With addition and multiplication defined by (5.6), R/K is a ring with zero element K:

$$K = K + 0, \qquad (K + x) + (K + 0) = K + (x + 0) = K + x \quad \forall\, x \in R.$$

R/K is called the *quotient ring* of R by K.[14]

The natural homomorphism $\nu: R \to R/K$ given by $\nu(x) = K + x$ is a ring epimorphism with kernel K, i.e. $\ker \nu = K$, because, for all $k \in K$, we have $\nu(k) = K + k = K$, the zero element of R/K. Suppose that $\varphi: R \to S$ is a ring

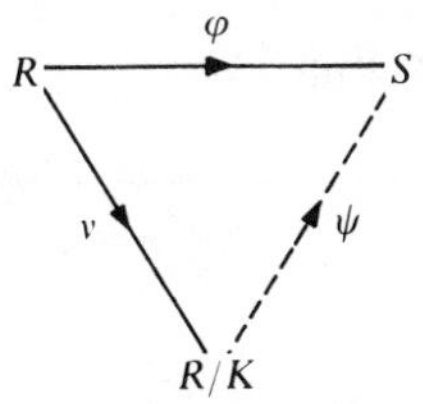

Fig. 5.3 $\varphi = \psi \circ \nu$

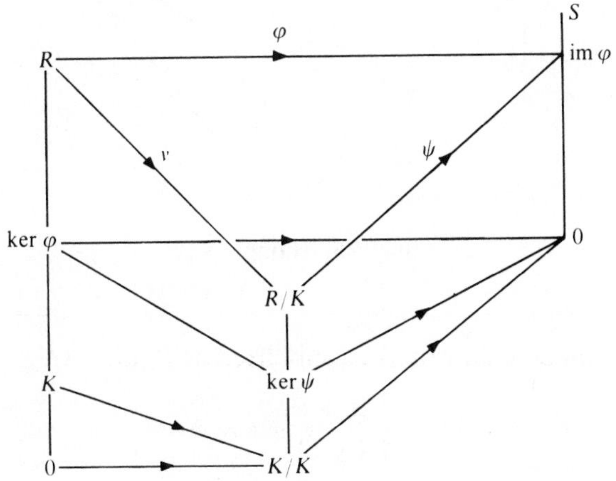

Fig. 5.4 First isomorphism theorem

homomorphism such that $\ker \varphi \supset K$. Then there exists a unique homomorphism $\psi: R/K \to S$ such that $\varphi = \psi \circ \nu$. Fig. 5.3 gives the commutative diagram illustrating this result. Furthermore, $\ker \psi = \ker \varphi/K$.

Just as for groups (Section 4.5), there is a first isomorphism theorem for rings. This states that, if $\varphi: R \to S$ is a ring homomorphism, then $R/\ker \varphi \cong \text{im } \varphi$, i.e. any two homomorphic images of R with the same kernel are isomorphic and $\ker \psi = \ker \varphi/K$ (Fig. 5.4).

Example 5.12

$R = \mathbb{Z}$, $K = 6\mathbb{Z}$, $\varphi: \mathbb{Z} \to \mathbb{Z}_3 = \text{im } \varphi$. We have

$$\mathbb{Z}/6\mathbb{Z} \cong \mathbb{Z}_6 = \{0, 1, 2, 3, 4, 5\}, \qquad \ker \varphi = 3\mathbb{Z} \supseteq K = 6\mathbb{Z}.$$

For $n \in \mathbb{Z}$, write $n = 6q + r$, where $r \in \{0, 1, 2, 3, 4, 5\}$. Then

$$\nu(n) = 6\mathbb{Z} + n = 6\mathbb{Z} + r,$$

and $\psi(6\mathbb{Z} + r) = 3\mathbb{Z} + s$, where $r = 3k + s$ with $s \in \{0, 1, 2\}$. Here,

$$\ker \psi = \{0, 3\}.$$

These results are illustrated in Fig. 5.5.

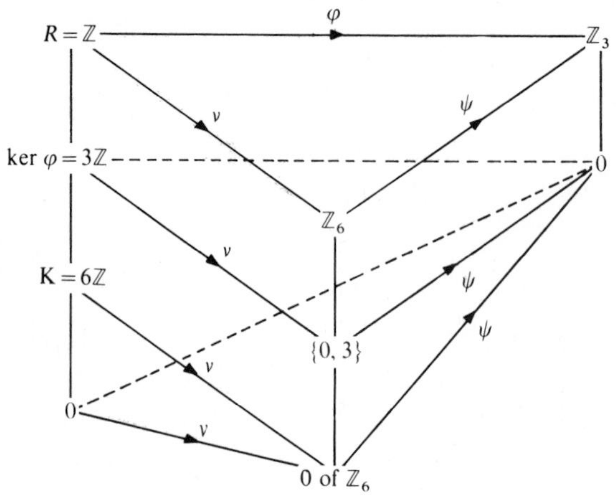

Fig. 5.5 Example 5.12

For an integral domain R and for any fixed $r \in R$, the set

$$rR = \left\{ \sum_{i=1}^{n} rx_i : x_i \in R,\ n \in \mathbb{N} \right\}$$
$$= \{rx : x \in R\}$$

is an ideal of R, called the ideal *generated* by r. [15] Such an ideal is called a *principal* ideal of R. In other words, a principal ideal K of an integral domain R is one that is generated by a single element $r \in R$, so that $K = rR$.

A ring R is called a *principal-ideal domain* (PID) if R is an integral domain and every ideal of R is principal. A PID here should not be confused with a proportional-integral-derivative controller well-known in classical control practice!

Example 5.13

As a consequence of the Euclidean division property described in Section 5.2, $\mathbb{R}[z]$ is a principal ideal domain. Certainly it was shown in Section 5.2 that $\mathbb{R}[z]$ is an integral domain, so it only remains to show that every ideal of $\mathbb{R}[z]$ is principal. Suppose that $K \lhd \mathbb{R}[z]$, and assume that $K \neq \{0\}$ (since $\{0\}$ is certainly a principal ideal). The set of degrees of the polynomial elements in K is a non-empty set of nonnegative integers, and so contains a smallest member. Let $g \in K$ be such that $g \neq 0$ and $\deg g \leqslant \deg k$ for all $k \in K$. The fact that $K = g\mathbb{R}[z]$ is now proved by showing that $g\mathbb{R}[z] \subseteq K$ and $K \subseteq g\mathbb{R}[z]$.

First $g \in K \lhd \mathbb{R}[z]$, so it follows immediately that $g\mathbb{R}[z] \subseteq K$. Now let $f \in K$. Then, for some $q,r \in \mathbb{R}[z]$, we have

$$f = qg + r,$$

with either $\deg r < \deg g$ or $r = 0$ by the Euclidean division property. Now

$$r = f - qg \in K.$$

If $r \neq 0$, then $\deg r < \deg g$, and this contradicts the assumption that $\deg g \leqslant \deg k$ for all $k \in K$. Therefore $r = 0$ and so $f = qg$, i.e. $f \in g\mathbb{R}[z]$. Thus, $K \subseteq g\mathbb{R}[z]$. It follows from the two results $g\mathbb{R}[z] \subseteq K$ and $K \subseteq g\mathbb{R}[z]$ that $K = g\mathbb{R}[z]$, and so $\mathbb{R}[z]$ is a PID.

Let R be a ring with identity but not necessarily a commutative ring. An element $u \in R$ is a *unit* of R if it has a multiplicative inverse in R.

Example 5.14

The only units of $\mathbb{Z}$ are 1 and -1, because $a \in \mathbb{Z}$ has an inverse $a^{-1} \in \mathbb{Z}$ iff $aa^{-1} = 1$, i.e. $a^{-1} = 1/a$. Now $1/a \in \mathbb{Z}$ only when $a = 1$ or -1, and so these are the only units of $\mathbb{Z}$.

If every nonzero element of R is a unit then R is a *skew field* (or *division ring*).

Let R be an integral domain. Then two elements $r,s \in R$ are called *associates* if $r|s$ and $s|r$. [7]

Example 5.15

Take $R = \mathbb{Q}$ and consider $3,6 \in \mathbb{Q}$. Now $3|6$ ($6 = 3 \cdot 2$ and $2 \in \mathbb{Q}$) and $6|3$ ($3 = 6 \cdot \frac{1}{2}$ and $\frac{1}{2} \in \mathbb{Q}$), so 3 and 6 are *associated in* $\mathbb{Q}$. Notice that they are not associates in $\mathbb{Z}$, because $3 = 6a$ does not hold for any $a \in \mathbb{Z}$.

5.4 Fields

In a ring, addition is very well behaved but multiplication can be difficult. For example, multiplication in $\mathbb{R}^{n \times n}$ is not commutative, nor do inverses necessarily exist. Algebraic systems known as fields are now to be described in which addition and multiplication are both nice. Indeed, they behave in a manner similar to those operations in $\mathbb{R}$ and $\mathbb{C}$.

A *field* is a commutative ring with identity in which the nonzero elements form a multiplicative group. Only the nonzero elements may belong to this group, because forming the multiplicative inverse is akin to division in $\mathbb{R}$, and division by the real number 0 is not permissible. A field can also be described as a commutative division ring; so, in a field, all nonzero elements are associates and have multiplicative inverses.

Example 5.16

$\mathbb{Q}$, $\mathbb{R}$, and $\mathbb{C}$ are all fields. $\mathbb{R}[z]$ is not a field because, as was seen in Section 5.2, z is not a unit in $\mathbb{R}[z]$, i.e. there does not exist a polynomial p such that $zp(z) = 1$. In other words, there is no inverse for z in multiplication.

Any integral domain D can be enlarged to (or embedded in) a field F such that every element of F can be expressed as a quotient of two elements of D. Such a field F is called a *field of quotients* of D.

Example 5.17

$D = \mathbb{Z}$, $F = \mathbb{Q}$. The rational numbers are formed by taking elements $a,b \in \mathbb{Z}$, with $b \neq 0$, to give the quotient $a/b \in \mathbb{Q}$.

Since $\mathbb{R}[z]$ is an integral domain, it is possible to construct the field of quotients $\mathbb{R}(z)$ of $\mathbb{R}[z]$. In other words, $\mathbb{R}(z)$ is the field of all rational functions in z. Any element of $\mathbb{R}[z]$ can be represented as a quotient $f(z) = p(z)/q(z)$ of two polynomials $p,q \in \mathbb{R}[z]$ with $q \neq 0$. We call f a *proper* rational function if $\deg p < \deg q$. The transfer functions g of single-input single-output systems, such as the one which arose in Example 1.5, are elements of $\mathbb{R}(s)$, the field of quotients of the integral domain $\mathbb{R}[s]$.[16]

Neither $(\mathbb{Z}, +, \cdot)$ nor $(\mathbb{R}^{n\times n}, +, \cdot)$ $(n \geqslant 2)$ is a field, but they are both rings. $\mathbb{Z}$ is nearer to being a field than $\mathbb{R}^{n\times n}$ $(n \geqslant 2)$ because multiplication is better behaved in $\mathbb{Z}$ than it is in $\mathbb{R}^{n\times n}$. For example, multiplication in $\mathbb{Z}$ is commutative whereas in $\mathbb{R}^{n\times n}$ $(n \geqslant 2)$ it is not. Again, $ab \neq 0$ for all nonzero $a,b \in \mathbb{Z}$, but this is not true in $\mathbb{R}^{n\times n}$, as the next example shows.

Example 5.18

$\boldsymbol{A},\boldsymbol{B} \in \mathbb{R}^{2\times 2}$, where

$$\boldsymbol{A} = \begin{bmatrix} 1 & 0 \\ 0 & 0 \end{bmatrix}, \qquad \boldsymbol{B} = \begin{bmatrix} 0 & 0 \\ 0 & 1 \end{bmatrix}.$$

Then $\boldsymbol{A} \neq \boldsymbol{0}$ and $\boldsymbol{B} \neq \boldsymbol{0}$. Nevertheless,

$$\boldsymbol{AB} = \begin{bmatrix} 0 & 0 \\ 0 & 0 \end{bmatrix} = \boldsymbol{BA}.$$

Generalizing the above discussion, the set of all $m \times n$ rational matrices (i.e. $m \times n$ matrices with entries from $\mathbb{R}(z)$) is denoted by $\mathbb{R}^{m\times n}$ (z). Transfer-function matrices $\boldsymbol{G}(s)$ with elements from $\mathbb{R}(s)$ (as in Example 1.6) belong to the space $\mathbb{R}^{m\times n}(s)$. A matrix $\boldsymbol{W} \in \mathbb{R}^{m\times n}(z)$ is called a proper rational matrix if all entries are proper rational functions.

Exercises

1. Show that
   (i) $3\mathbb{Z}$ is a commutative ring but without an identity,
   (ii) $\mathbb{Z}_2$ is a commutative ring with identity.
2. Prove that $\mathbb{Z}_6$ is not an integral domain.
3. If R is an integral domain and $a,x,y \in R$, then prove that
   (i) $-(xy) = x(-y)$,
   (ii) $ax = ay \;\Rightarrow\; a(x-y) = 0$.
4. Given that $S = 4\mathbb{Z}$ and $T = 6\mathbb{Z}$, calculate $S + T$ and ST.
5. A ring $R = \{0, a, b, c\}$ is given by the following tables:

$+$	0	a	b	c
0	0	a	b	c
a	a	0	c	b
b	b	c	0	a
c	c	b	a	0

$\cdot$	0	a	b	c
0	0	0	0	0
a	0	a	0	a
b	0	0	b	b
c	0	a	b	c.

Show that $S = \{0, a\}$ is a subring of R and that R and S have different identities.

6. Show that addition in $\mathbb{R}[z]$ is associative.

7. Demonstrate that multiplication in $\mathbb{R}[z]$ is associative.

8. $R = \left\{ \begin{bmatrix} a & b \\ -\bar{b} & \bar{a} \end{bmatrix} \in \mathbb{C}^{2\times 2} : a,b \in \mathbb{C} \right\}$. Show that R is a division ring which is not commutative.

9. Prove that every field is an integral domain. Explain why $\mathbb{Z}$ is an integral domain but not a field.

Notes for Chapter 5

1. As in Chapter 4, the addition of two elements $a,b \in R$ is denoted by $a+b$ and their multiplication by ab. Note that these operations are to be interpreted fairly liberally. For example, multiplication could mean composition of functions.
2. The proofs of these results are very instructive in that they use several properties of rings. The reader is advised to study each step in the proofs.

(i) $x+0=x$ so that $x+(-x)=0$. Then

$$x0 = x[x+(-x)] = xx + x(-x) = xx - xx = 0,$$

and similarly for $0x = 0$.

(ii) $xy + (-x)y = [x+(-x)]y = 0y = 0$ so that

$$(-x)y = -(xy),$$

and similarly for $x(-y) = -(xy)$.

(iii) From (ii),

$$(-x)(-y) = -[x(-y)] = -[-(xy)] = xy.$$

3. It is worth remarking that, in a commutative ring R, only one of the equations in (5.1) is required, because one implies the other. For example, take the first equation in (5.1), namely

$$a(b+c) = ab + ac.$$

Now, $ab = ba$ and $ac = ca$ for all $a,b,c \in R$. Therefore

$$a(b+c) = ba + ca.$$

But $a(b+c) = (b+c)a$, so that

$$(b+c)a = ba + ca,$$

and this is the second equation of (5.1). Similarly, this second equation implies the first. On the other hand, if R is not a commutative ring, then both equations are required.
4. If $ax = ay$, then the properties of a ring give $a(x-y) = 0$ (Exercise 3(ii)). Since the ring is an integral domain and $a \neq 0$, it follows that $x - y = 0$, whence $x = y$. Of course, if a ring has zero-divisors, then cancellation is not allowed, because $ax = ay$ $(a \neq 0) \not\Rightarrow x = y$.
5. Subsets S and T of a ring R can also be subtracted under the definition

$$S - T \triangleq \{s + (-t) : s \in S, t \in T\}.$$

Note that $S+T$ is defined as for an additive group (Table 4.5). Also, ST consists of all sums of a finite number of products st of elements $s \in S$ and $t \in T$.
6. Any ring R is an additive group with respect to addition, so that every $r \in R$ has an additive inverse $-r$. For all $r,s \in R$, subtraction is then defined by $r-s \triangleq r+(-s)$. The set R is closed under addition and so is also closed under subtraction. A non-empty subset S of an additive group is a subgroup if it is closed under subtraction. To see this, suppose that $S \neq \varnothing$ and S is closed under subtraction. Then there exists $a \in S$ for which $a-a \in S$ so that $O \in S$. Also, $0-x=-x \in S$ and

$$y+x=y-(-x) \in S, \quad \text{for all } x,y \in S.$$

7. $4|a$ means '4 divides a', i.e. there exists $b \in \mathbb{Z}$ such that $a=4b$ (see also Note 2.12). In general, for two elements a and b belonging to a set S, the notation $a|b$ means 'a divides b' which in turn implies that there exists an element $c \in S$ such that $b=ac$. In the discussion of $\mathbb{R}[z]$ in Section 5.2, if $f,g \in \mathbb{R}[z]$, then $g|f$ implies that $\deg g \leqslant \deg f$.
8. The degree of the zero element in $\mathbb{R}[z]$ is not defined.
9. The multiplication of $q \in \mathbb{R}[z]$ by any polynomial $p \in \mathbb{R}[z]$ of the form $p=(\alpha, 0, 0, \dots)$, with $\alpha \in \mathbb{R}$, is such that $pq=(\alpha q_i)$ which in turn can be written as αq. This can then be interpreted as multiplication of elements from different spaces: $\alpha \in \mathbb{R}$ and $q \in \mathbb{R}[z]$. The same situation occurs in vector spaces (Chapter 6).
10. If R is a ring, with identity 1, then so is $\varphi(R)$, and $\varphi(1)$ is its identity, but this may not be the identity in S.
11. Of course, given a ring homomorphism $\varphi: R \to S$, the map φ is also a semigroup homomorphism from $(R, \cdot)$ to $(S, \cdot)$. If R has an identity 1_R of multiplication, then the semigroup homomorphism could be defined as

$$\{r \in R: \varphi(r)=\varphi(1_R)\}.$$

One difficulty is that, although $\varphi(0_R)=0_S$, it may not be true that $\varphi(1_R)=1_S$.

12. The last 'if' is obvious because, if $\ker \varphi=\{0_R\}$, then

$$\varphi(a-b)=0_S \quad \Rightarrow \quad a-b \in \ker \varphi,$$

so that $a-b=0_R$. The last 'only if' is not quite so obvious. To prove that $[\varphi(a-b)=0_S \Rightarrow a-b=0_R] \Rightarrow \ker \varphi=\{0_R\}$, choose any $a \in \ker \varphi$ and $b=0$. Then, from the left-hand side, $\varphi(a)=0 \Rightarrow a=0$ and hence $\ker \varphi=\{0\}$.
13. The cosets $K+x$ are elements of the additive group $(R/K, +)$, so one really only has to deal with multiplication.
14. Other names for a quotient ring are *factor ring* and *residue class ring*. The quotient ring of R by K is sometimes called the quotient ring of R modulo K.
15. One needs commutativity so that $rx=xr$, and an identity is required in R so that $r=r \cdot 1 \in rR$.
16. In a field of quotients, equivalent fractions have to be identified. For example, define $\mathbb{Q}$ as a set of equivalence classes of pairs $(p,q) \in \mathbb{Z} \times (\mathbb{Z}\setminus\{0\})$ with $(p,q) \sim (p_1, q_1)$ iff $pq_1=qp_1$ in $\mathbb{Z}$. By denoting the equivalence class $[(p,q)]$ by $p/q \in \mathbb{Q}$ it is clear that 'p/q' has the same meaning in $\mathbb{Z}$ and $\mathbb{Q}$ under the embedding $p \mapsto [(p,1)]$ $(p \in \mathbb{Z})$.
Similarly, in $\mathbb{R}(z)$, we have $(p(z), q(z)) \sim (p_1(z), q_1(z))$ iff $p(z)q_1(z)=q(z)p_1(z)$ in $\mathbb{R}[z]$, and the elements of $\mathbb{R}(z)$ are the equivalence classes.

6

VECTOR SPACES AND MODULES

From the previous chapters, it is evident that mathematical objects such as vectors, matrices, polynomials, and mappings, all of which occur in dynamical system theory, can be added together and multiplied by real (or complex) numbers. The results of these mathematical operations of addition and scalar multiplication are vectors when acting on vectors, polynomials when acting on polynomials, and so on. Scalars need not necessarily be from $\mathbb{R}$ or $\mathbb{C}$; they can be elements from any field. Scalar multiplication by elements from a field gives rise to a *vector space* (also called a *linear space*). If, on the other hand, the scalars are chosen from a ring rather than from a field, then the algebraic system so formed is known as a *module*. The difference between a field and a module lies in the set of scalars; in a module, division by, and cancellation of, scalars is not always possible, whereas in a field it is.

6.1 Vector spaces

In the linear time-invariant second-order system (1.5) of Example 1.1,

$$\dot{\boldsymbol{x}}(t) = \boldsymbol{A}\boldsymbol{x}(t) + \boldsymbol{b}u(t), \tag{6.1}$$

the term $\boldsymbol{b}u(t)$ is the multiplication of the vector $\boldsymbol{b} = (0, 1) \in \mathbb{R}^2$ by the real number $u(t) = 1 - e^{-t}$. The result of this operation,

$$\boldsymbol{b}u(t) = \begin{bmatrix} 0 \\ 1 \end{bmatrix}(1 - e^{-t}) = \begin{bmatrix} 0 \\ 1 - e^{-t} \end{bmatrix},$$

is obtained by the usual definition of multiplication of a vector or matrix by a scalar. Furthermore, the result of this multiplication, being the vector $(0, 1 - e^{-t})$, also belongs to $\mathbb{R}^2$.

Again, in (6.1), the term $\boldsymbol{A}\boldsymbol{x}(t)$ is the matrix

$$\boldsymbol{A} = \begin{bmatrix} 0 & 1 \\ 0 & 0 \end{bmatrix} \in \mathbb{R}^{2 \times 2}$$

post-multiplied by the column vector $\boldsymbol{x} \in \mathbb{R}^2$. This results in the vector

$$\boldsymbol{A}\boldsymbol{x}(t) = \begin{bmatrix} 0 & 1 \\ 0 & 0 \end{bmatrix}\begin{bmatrix} x_1(t) \\ x_2(t) \end{bmatrix} = \begin{bmatrix} x_2(t) \\ 0 \end{bmatrix},$$

another vector belonging to $\mathbb{R}^2$. Hence, the right-hand side of (6.1) is the sum of two vectors in $\mathbb{R}^2$. Addition of vectors is defined as the addition of corresponding elements of the vectors (componentwise addition). Thus, (6.1) can be written as

$$\begin{bmatrix} \dot{x}_1(t) \\ \dot{x}_2(t) \end{bmatrix} = \begin{bmatrix} x_2(t) \\ 0 \end{bmatrix} + \begin{bmatrix} 0 \\ 1 - e^{-t} \end{bmatrix} = \begin{bmatrix} x_2(t) \\ 1 - e^{-t} \end{bmatrix} \tag{6.2}$$

and the result of the addition is seen to be another vector in $\mathbb{R}^2$. Equality between vectors is defined as equality between corresponding elements of the two vectors so that, from the first and last vectors in (6.2),

$$\dot{x}_1(t) = x_2(t), \qquad \dot{x}_2(t) = 1 - e^{-t} = u(t),$$

which are the original differential equations (1.4).

In the above discussion, vectors in $\mathbb{R}^2$ were added together and multiplied by scalars from $\mathbb{R}$. These ideas can be generalized to objects other than vectors, and scalars from fields other than $\mathbb{R}$. This then leads to the definition of a vector space. Let F be a field, and V an additive abelian group on which there is defined scalar multiplication of an element in V by an element from F to give an element in V. Then V is called a *vector space over* F (or an F-*vector space*) provided that, with 1 as the unity in F, the following conditions hold for all $m,n \in F$ and $u,v \in V$:

$$\begin{aligned} m(u+v) &= mu + mv \qquad \text{(linearity)}, \\ (m+n)u &= mu + nu, \\ m(nu) &= (mn)u \quad \text{and} \quad 1u = u. \end{aligned} \tag{6.3}$$

Example 6.1

The space $\mathbb{R}^n$ discussed in Section 1.5 is an abelian group with group operation given by componentwise addition as shown in Example 4.10 for $n = 2$. The group $\mathbb{R}^n$ becomes a vector space over $\mathbb{R}$ by defining scalar multiplication in the following way. For any $\alpha \in \mathbb{R}$ and for all $\boldsymbol{x} = (x_1, \ldots, x_n) \in \mathbb{R}^n$, [1] we define

$$\alpha \boldsymbol{x} = \alpha(x_1, \ldots, x_n) \triangleq (\alpha x_1, \ldots, \alpha x_n).$$

$\mathbb{R}^n$ is then called a vector space over the reals (i.e. the real numbers), or an $\mathbb{R}$-vector space.

Because of linearity in a vector space,

$$\alpha(\boldsymbol{x} + \boldsymbol{y}) = \alpha \boldsymbol{x} + \alpha \boldsymbol{y} \quad \forall\, \alpha \in \mathbb{R} \quad \forall\, \boldsymbol{x},\boldsymbol{y} \in \mathbb{R}^n. \tag{6.4}$$

Equation (6.4) states that the scalar multiple of the vector $\boldsymbol{x} + \boldsymbol{y}$ yields the

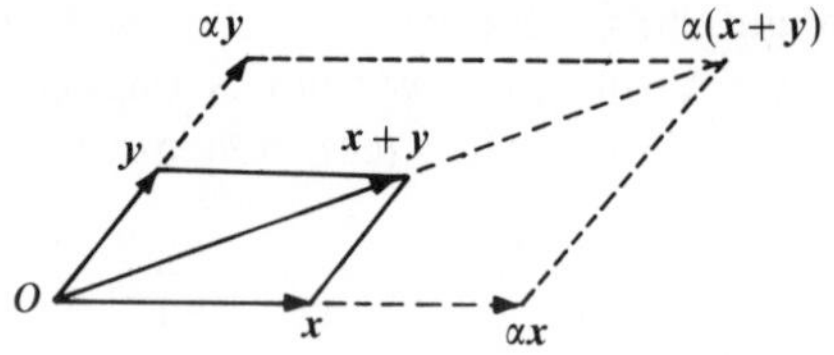

Fig. 6.1 Linearity of $\mathbb{R}^2$

same vector as the sum of the scalar multiples of $\boldsymbol{x}$ and $\boldsymbol{y}$ separately. This is illustrated for $\mathbb{R}^2$ in Fig. 6.1.

It is important to realize that elements of a vector space need not be vectors in the usual sense of the word. Nevertheless, elements in a vector space are called vectors even though they may actually be matrices, mappings, or other mathematical objects. An example where the 'vectors' of a vector space are matrices is now given.

Example 6.2

The space $\mathbb{R}^{2\times 2}$ is a vector space over $\mathbb{R}$, with scalar multiplication of a matrix

$$\boldsymbol{A} = \begin{bmatrix} a_{11} & a_{12} \\ a_{21} & a_{22} \end{bmatrix} \in \mathbb{R}^{2\times 2}$$

by a scalar $\alpha \in \mathbb{R}$ defined in the usual way [1] as

$$\alpha\boldsymbol{A} = \begin{bmatrix} \alpha a_{11} & \alpha a_{12} \\ \alpha a_{21} & \alpha a_{22} \end{bmatrix} \in \mathbb{R}^{2\times 2},$$

and addition of two matrices $\boldsymbol{A},\boldsymbol{B} \in \mathbb{R}^{2\times 2}$ defined by componentwise addition as

$$\boldsymbol{A}+\boldsymbol{B} = \begin{bmatrix} a_{11} & a_{12} \\ a_{21} & a_{22} \end{bmatrix} + \begin{bmatrix} b_{11} & b_{12} \\ b_{21} & b_{22} \end{bmatrix}$$
$$\triangleq \begin{bmatrix} a_{11}+b_{11} & a_{12}+b_{12} \\ a_{21}+b_{21} & a_{22}+b_{22} \end{bmatrix} \in \mathbb{R}^{2\times 2}.$$

Similarly, $\mathbb{R}^{m\times n}$ is an additive abelian group by Exercise 4.3(ii); so, by generalizing the above definitions, $\mathbb{R}^{m\times n}$ is a linear vector space over $\mathbb{R}$.

Within vector spaces there can exist other vector spaces called subspaces. A *subspace* of a vector space V over $\mathbb{R}$ is a subset of V that is also a vector space

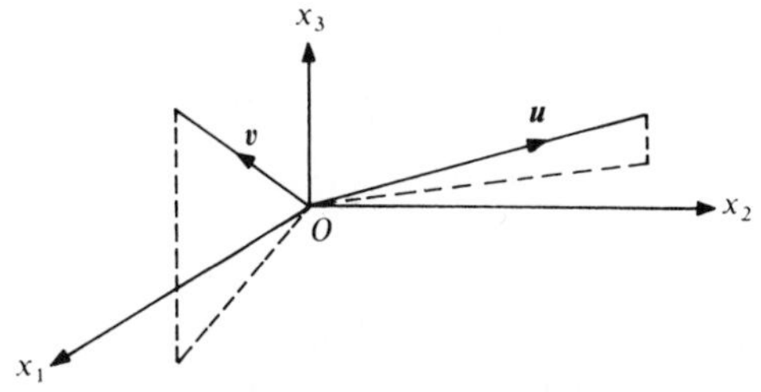

Fig. 6.2 A subspace of $\mathbb{R}^3$

over $\mathbb{R}$. Such a subset is therefore a subgroup of $(V, +)$. Furthermore, any subgroup of $(V, +)$ is naturally a subspace if it is closed under scalar multiplication. Consider the vector space $\mathbb{R}^3$ as an example. The elements of this space can be considered geometrically as the set of all vectors in three-dimensional space with origin O (Fig. 6.2). Choose any plane through O (e.g. the one defined by vectors $\boldsymbol{u}$ and $\boldsymbol{v}$ in Fig. 6.2). Then that plane is a vector space in its own right and so is a subspace of $\mathbb{R}^3$. It is a vector space over $\mathbb{R}$, because the sum of any two vectors in the plane is also a vector in the same plane, and similarly a scalar multiple of any vector in the plane also lies in the plane, the scalar belonging to $\mathbb{R}$. Since all subspaces are vector spaces, every subspace contains the null element, and so every subspace of $\mathbb{R}^3$ must pass through the origin.

Example 6.3

A subset of $\mathbb{R}^2$ is

$$H^2 = \{(x_1, x_2) \in \mathbb{R}^2 : x_2 \geqslant 0\}, \qquad \text{(Fig. 6.3)}$$

but H^2 is not a subspace (vector space) since it is not an additive group—it is not closed on taking negatives.

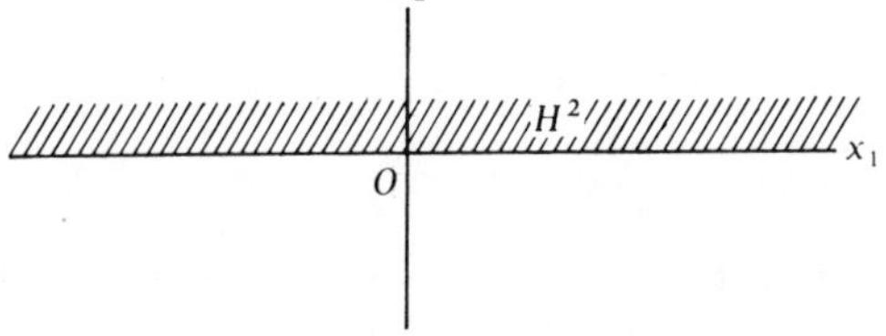

Fig. 6.3. Subset H^2 of $\mathbb{R}^2$

The set of points $\{(x_1, 0) : x_1 \in \mathbb{R}\}$ is also a subset of $\mathbb{R}^2$ and is the boundary of H^2, usually denoted by ∂H^2. Unlike H^2, the boundary ∂H^2 is a subspace of $\mathbb{R}^2$.

A mapping $f:X \to Y$ is called a *linear map* if

$$\begin{aligned} f(x+y) &= f(x)+f(y), \\ f(\alpha x) &= \alpha f(x), \end{aligned} \qquad (6.5)$$ [2]

for all $x,y \in X$ and $\alpha \in \mathbb{R}$.

The graph of a linear map $f:\mathbb{R} \to \mathbb{R}$ is a line $f(x) = ax$ through the origin, and the graph of a linear map $f:\mathbb{R}^2 \to \mathbb{R}$ is a plane $f(x, y) = ax + by$ through the origin. For a linear map $f: X \to Y$, with 0_X and 0_Y the zero elements in X and Y respectively, given any $x \in X$, we have

$$\begin{aligned} f(0_X) &= f(x-x) = f(x+(-x)) \\ &= f(x)+f(-x) && \text{(by (6.5))} \\ &= f(x)-f(x) && \text{(by (6.5))} \\ &= 0_Y \qquad \text{(by definition of } 0_Y). \end{aligned}$$

Thus, all linear maps $f:X \to Y$ take the zero element of X into the zero element of Y.

If X and Y are vector spaces, then (6.5) defines a vector-space homomorphism f from X to Y. It preserves the operations of vector addition and scalar multiplication, for the following reasons. On the left-hand sides of (6.5), these operations are applied in the vector space X and the resulting vectors are then mapped into Y; on the right-hand sides of (6.5), x and y are first mapped into Y and then the vector space operations are applied in the vector space Y. Equations (6.5) simply state that the resulting vectors on the left-hand and right-hand sides are the same. This preservation of vector-space operations under a linear map makes such maps on vector spaces particularly important. [3] The kernel

$$\{x : x \in X, f(x) = 0_Y\}$$

of a vector space homomorphism $f: X \to Y$ is a subspace of X, because

$$f(\alpha x + \beta y) = \alpha f(x) + \beta f(y) \quad \forall\ x,y \in X \quad \forall\ \alpha,\beta \in \mathbb{R},$$

so that, if $x,y \in \ker f$, then $\alpha x + \beta y \in \ker f$. Of course, $\ker f$ is the kernel of the corresponding additive group homomorphism. Furthermore, if $K = \ker f$, then

$$f(K+r) = f(r),$$

and the additive cosets modulo K turn out to be inverse images of elements in the image. With a vector space structure, a homomorphism has to preserve linearity, so the kernel is important.

6.2 Linear independence and bases

In Fig. 6.1, the addition of geometrical vectors in $\mathbb{R}^2$ by the parallelogram law was used to illustrate linearity as given by (6.4). The reader will no doubt be familiar with this geometrical interpretation, so the next example is simply to refresh the memory.

Example 6.4

Two geometrical vectors $\overrightarrow{OA}$ and $\overrightarrow{OC}$ are the representations of two points A and C in $\mathbb{R}^2$ (Fig. 6.4). These two points are given by the vectors (a, b) and (h, k). The parallelogram law then gives the geometical vector $\overrightarrow{OB}$ as the result of the vector addition $\overrightarrow{OA} + \overrightarrow{OC}$. Vector $\overrightarrow{OB}$ is the representation of the point B in $\mathbb{R}^2$ and this point is given by a vector (p, q), say. From Fig. 6.4, it is easy to deduce that $p = a + h$ and $q = b + k$. In vector notation, these two equations can be written in the form

$$\begin{bmatrix} p \\ q \end{bmatrix} = \begin{bmatrix} a \\ b \end{bmatrix} + \begin{bmatrix} h \\ k \end{bmatrix}.$$

Thus, the componentwise addition of the two vectors (a, b) and (h, k) yields the vector (p, q). Alternatively, this result can be interpreted as the sum of the two geometrical vectors $\overrightarrow{OA}$ and $\overrightarrow{OC}$ by the parallelogram law to give vector $\overrightarrow{OB}$. Either way the result is the same.

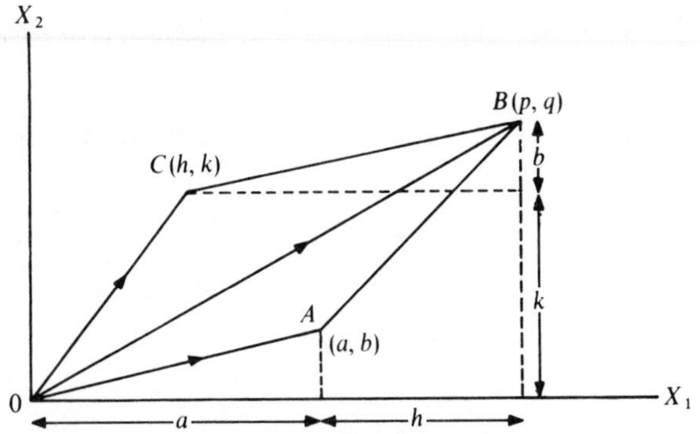

Fig. 6.4 Addition in $\mathbb{R}^2$

Two special vectors in $\mathbb{R}^2$ are (1, 0) and (0, 1). Looked upon as geometrical vectors, they are both of unit length, the former having the direction OX_1 and the latter OX_2 in Fig. 6.4. Clearly they are the same vectors as those usually denoted by $\boldsymbol{i}$ and $\boldsymbol{j}$ in vector analysis, but here the notation $\boldsymbol{e}_1$ and $\boldsymbol{e}_2$ will be used:

$$e_1 = \begin{bmatrix} 1 \\ 0 \end{bmatrix}, \quad e_2 = \begin{bmatrix} 0 \\ 1 \end{bmatrix}.$$

The importance of these unit vectors e_1 and e_2 is that they are orthogonal, and any vector x in $\mathbb{R}^2$ can be expressed as a sum of multiples of e_1 and e_2 (called a *linear combination* of e_1 and e_2) in the form

$$\begin{bmatrix} x_1 \\ x_2 \end{bmatrix} = x_1 \begin{bmatrix} 1 \\ 0 \end{bmatrix} + x_2 \begin{bmatrix} 0 \\ 1 \end{bmatrix}$$

i.e., with $e_1, e_2 \in \mathbb{R}^2$, we may write

$$x = x_1 e_1 + x_2 e_2.$$

If axes different from OX_1 and OX_2 are chosen, then the coordinates (x_1, x_2) of a given point relative to these new axes will be different from those relative to OX_1 and OX_2.

Example 6.5

Relative to e_1 and e_2, the point P has coordinates (1, 1) (Fig. 6.5). Taking axes OY_1 and OY_2 through the points (2, 1) and (1, 2) respectively, the unit vectors [4] in the directions OY_1 and OY_2 are $(2/\sqrt{5}, 1/\sqrt{5})$ and $(1/\sqrt{5}, 2/\sqrt{5})$ respectively. Then, if (α, β) are the coordinates of P relative to these unit vectors,

$$\begin{bmatrix} 1 \\ 1 \end{bmatrix} = 1 \begin{bmatrix} 1 \\ 0 \end{bmatrix} + 1 \begin{bmatrix} 0 \\ 1 \end{bmatrix} \qquad \text{(relative to axes } OX_1X_2\text{)}$$

$$= \alpha \begin{bmatrix} 2/\sqrt{5} \\ 1/\sqrt{5} \end{bmatrix} + \beta \begin{bmatrix} 1/\sqrt{5} \\ 2/\sqrt{5} \end{bmatrix} \qquad \text{(relative to axes } OY_1Y_2\text{)}.$$

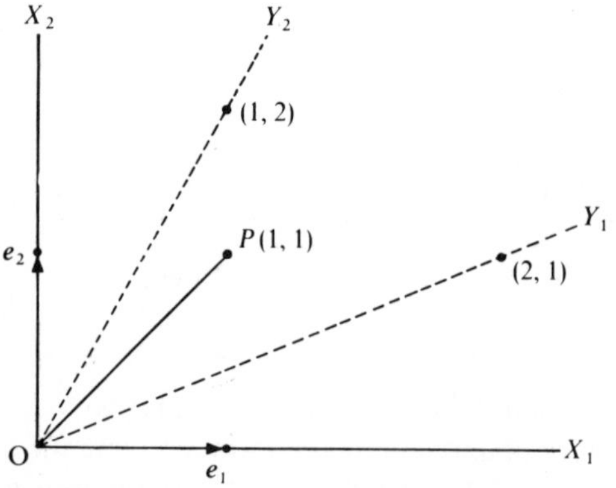

Fig. 6.5 Change of axes in $\mathbb{R}^2$

It follows that

$$\frac{2}{\sqrt{5}}\alpha + \frac{1}{\sqrt{5}}\beta = 1, \qquad \frac{1}{\sqrt{5}}\alpha + \frac{2}{\sqrt{5}}\beta = 1,$$

whence $\alpha = \sqrt{5}/3 = \beta$. Thus, the point P has coordinates $(\sqrt{5}/3, \sqrt{5}/3)$ relative to axes OY_1 and OY_2.

Unit vectors $\boldsymbol{e}_1$ and $\boldsymbol{e}_2$ are not the only two vectors that will generate all vectors in $\mathbb{R}^2$ through linear combinations, although they are the most useful since they are so simple and since they refer to orthogonal axes. In order to generate all vectors in $\mathbb{R}^2$, linear combinations of at least two vectors are required. Any set S of vectors in $\mathbb{R}^2$ such that every vector in $\mathbb{R}^2$ is a linear combination of the vectors in S is said to *span* the space $\mathbb{R}^2$. The minimum number of vectors which will span $\mathbb{R}^2$ is two but they must be linearly independent vectors, a term which is now defined. Two nonzero vectors $\boldsymbol{x} = (x_1, x_2)$ and $\boldsymbol{y} = (y_1, y_2)$ are said to be *linearly independent* if, for $c_1, c_2 \in \mathbb{R}$,

$$c_1\boldsymbol{x} + c_2\boldsymbol{y} = 0 \quad \Rightarrow \quad c_1 = 0 \text{ and } c_2 = 0.$$

In other words, $\boldsymbol{x}$ and $\boldsymbol{y}$ are linearly independent if neither is a multiple of the other. Geometrically, this means $\boldsymbol{x}$ and $\boldsymbol{y}$, interpreted as geometrical vectors, must not have the same (or opposite) direction. Two vectors that are not linearly independent are called *linearly dependent*.

Example 6.6

Two linearly independent vectors $\boldsymbol{x}$ and $\boldsymbol{y}$ in $\mathbb{R}^2$ must both be different from the null vector. Otherwise, if $\boldsymbol{x} = (0, 0)$, then

$$c_1(0, 0) + c_2(y_1, y_2) = (0, 0),$$

provided that $c_2 = 0$; the scalar c_1 can be chosen nonzero, say $c_1 = 1$. Thus, the null vector and any other vector in $\mathbb{R}^2$ form a linearly dependent set.

Example 6.7

Consider the vectors (2, 1) and (1, 2) in $\mathbb{R}^2$. For $c_1, c_2 \in \mathbb{R}$, the equation

$$c_1\begin{bmatrix} 2 \\ 1 \end{bmatrix} + c_2\begin{bmatrix} 1 \\ 2 \end{bmatrix} = \begin{bmatrix} 0 \\ 0 \end{bmatrix}$$

implies the equation pair

$$2c_1 + c_2 = 0,$$

$$c_1 + 2c_2 = 0.$$

In matrix notation, these two homogeneous linear equations in c_1 and c_2 can be written as

$$\begin{bmatrix} 2 & 1 \\ 1 & 2 \end{bmatrix}\begin{bmatrix} c_1 \\ c_2 \end{bmatrix} = \begin{bmatrix} 0 \\ 0 \end{bmatrix}. \tag{6.6}$$

The determinant of the 2×2 matrix in (6.6) is nonzero (actually 3) and so the only solution to (6.6) is $c_1 = c_2 = 0$ (Strang 1980). The two vectors (2, 1) and (1, 2) are therefore linearly independent. Any vector $(x_1, x_2) \in \mathbb{R}^2$ could then be written as a linear combination of vectors (2, 1) and (1, 2) in the form

$$c_1\begin{bmatrix} 2 \\ 1 \end{bmatrix} + c_2\begin{bmatrix} 1 \\ 2 \end{bmatrix} = \begin{bmatrix} x_1 \\ x_2 \end{bmatrix},$$

by choosing $c_1 = \frac{1}{3}(2x_1 - x_2)$ and $c_2 = \frac{1}{3}(2x_2 - x_1)$ (cf. Example 6.5).

Any set of two linearly independent vectors $\boldsymbol{x}, \boldsymbol{y} \in \mathbb{R}^2$ generate all vectors in $\mathbb{R}^2$ by linear combinations of $\boldsymbol{x}$ and $\boldsymbol{y}$. Such a set is called a *basis* for $\mathbb{R}^2$. [5] Any basis of $\mathbb{R}^2$ contains the minimum number (i.e. 2) of vectors which span $\mathbb{R}^2$ and this number 2 is called the *dimension* of the vector space $\mathbb{R}^2$. The particular basis $(\boldsymbol{e}_1, \boldsymbol{e}_2)$ is called the *standard* (or *canonical*) basis for $\mathbb{R}^2$. From Example 6.5, the reader will deduce that $\mathbb{R}^2$ does not have a unique basis. However, given a basis, $(\boldsymbol{x}, \boldsymbol{y})$, then the components v_1 and v_2 of any vector $\boldsymbol{v} \in \mathbb{R}^2$ are unique relative to this basis. So in Example 6.7, the vector (3, 3) (i.e. $x_1 = 3, x_2 = 3$), with coordinates relative to the basis $(\boldsymbol{e}_1, \boldsymbol{e}_2)$, is the vector (1, 1) (i.e. $c_1 = 1, c_2 = 1$) with coordinates relative to the basis ((2, 1), (1, 2)).

More than two (say $k > 2$) vectors in $\mathbb{R}^2$ are always linearly dependent. This is because the matrix $\boldsymbol{A}$, whose columns are the given vectors, is of order $2 \times k$ with $k > 2$. The rank of $\boldsymbol{A}$ is at most 2 and therefore less than k. The equation $\boldsymbol{Ac} = \boldsymbol{0}$, with $\boldsymbol{c} = (c_1, \ldots, c_k)$, therefore has more unknowns than equations and so there is a solution $\boldsymbol{c} \neq \boldsymbol{0}$ (Strang 1980). Thus, the k vectors are linearly dependent. This is illustrated in the next example.

Example 6.8

The vectors $\boldsymbol{u} = (2, 1)$, $\boldsymbol{v} = (1, 2)$ and $\boldsymbol{w} = (1, 1)$ in $\mathbb{R}^2$ are linearly dependent. To see this, consider the vector equation

$$c_1\boldsymbol{u} + c_2\boldsymbol{v} + c_3\boldsymbol{w} = \boldsymbol{0},$$

which can be written in matrix form as

$$\begin{bmatrix} 2 & 1 & 1 \\ 1 & 2 & 1 \end{bmatrix}\begin{bmatrix} c_1 \\ c_2 \\ c_3 \end{bmatrix} = \begin{bmatrix} 0 \\ 0 \end{bmatrix}.$$

This equation represents two equations in three unknowns. The solution is $c_1 = \lambda, c_2 = \lambda, c_3 = -3\lambda$, where $\lambda \in \mathbb{R}$ is arbitrary. Thus, a nonzero choice of λ illustrates the fact that $\boldsymbol{u}, \boldsymbol{v}, \boldsymbol{w}$ are linearly dependent. [6]

The ideas discussed above for the space $\mathbb{R}^2$ can be generalized immediately to vectors in $\mathbb{R}^n$ although, for $n > 3$, the generalizations are understandable only algebraically and not geometrically. Nonzero vectors $\boldsymbol{x}_1, \ldots, \boldsymbol{x}_n \in \mathbb{R}^n$ are said to be linearly independent if, for $c_1, \ldots, c_n \in \mathbb{R}$, the equation

$$c_1\boldsymbol{x}_1 + \cdots + c_n\boldsymbol{x}_n = \boldsymbol{0} \tag{6.7}$$

implies that $c_1 = \cdots = c_n = 0$. This means that no element of the set of vectors $\{\boldsymbol{x}_1, \ldots, \boldsymbol{x}_n\}$ can be written as a linear combination of the others. If any c, say c_k, can be chosen nonzero and (6.7) is valid, then the vectors $\boldsymbol{x}_1, \ldots, \boldsymbol{x}_n$ are linearly dependent. In this case, $\boldsymbol{x}_k$ can be written as a linear combination of $\boldsymbol{x}_1, \ldots, \boldsymbol{x}_{k-1}, \boldsymbol{x}_{k+1}, \ldots, \boldsymbol{x}_n$ in the form

$$\boldsymbol{x}_k = -\frac{c_1}{c_k}\boldsymbol{x}_1 - \cdots - \frac{c_{k-1}}{c_k}\boldsymbol{x}_{k-1} - \frac{c_{k+1}}{c_k}\boldsymbol{x}_{k+1} - \cdots - \frac{c_n}{c_k}\boldsymbol{x}_n.$$

Example 6.9

The vectors $\boldsymbol{e}_1, \ldots, \boldsymbol{e}_i, \ldots, \boldsymbol{e}_n$ in $\mathbb{R}^n$, where $\boldsymbol{e}_i = (0, 0, \ldots, 0, 1, 0, \ldots, 0)$, the unit component being in the ith position, form a set of n linearly independent vectors. To see this, consider the equation

$$c_1\boldsymbol{e}_1 + \cdots + c_n\boldsymbol{e}_n = \boldsymbol{0},$$

which can be written fully as

$$c_1\begin{bmatrix}1\\0\\\vdots\\0\\0\end{bmatrix} + c_2\begin{bmatrix}0\\1\\0\\\vdots\\0\end{bmatrix} + \cdots + c_n\begin{bmatrix}0\\0\\\vdots\\0\\1\end{bmatrix} = \begin{bmatrix}0\\0\\\vdots\\0\\0\end{bmatrix}. \tag{6.8}$$ [7]

Since $c_i = 0$ $(i = 1, \ldots, n)$, this implies that the vectors $\boldsymbol{e}_1, \ldots, \boldsymbol{e}_n$ are linearly independent.

Any set of n linearly independent vectors $\boldsymbol{x}_1, \ldots, \boldsymbol{x}_n \in \mathbb{R}^n$ is a basis for $\mathbb{R}^n$, and the set of vectors $\boldsymbol{e}_1, \ldots, \boldsymbol{e}_n$ of Example 6.9 is called the standard (or canonical) basis for $\mathbb{R}^n$. It follows from what has been said earlier for $\mathbb{R}^2$ that any set of vectors in $\mathbb{R}^n$ such that all linear combinations of them will generate all vectors in $\mathbb{R}^n$ is said to be a spanning set (or that the set spans $\mathbb{R}^n$). Again, n

vectors in $\mathbb{R}^n$ is the minimum number that will span the space, and n is the dimension of $\mathbb{R}^n$. As before, the dimension is the number of vectors in a basis.

The concepts of spanning set, basis, and dimension extend to all vector spaces, not just $\mathbb{R}^n$. This is illustrated in the next example.

Example 6.10

Any matrix

$$A = \begin{bmatrix} a_{11} & a_{12} \\ a_{21} & a_{22} \end{bmatrix}$$

belonging to the vector space $\mathbb{R}^{2\times 2}$ (Example 6.2) can be expressed as a linear combination

$$a_{11}\begin{bmatrix} 1 & 0 \\ 0 & 0 \end{bmatrix} + a_{12}\begin{bmatrix} 0 & 1 \\ 0 & 0 \end{bmatrix} + a_{21}\begin{bmatrix} 0 & 0 \\ 1 & 0 \end{bmatrix} + a_{22}\begin{bmatrix} 0 & 0 \\ 0 & 1 \end{bmatrix}.$$

The basis chosen here (the standard basis) is the set of four matrices

$$\left\{\begin{bmatrix} 1 & 0 \\ 0 & 0 \end{bmatrix}, \begin{bmatrix} 0 & 1 \\ 0 & 0 \end{bmatrix}, \begin{bmatrix} 0 & 0 \\ 1 & 0 \end{bmatrix}, \begin{bmatrix} 0 & 0 \\ 0 & 1 \end{bmatrix}\right\}. \tag{6.9}$$

Thus, the vector space $\mathbb{R}^{2\times 2}$ is of dimension 4 because there are four elements in this basis. The standard basis here is akin to the standard basis $\{e_1, e_2\}$ for the 2-dimensional vector space $\mathbb{R}^2$ and the reader should check that the matrices in (6.9) are linearly independent. These matrices are often denoted by E_{ij} with 1 in the (i, j)th position and zeros elsewhere, so that the set (6.9) becomes $\{E_{11}, E_{12}, E_{21}, E_{22}\}$. It follows that

$$A = \sum_{i=1}^{2}\sum_{j=1}^{2} a_{ij}E_{ij},$$

a form which is most useful for the Kronecker product (Graham 1981).

Generalization of Example 6.10 leads to the fact that the vector space $\mathbb{R}^{m\times n}$ is of dimension mn.

For any vector space V over a field F, a set of n linearly independent vectors in V is a basis for V if all sets of $n+1$ vectors in V are linearly dependent (i.e. all vectors can be expressed as linear combinations of the n independent ones). The vector space V is then said to be finite-dimensional of dimension n. Any basis for V is a linearly independent set of n vectors that span V. [8] If no such $n \in \mathbb{N}$ exists for V, then V is not finite-dimensional, in which case it is said to be infinite-dimensional.

Example 6.11

Let P be the set of all polynomials in $\mathbb{R}[x]$ of degree not greater than $n \in \mathbb{N}$, together with the zero polynomial 0. This set P, with the usual addition and multiplication by real numbers,. is a vector space over $\mathbb{R}$ (check!) Any polynomial $p(x) \in P$ such that

$$p(x) = p_0 + p_1 x + p_2 x^2 + \cdots + p_n x^n$$

is a linear combination of the polynomials $1, x, x^2, \dots, x^n$ with $p_i \in \mathbb{R}$. Furthermore, $p(x) = 0$ (the zero polynomial) iff $p_i = 0 \ (i = 0, \dots, n)$. Thus, the vectors in the set $\{1, x, x^2, \dots, x^n\}$ are linearly independent and the set spans P, so forming a finite basis for P. The dimension of P is therefore $n + 1$. Examples of infinite-dimensional vector spaces are $C[0, 1]$ and ℓ^2 (Kreyszig 1978).

If V_1 and V_2 are two vector spaces over the same field F, then V_1 and V_2 are said to be isomorphic if there is a bijective mapping $f: V_1 \to V_2$ such that

$$f(x + y) = f(x) + f(y), \qquad f(\alpha x) = \alpha f(x),$$

for all $x, y \in V_1$ and all $\alpha \in F$. The mapping f is an isomorphism of V_1 onto V_2.

6.3 Modules

The set of all multi-input real vector sequences

$$\boldsymbol{u} = (\dots, \mathbf{0}, \mathbf{0}, \boldsymbol{u}(-q), \boldsymbol{u}(-q+1), \dots, \boldsymbol{u}(-1), \boldsymbol{u}(0); \mathbf{0}, \mathbf{0}, \dots), \tag{6.10}$$

in which $\boldsymbol{u}(t) \in U \subseteq \mathbb{R}^m$ for $t \in \mathbb{Z}$, has been discussed in Section 4.2 and Example 1.3.[9] This set forms an input space, usually denoted by Ω. The m-vector $\bar{\boldsymbol{u}}(z)$ of (1.7) contains all the information on such an input sequence $\boldsymbol{u} \in \Omega$ and thus is a valid representation of the sequence. Elements of Ω may thus be represented by m-vectors such as $\bar{\boldsymbol{u}}(z)$ with components from the ring $\mathbb{R}[z]$, i.e. $\bar{\boldsymbol{u}}(z) \in \mathbb{R}^m[z]$. Since $\mathbb{R}[z]$ is an additive abelian group (because $\mathbb{R}[z]$ is a ring, cf. Section 5.2), so is $\mathbb{R}^m[z]$ under componentwise addition (cf. $\mathbb{R}$ and $\mathbb{R}^n$ in Example 4.10). Thus, $\mathbb{R}^m[z]$ can be given the structure of an $\mathbb{R}$-vector space by defining scalar multiplication, for all $\alpha \in \mathbb{R}$, as

$$\alpha(\bar{u}_1(z), \dots, \bar{u}_m(z)) \triangleq (\alpha \bar{u}_1(z), \dots, \alpha \bar{u}_m(z)), \tag{6.11}$$

just as for $\mathbb{R}^n$ in Example 6.1. An isomorphism is therefore established between all input sequences in Ω of the form (6.10) and the $\mathbb{R}$-vector space $\mathbb{R}^m[z]$, i.e. $\Omega \cong \mathbb{R}^m[z]$. Elements of the input space Ω may therefore be thought of as m-vectors with components from $\mathbb{R}[z]$.

Alternatively, the m-vectors of $\mathbb{R}^m[z]$ may be multiplied by elements from the ring $\mathbb{R}[z]$ instead of by elements from the field $\mathbb{R}$.

Example 6.12

$$\bar{\boldsymbol{u}}(z) = \boldsymbol{u}(0) + \boldsymbol{u}(-1)z + \boldsymbol{u}(-2)z^2 + \cdots + \boldsymbol{u}(-q)z^q$$

represents the input sequence (6.10). Multiplication of $\bar{\boldsymbol{u}}(z) \in \mathbb{R}^m[z]$ by the particular polynomial $z \in \mathbb{R}[z]$ yields

$$z\bar{\boldsymbol{u}}(z) = \boldsymbol{u}(0)z + \boldsymbol{u}(-1)z^2 + \boldsymbol{u}(-2)z^3 + \cdots + \boldsymbol{u}(-q)z^{q+1}.$$

This new polynomial in $\mathbb{R}^m[z]$ represents the sequence

$$(\ldots\ \mathbf{0},\ \mathbf{0},\ \boldsymbol{u}(-q),\ \boldsymbol{u}(-q+1)\ ,\ldots,\ \boldsymbol{u}(0),\ \mathbf{0};\ \mathbf{0},\ \mathbf{0}\ ,\ldots,\ ). \tag{6.12}$$

Thus, under multiplication by z, a mapping is defined by (6.10)→(6.12) and corresponds to a backward shift one place (right to left) together with the addition of a zero at $t = 0$.

Scalar multiplication of $\mathbb{R}^m[z]$ over $\mathbb{R}[z]$ is given, for all $\pi(z) \in \mathbb{R}[z]$ and $\bar{\boldsymbol{u}}(z) \in \mathbb{R}^m[z]$, by $\pi(z)\bar{\boldsymbol{u}}(z)$ as in Example 6.12. $\mathbb{R}^m[z]$ then becomes an algebraic system known as an $\mathbb{R}[z]$-*module* if $\alpha \in \mathbb{R}$ is replaced by $\pi(z) \in \mathbb{R}[z]$ in (6.11). Elements of this $\mathbb{R}[z]$-module can be added together and multiplied by scalars from $\mathbb{R}[z]$, and so the structure of the module is very similar to that of a vector space. The basic difference is that a vector space is defined over a field ($\mathbb{R}$ for an $\mathbb{R}$-vector space), whereas a module is defined over a ring with identity (such as $\mathbb{R}[z]$ for an $\mathbb{R}[z]$-module). [10] The formal definition of a module is as follows.

Given a ring R with identity 1 and an abelian group M, then M is an R-module if products of elements of M by scalars in R can be defined, satisfying the properties [11]

$$\text{(i)}\quad r_1 m_1 \in M,$$

$$\text{(ii)}\quad r_1(m_1 + m_2) = r_1 m_1 + r_1 m_2, \qquad \text{(iii)}\quad (r_1 + r_2)m_1 = r_1 m_1 + r_2 m_1,$$

$$\text{(iv)}\quad r_1(r_2 m_1) = (r_1 r_2)m_1, \qquad \text{(v)}\quad 1m_1 = m_1, \tag{6.13}$$

for all $r_1, r_2 \in R$ and all $m_1, m_2 \in M$. One may obviously associate with such an R-module a map $\varphi: R \times M \to M$ defined by $\varphi(r, m) = rm$ and satisfying (6.13). It follows from these properties that $O_R m_1 = O_M = r_1 O_M$ and [12]

$$(-r_1)m_1 = -(r_1 m_1) = r_1(-m_1). \tag{6.14}$$

Example 6.13

It is possible to make $\mathbb{Z}$ into a module over itself, i.e. a $\mathbb{Z}$-module. The set of integers $\mathbb{Z}$ is a ring with a 1. As part of its ring structure, $\mathbb{Z}$ is also an additive abelian group $(\mathbb{Z}, +)$ (see Example 4.9). Using the ordinary multiplication in $\mathbb{Z}$ as the product rm of a scalar $r \in \mathbb{Z}$ with $m \in (\mathbb{Z}, +)$, (i)–(iv) of (6.13) are satisfied by the closure, distributive, and associative laws for integers. Condition (v) just declares the property of the identity $1 \in \mathbb{Z}$. Hence, the set (group) of integers $(\mathbb{Z}, +)$ (or $\mathbb{Z}$, see Example 5.1) may be thought of as a module over the ring $\mathbb{Z}$, i.e. as a $\mathbb{Z}$-module. Again a map $\varphi: \mathbb{Z} \times (\mathbb{Z}, +) \to (\mathbb{Z}, +)$ can be defined as $(r, m) \mapsto rm$.

The $\mathbb{R}[z]$-module representing the input space Ω described above is illustrated in Fig. 6.6. The module mapping $\varphi: \mathbb{R}[z] \times \mathbb{R}^m[z] \to \mathbb{R}^m[z]$ is defined by $\varphi(\pi, \bar{\boldsymbol{u}}) = \pi(z)\bar{\boldsymbol{u}}(z)$.

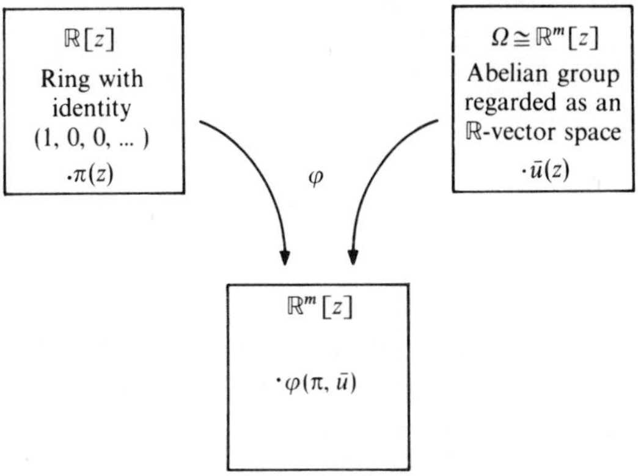

Fig. 6.6 $\mathbb{R}^m[z]$ as an $\mathbb{R}[z]$-module

From (1.7) it is easy to see that an element $\bar{\boldsymbol{u}}(z) \in \mathbb{R}^m[z]$ can be written as

$$\bar{\boldsymbol{u}}(z) = \bar{u}_1(z)\boldsymbol{e}_1 + \dots + \bar{u}_m(z)\boldsymbol{e}_m,$$

where $\boldsymbol{e}_j$ is the jth column of $\mathbf{I}_m$ and $\bar{u}_j(z) \in \mathbb{R}[z]$ $(j = 1, \dots, m)$. Because of this, $\mathbb{R}^m[z]$ is called a *finitely generated free* $\mathbb{R}[z]$-module with *generators* $\boldsymbol{e}_1, \dots, \boldsymbol{e}_m \in \mathbb{R}^m[z]$. The module $\mathbb{R}^m[z]$ is called *free* because the $\bar{u}_j(z)$ are unique for each $\bar{\boldsymbol{u}}(z)$ and the set of generators $\{\boldsymbol{e}_1, \dots, \boldsymbol{e}_m\}$ is a *basis* for $\mathbb{R}^m[z]$. This is why $\mathbb{R}^m[z]$ can be regarded as an $\mathbb{R}$-vector space simply by restricting the scalars from $\mathbb{R}[z]$ to polynomials of degree zero. [13]

Example 6.14

The finite module $\mathbb{Z}_n$ is not free as a module over $\mathbb{Z}$ since, if g is any generator, then the set $\{g\}$ is not linearly independent (e.g. $ng = 0$). However, the module $\mathbb{Z}_n$ is free as a $\mathbb{Z}_n$-module since then the equation $[n]g = 0$ does not mean that $\{g\}$ is linearly dependent.

Just as the input module Ω was defined above, so an output module Γ can be defined in a similar manner. Restricting outputs to be nonzero only when $t > 0$, for causality,[14] but allowing sequences $\boldsymbol{y} = (\ldots, \mathbf{0}, \mathbf{0}; \boldsymbol{y}(1), \boldsymbol{y}(2), \ldots)$ with an infinite number of nonzero elements, z is again used as a time shift operator. Let $\bar{\boldsymbol{y}}(z^{-1}) = \boldsymbol{y}(1)z^{-1} + \boldsymbol{y}(2)z^{-2} + \cdots$, with $\boldsymbol{y}(i) \in Y \subseteq \mathbb{R}^l$. Then

$$\bar{\boldsymbol{y}}(z^{-1}) = \begin{bmatrix} y_1(1) \\ \vdots \\ y_l(1) \end{bmatrix} z^{-1} + \begin{bmatrix} y_1(2) \\ \vdots \\ y_l(2) \end{bmatrix} z^{-2} + \cdots = \begin{bmatrix} \bar{y}_1(z^{-1}) \\ \vdots \\ \bar{y}_l(z^{-1}) \end{bmatrix}$$

where $\bar{y}_j(z^{-1}) = y_j(1)z^{-1} + y_j(2)z^{-2} + \cdots$ $(j = 1, \ldots, l)$ and $\bar{y}_j(z^{-1}) \in \mathbb{R}[\![z^{-1}]\!]$, the set of all formal power series in z^{-1}.[15] The set Γ of all such output series $\bar{\boldsymbol{y}}(z^{-1})$ becomes an $\mathbb{R}[z]$-module under multiplication of any $\bar{\boldsymbol{y}}(z^{-1}) \in \Gamma$ by any polynomial $p(z) \in \mathbb{R}[z]$ and by deleting all terms in z^t for $t \geqslant 0$. Then Γ is an $\mathbb{R}[z]$-module under the ring operation $p(z)\bar{\boldsymbol{y}}(z^{-1})$.

For the $\mathbb{R}[z]$-module Γ, the time shift operator is $\boldsymbol{y} \to z\boldsymbol{y}$ but Γ is not a free module as the next example shows.

Example 6.15

Suppose $\bar{y}$ is represented by $2z^{-1} + z^{-2} + z^{-3}$ and $\bar{y}'$ by $z^{-1} + z^{-2} + z^{-3}$. Then $z\bar{y}$ is given by $z^{-1} + z^{-2}$ and $z\bar{y}'$ is also $z^{-1} + z^{-2}$, i.e. $z\bar{y} = z\bar{y}'$.

Nor is Γ finitely generated. To see this, consider a scalar output ($m = 1$) so that $\Gamma \subseteq \mathbb{R}[\![z^{-1}]\!]$. Suppose there exists a finite set of generators. Let

$$\boldsymbol{y}(1)z^{-1} + \cdots + \boldsymbol{y}(k)z^{-k}$$

be the generator with maximum degree k in z^{-1}. Then, on multiplication by $p(z) \in \mathbb{R}[z]$, no output of degree $k+1$ in z^{-1} can be produced. Hence it is impossible to generate $R\,[\![z^{-1}]\!]$ finitely as an $\mathbb{R}[z]$-module.

An input–output linear map $\boldsymbol{f}: \Omega \to \Gamma$ can now be defined by $\boldsymbol{f}(\boldsymbol{u}) = \boldsymbol{y}$ and shown to be an $\mathbb{R}[z]$-module homomorphism. In both the input module Ω and the output module Γ, the effect of multiplication by z is a backwards time shift. Hence, if an input $\boldsymbol{u}$ produces an output $\boldsymbol{y}$, then the output from $z\boldsymbol{u}$ is $z\boldsymbol{y}$. Time invariance is expressed by $\boldsymbol{f}(z\boldsymbol{u}) = z\boldsymbol{y}$, an equation which simply implies that shifting the input $\boldsymbol{u}$ shifts the output but otherwise leaves it unchanged. By linearity for $p(z) \in \mathbb{R}[z]$, it follows that

$$\boldsymbol{f}(p(z)\boldsymbol{u}) = p(z)\boldsymbol{f}(\boldsymbol{u}) = p(z)\boldsymbol{y},$$

so that $\boldsymbol{f}$ is an $\mathbb{R}[z]$-module homomorphism (see Section 6.4). Time invariance is illustrated by the commutative diagram in Fig. 6.7.

Leaving the special case of discrete dynamical systems, suppose M is an R-module over a ring R. Define a map $\varphi(r): M \to M$ for a specified $r \in R$ by

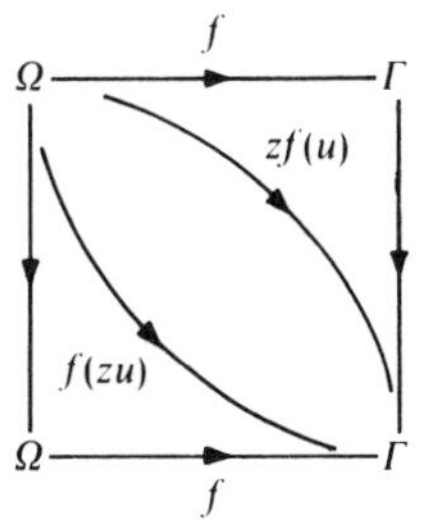

Fig. 6.7 Time invariance

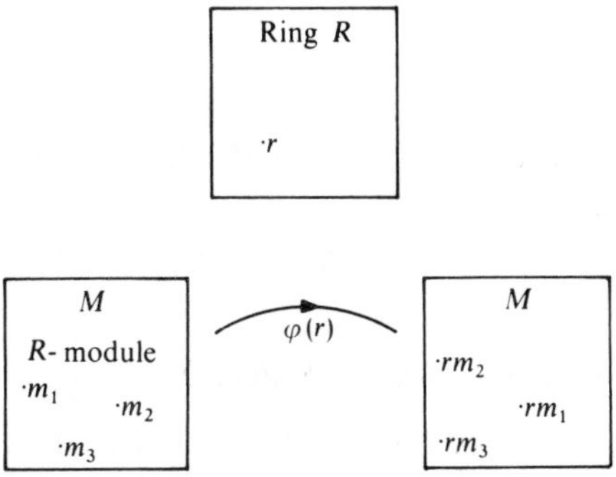

Fig. 6.8. Group homomorphism $\varphi(r)$ from M to M

$\varphi(r)(m) = rm$ (Fig. 6.8). Clearly $\varphi(r)$ is a group homomorphism of the abelian group M into itself, i.e. an endomorphism, because $r(m_1 + m_2) = rm_1 + rm_2$: see (6.13ii). Let End M be the set of all group endomorphisms of M into M, so that $\varphi(r) \in \text{End}\, M$. The set End M can be given a ring structure [16] by defining addition and multiplication as follows:

$$(\alpha + \beta)(m) \triangleq \alpha(m) + \beta(m), \qquad (\alpha\beta)(m) \triangleq \alpha(\beta(m)), \tag{6.15}$$

for all $\alpha, \beta \in \text{End}\, M$ and $m \in M$.

Another mapping $\varphi : R \to \text{End}\, M$ can now be defined by $r \mapsto \varphi(r)$. If $r, s \in R$, and $m \in M$, then

$$\varphi(r + s)m = (r + s)m = rm + sm = [\varphi(r) + \varphi(s)]m,$$

$$\varphi(rs)m = (rs)m = r(sm) = \varphi(r)\varphi(s)m.$$

Since both R and End M are rings, φ is a ring homomorphism of R into the ring of endomorphisms of the abelian group M. Conversely, given any ring homomorphism φ of a ring R into the ring End M, with $\varphi(1)$ equal to the identity map on M, then M is an R-module under the definition $rm \triangleq \varphi(r)(m)$.

Example 6.16

It is possible to make a vector space V over a field F into an $F[x]$-module in the following way. Let α be a linear operator on V, i.e. $\alpha \in \text{End } V$. Suppose that

$$f(x) = a_0 + a_1 x + a_2 x^2 + \cdots + a_n x^n \in F[x].$$

Then

$$f(\alpha) = a_0 \iota + a_1 \alpha + \ldots + a_n \alpha^n \in \text{End } V,$$

where $\iota = \text{id}_{\text{End } V}$, the identity of End V. For any $v \in V$, this endomorphism $f(\alpha)$ operates on v to give

$$f(\alpha)(v) = a_0 v + a_1 \alpha(v) + \cdots + a_n \alpha^n(v),$$

where $\alpha^n(v) = \alpha(\alpha(\ldots(\alpha(v))\ldots))$ with n α's. For a fixed $\alpha \in \text{End } V$, a map ψ: $F[x] \times V \to V$, defined by $\psi(f, v) = f(\alpha)(v)$, makes V into an $F[x]$-module [17] via α (see Exercise 8).

6.4 Submodules and module homomorphisms

If S is a subset of a ring R and X and Y are subsets of an R-module M, then

$$X + Y \triangleq \{x + y \colon x \in X, y \in Y\},$$

$$SX \triangleq \left\{ \sum_{i=1}^{n} s_i x_i \colon s_i \in S, x_i \in X, n \in \mathbb{N} \right\}.$$

A *submodule* N of an R-module M is a subset of M which is also an R-module. Thus, a nonempty subset N of M is a submodule of M if it is closed under subtraction and multiplication by ring elements. That is,

$$x - y \in N \quad \text{and} \quad rx \in N \quad \forall\, r \in R \quad \forall\, x,y \in N,$$

i.e. $N - N = N$ and $RN = N$.

Example 6.17

$\mathbb{Z}$ is a $\mathbb{Z}$-module by Example 6.13. $2\mathbb{Z}$ is the set of all even integers $\{0, \pm 2, \pm 4, \pm 6, \ldots\}$ and is a subgroup of $(\mathbb{Z}, +)$. The difference between any two even integers is an even integer and multiplying any even integer by any integer gives an even integer. That is to say,

$$x - y \in 2\mathbb{Z} \quad \text{and} \quad rx \in 2\mathbb{Z}$$

for all $r \in \mathbb{Z}$ and $x,y \in 2\mathbb{Z}$. Thus, $2\mathbb{Z}$ is a submodule of the $\mathbb{Z}$-module $\mathbb{Z}$.

For any R-module M, $\{0\}$ and M itself are submodules. A submodule N of M is called *proper* if it is not $\{0\}$ or M. There are a number of results and definitions concerning submodules which are of importance in the theory of dynamical systems. Some of these facts are presented below without proof; for more details, the interested reader is referred to Hartley and Hawkes (1970). The results on module homomorphisms are parallel to those on group homomorphisms (regarding the module as an abelian group) together with the module operation of product by scalar in the ring superimposed.

A finite sum of submodules of an R-module M is also a submodule. If X is a nonempty subset of M, then the submodule of M generated by X is RX. Furthermore, if $X = \{x ,..., x_n\}$, then $RX = Rx_1 + \dots + Rx_n$. An R-module M is finitely generated if $M = RX$ for some finite set X; and M is *cyclic* if $M = Rx$ for some $x \in M$.

An R-module homomorphism of an R-module M into an R-module N [18] is a map $\varphi: M \to N$ such that

$$\begin{aligned} \varphi(m_1 + m_2) &= \varphi(m_1) + \varphi(m_2) \quad &\text{(abelian group homomorphism)} \\ \varphi(rm) &= r\varphi(m) \quad &\text{(module structure)} \end{aligned} \tag{6.16}$$

for all $m, m_1, m_2 \in M$ and all $r \in R$. It is important to note the difference between the module structure of an R-module homomorphism in (6.16) and the ring structure of a ring homomorphism in (5.5), a difference now illustrated.

Example 6.18

Suppose $R = \mathbb{Z}$ and consider the map $\varphi: \mathbb{Z} \to \mathbb{Z}$ defined by $n \mapsto 2n$. Then φ is a $\mathbb{Z}$-module homomorphism (actually an endomorphism), but it is not a ring homomorphism. For all $n_1, n_2 \in \mathbb{Z}$, we have $\varphi(n_1 n_2) = 2n_1 n_2$ and $n_1 \varphi(n_2) = n_1 \cdot 2n_2$ so that $\varphi(n_1 n_2) = n_1 \varphi(n_2)$. However,

$$\varphi(n_1)\varphi(n_2) = 2n_1 \cdot 2n_2 = 4n_1 n_2,$$

and this is equal to $\varphi(n_1 n_2)$ only if $n_1 = 0$ or $n_2 = 0$ (or both).

In the discussion in Section 3.4 on the solution of the linear algebraic equation $\boldsymbol{Ax} = \boldsymbol{0}$ when $\boldsymbol{A}$ is an $n \times n$ singular matrix, the vector space $\mathbb{R}^n$ is an $\mathbb{R}$-module. Because the matrix transformation $\boldsymbol{x} \mapsto \boldsymbol{Ax}$ is linear, it is

(i) an $\mathbb{R}$-module homomorphism of $\mathbb{R}^n$ to $\mathbb{R}^n$ and hence

(ii) a group homomorphism of $(\mathbb{R}^n, +)$ to $(\mathbb{R}^n, +)$. Both homomorphisms (i) and (ii) have the same kernel.

The image $\varphi(M)$ of a module homomorphism $\varphi: M \to N$ is a submodule of N and $\ker \varphi$ is a submodule of M. The homomorphism φ is a monomorphism iff $\ker \varphi = \{0\}$. If $\varphi: M \to N$ is an R-module isomorphism, then so is φ^{-1}, and $M = N$. Just as for groups and rings, if K is a submodule of a module M, then the set of additive cosets of M (modulo K) is given by

$$M/K = \{K + m : m \in M\}.$$

M/K can be made into an R-module by defining

$$\begin{aligned} (K+m) + (K+m') &\triangleq K + (m+m'), \\ r(K+m) &\triangleq K + (rm), \end{aligned} \tag{6.17}$$

for all $m, m' \in M$ and all $r \in R$. The R-module M/K is called the *quotient module* of M by K.

A natural homomorphism $\nu : M \to M/K$ is an R-epimorphism with kernel K, defined by $\nu(m) = K + m$, so that $\nu(M) = M/K$. Let K be a submodule of the R-module M, and let $\varphi : M \to N$ be an R-homomorphism for which $\ker \varphi \supset K$. Then there exists a unique R-homomorphism $\psi : M/K \to N$ such that $\varphi = \psi \circ \nu$ (Fig. 6.9). Furthermore,

$$M/\ker \varphi \cong \varphi(M). \tag{6.18}$$

This is an exact parallel to the theory of rings except for the difference between the second definitions in (6.17) and (5.6).

Now consider again the input–output map $\boldsymbol{f} : \Omega \to \Gamma$, defined by $\boldsymbol{u} \mapsto \boldsymbol{f}(\boldsymbol{u}) = \boldsymbol{y}$, associated with a discrete dynamical system. The image module $\boldsymbol{f}(\Omega)$ is a submodule of Γ, and $\boldsymbol{f}(\Omega) \cong \Omega/\ker \boldsymbol{f}$ from (6.18). The quotient $\mathbb{R}[z]$-module $\Omega/\ker \boldsymbol{f}$ is associated in a natural way with the input–output map $\boldsymbol{f}$. With this quotient module $\Omega/\ker \boldsymbol{f}$ denoted by X_f, the module X_f is called the *natural state set*. Since Ω is finitely generated, X_f is a finitely generated $\mathbb{R}[z]$-module, because X_f is generated (as an $\mathbb{R}[z]$-module) by the images of the generators of Ω. The elements of X_f are the additive cosets $\ker \boldsymbol{f} + \boldsymbol{u}$ which may be regarded as equivalence classes $(\boldsymbol{u})_f$ (the Nerode equivalence classes) of Ω consisting of classes of inputs that give the same output. That is to say, the Nerode equivalence relation induced by $\boldsymbol{f}$ is defined as follows:

$$\boldsymbol{u} \sim \boldsymbol{u}' \quad \text{iff} \quad \boldsymbol{f}(\boldsymbol{u} + \boldsymbol{v}) = \boldsymbol{f}(\boldsymbol{u}' + \boldsymbol{v}) \ \forall \ \boldsymbol{v} \in \Omega \tag{6.19}$$

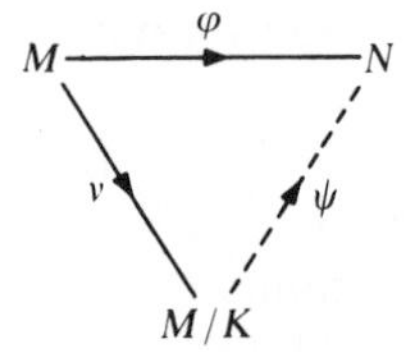

Fig. 6.9 $\varphi = \psi \circ \nu$

where + denotes concatenation ((4.1) and Section 4.2). Alternatively, $\forall \boldsymbol{u} \in \Omega$,

$$f(\boldsymbol{u}+\boldsymbol{v}) = f(\boldsymbol{u})$$

for all $\boldsymbol{u} \in \Omega$ and $\boldsymbol{v} \in \ker f$. The set $\{(\boldsymbol{u})_f : \boldsymbol{u} \in \Omega\}$ of all Nerode equivalence classes of the inputs is the natural state set X_f, i.e.

$$X_f = \{\ker f + \boldsymbol{u} : \boldsymbol{u} \in \Omega\} = \{(\boldsymbol{u})_f : \boldsymbol{u} \in \Omega\}.$$

In the isomorphism $f(\Omega) \cong X_f$, the correspondence is $f(\boldsymbol{u}) \leftrightarrow (\boldsymbol{u})_f$.

The set of equivalence classes $\{[\boldsymbol{u}] : \boldsymbol{u} \in \Omega\}$ of Example 4.22 can be given the structure of an $\mathbb{R}[z]$-module by operations defined by

$$[\boldsymbol{u}] + [\boldsymbol{v}] = [\boldsymbol{u} + \boldsymbol{v}], \qquad z[\boldsymbol{u}] = [z\boldsymbol{u}]. \tag{6.20}$$

The set $\{[\boldsymbol{u}] : \boldsymbol{u} \in \Omega\}$ is also equal to the natural state set X_f (Exercise 12).

6.5 Torsion modules and free modules

In an R-module M, an element $m \in M$ is a *torsion element* if $rm = 0$ for some nonzero $r \in R$. A *torsion module* is a module consisting only of torsion elements. A *torsion-free module* is a module with no nonzero torsion elements. A *torsion-free element* is any element that is not a torsion element.

The *annihilator* A_X of a subset X of an R-module M is the ideal of R given by

$$A_X = \{r \in R : rm = 0_M, m \in X\}.$$

The *order ideal* $o(m)$ of an element $m \in M$ is the annihilator of $\{m\}$, i.e.

$$o(m) = A_{\{m\}} = \{r \in R : rm = 0_M\}.$$

Example 6.19

$\mathbb{Z}_4 = \{0, 1, 2, 3\}$ as a $\mathbb{Z}_4$-module.
The element 1 is torsion-free because, apart from the zero element 0, there is no r in the ring $\mathbb{Z}_4$ for which $r0 = 0$. However, with $\mathbb{Z}_4$ as a $\mathbb{Z}$-module, $4 \cdot 1 = 4 = 0 \pmod 4$, and so 1 is not a torsion-free element. Hence, torsion depends upon the ring R. Again, with $\mathbb{Z}_4$ as a $\mathbb{Z}$-module, the order ideal of 1 is given by

$$o(1) = \{n \in \mathbb{Z} : n1 = 0\}$$

so that $n \in o(1)$ iff $4 | n$. Thus $o(1) = 4\mathbb{Z}$.

Example 6.20

Suppose V is a vector space over a field F and that it is considered as an F-module. Then V is a torsion-free module because, for any nonzero $a \in F$

and any $v \in V$, we have

$$av = 0 \quad \Rightarrow \quad (a^{-1}a)v = 0.$$

But $(a^{-1}a)v = v$, so that $v = 0$.

An R-module M is free on the generating set $X \subseteq M$ if (i) $M = RX$ and (ii) for any R-module N and any map $\varphi: X \to N$, there exists an R-homomorphism $\psi: M \to N$ such that $\psi(x) = \varphi(x)$ for all $x \in X$. X is called a *free basis* of M. [19]

Linear dependence and independence of subsets of an R-module M are defined as for a vector space: $\{m_1, ..., m_n\}$ is a linearly independent set iff there exist $r_i \in R$, not all zero, such that

$$\sum_{i=1}^{n} r_i m_i = 0 \quad \Rightarrow \quad r_i = 0 \ (1 \leqslant i \leqslant n).$$

Example 6.21

$M = 2\mathbb{Z} = \{0, \pm 2, \pm 4, \pm 6, ...\}$ is a $\mathbb{Z}$-module (Example 6.17) and is freely generated by $X = \{2\}$, because $2\mathbb{Z} = \mathbb{Z}\{2\}$. The elements $4, 6 \in 2\mathbb{Z}$ span $2\mathbb{Z}$, because $6 - 4 = 2$ and 2 certainly spans $2\mathbb{Z}$. But elements 4 and 6 are linearly dependent over $\mathbb{Z}$, because $3 \cdot 4 - 2 \cdot 6 = 0$ and $3, -2 \in \mathbb{Z}$, with $4, 6 \in 2\mathbb{Z}$. So $2\mathbb{Z}$ is generated by $\{4, 6\}$ but not freely, i.e. $\{4, 6\}$ is a basis of $2\mathbb{Z}$ but not a free basis.

Thus, in a module, a basis may not be free (i.e. not linearly independent). In other words, a spanning set may not contain a free basis as a subset. Compare this with the situation in a vector space. There the elements in a spanning set may be linearly dependent, but the elements in a basis are linearly independent.

Any finite free basis of a module having such a basis has the same number of elements. To see this, let M be an R-module with free basis $\{x_1, ..., x_m\}$. Suppose that M has a second free basis $\{y_1, ..., y_n\}$ with $n > m$. Let $\varphi: M \to M$ be defined by $\varphi(x_i) = y_i \quad (1 \leqslant i \leqslant m)$ and

$$\varphi\left(\sum r_i x_i\right) = \sum r_i y_i \quad (r_i \in R).$$

Then φ is a module isomorphism. It follows that $y_1, ..., y_m$ generate M, and hence $m = n$.

Some of the above topics can now be injected into the module approach to linear time-invariant discrete dynamical systems.

6.6 State-space module X_Σ.

Instead of the input–output map $f: \Omega \to \Gamma$, consider the internal state description of a linear time-invariant discrete multivariable system Σ which, working over the field $\mathbb{R}$, involves three vector spaces:

(i) input space $U \subseteq \mathbb{R}^m$ with canonical basis $\{\boldsymbol{m}_1, ..., \boldsymbol{m}_m\}$,
(ii) output space $Y \subseteq \mathbb{R}^l$ with canonical basis $\{\boldsymbol{l}_1, ..., \boldsymbol{l}_l\}$,
(iii) state space $X \subseteq \mathbb{R}^n$ with canonical basis $\{\boldsymbol{e}_1, ..., \boldsymbol{e}_n\}$.

The equations

$$\begin{aligned} \boldsymbol{x}(t+1) &= \boldsymbol{A}\boldsymbol{x}(t) + \boldsymbol{B}\boldsymbol{u}(t) \\ \boldsymbol{y}(t) &= \boldsymbol{C}\boldsymbol{x}(t) \end{aligned} \tag{6.21}$$

can be regarded as a sequence of maps B, A, C (Fig. 6.10), where $B: U \to X$, $A: X \to X$, and $C: X \to Y$ are vector space homomorphisms represented in the usual way by $n \times m$, $n \times n$, and $l \times n$ matrices respectively. Vector spaces U, Y, and X can be regarded as R-modules, so that the system Σ is described by R-module homomorphisms.

Given $A: X \to X$, it is possible to make X into an $\mathbb{R}[z]$-module in the following way (cf. Example 6.16). Denote by End X the set of all linear maps $X \to X$. Then End X is an $\mathbb{R}$-algebra; that is to say End X is both a ring and an $\mathbb{R}$-vector space with the same additive group structure and, for all $A, B \in \text{End}\, X$ and all $\lambda \in \mathbb{R}$, we have

$$\lambda(AB) = (\lambda A)B = A(\lambda B)$$

(see also Section 11.5). Given $A \in \text{End}\, X$ and an element $p \in \mathbb{R}[z]$, say

$$p(z) = p_0 + p_1 z + p_2 z^2 + \cdots + p_n z^n,$$

then, with I denoting the identity map in End X,

$$p(A) \triangleq p_0 I + p_1 A + p_2 A^2 + \cdots + p_n A^n$$

is a well-defined element of End X. The effect of $p(A)$ on an arbitrary element $x \in X$ is to produce another element in X, namely

$$\begin{aligned} p(A)(\boldsymbol{x}) &= (p_0 I + p_1 A + \cdots + p_n A^n)\boldsymbol{x} \\ &= p_0 x + p_1 Ax + \cdots + p_n A^n x. \end{aligned}$$

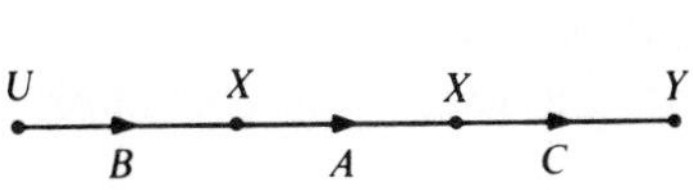

Fig. 6.10. Vector space homomorphisms A, B, C

Now suppose that A is a fixed element of End X so that a map $\mathbb{R}[z] \times X \to X$ can be defined by $(p, \boldsymbol{x}) \mapsto p(A)\boldsymbol{x}$. This map is easily seen to make X into an $\mathbb{R}[z]$-module (just check with (6.13)), which is the state-space module X_Σ via A. [20] The module operation is defined as

$$p(z)\boldsymbol{x} = p(A)\boldsymbol{x},$$

for any $p \in \mathbb{R}[z]$ and $\boldsymbol{x} \in X$, in which $p(z)\boldsymbol{x} \in X_\Sigma$, and $p(A)$ is a polynomial operator on X.

Under the above operation, X_Σ as an $\mathbb{R}[z]$-module is finitely generated, because a basis $\{\boldsymbol{x}_1, \ldots, \boldsymbol{x}_n\}$ for X is such that each $\boldsymbol{x} \in X$ may be written

$$\boldsymbol{x} = p_1\boldsymbol{x}_1 + \cdots. + p_n\boldsymbol{x}_n$$

where $p_1, \ldots, p_n \in \mathbb{R}$ belong to $\mathbb{R}[z]$ as polynomials of degree 0.

X_Σ is a torsion module. To see this, let $\boldsymbol{x}$ be any element of X_Σ and choose a polynomial

$$p(z) = b_0 + b_1 z + b_2 z^2 + \cdots + b_n z^n.$$

Then under the module operation,

$$\begin{aligned} p(z)\boldsymbol{x} &= p(A)\boldsymbol{x} \\ &= b_0\boldsymbol{x} + b_1(A\boldsymbol{x}) + b_2(A^2\boldsymbol{x}) + \cdots + b_n(A^n\boldsymbol{x}). \end{aligned}$$

But the $n + 1$ elements $\boldsymbol{x}$, $A\boldsymbol{x}$, $A^2\boldsymbol{x}, \ldots, A^n\boldsymbol{x}$ must be linearly dependent in X since the state space X is of dimension n. Thus, there exist $b_0, \ldots, b_n \in \mathbb{R}$, not all zero, such that

$$b_0\boldsymbol{x} + b_1(A\boldsymbol{x}) + \cdots + b_n(A^n\boldsymbol{x}) = 0$$

i.e. $p(z)\boldsymbol{x} = \boldsymbol{0}$ with $p(z) \neq 0$. Hence $\boldsymbol{x}$ is a torsion element and, since $\boldsymbol{x}$ is arbitrary, X_Σ is a torsion module.

For $1 \leqslant j \leqslant m$, the pulse input $\boldsymbol{u} = \boldsymbol{m}_j$, where $\{\boldsymbol{m}_1, \ldots, \boldsymbol{m}_m\}$ is the canonical basis for the input space U, gives a state-space sequence from the state equation (6.21):

$$\boldsymbol{x}(0) = \boldsymbol{0}, \quad \boldsymbol{x}(1) = \boldsymbol{B}\boldsymbol{m}_j = \boldsymbol{b}_j, \quad \boldsymbol{x}(2) = \boldsymbol{A}\boldsymbol{b}_j, \quad \boldsymbol{x}(3) = \boldsymbol{A}^2\boldsymbol{b}_j, \ldots$$

and an output sequence

$$\boldsymbol{y}(1) = \boldsymbol{C}\boldsymbol{b}_j, \quad \boldsymbol{y}(2) = \boldsymbol{C}\boldsymbol{A}\boldsymbol{b}_j, \quad \boldsymbol{y}(3) = \boldsymbol{C}\boldsymbol{A}^2\boldsymbol{b}_j, \ldots,$$

where $\boldsymbol{b}_j$ is the jth column of $\boldsymbol{B}$. Then the ith output from the input $\boldsymbol{m}_j$ is the (i, j) entries in the sequence $(\boldsymbol{CB}, \boldsymbol{CAB}, \boldsymbol{CA}^2\boldsymbol{B}, \ldots)$. In this case, Σ is a realization (Kalman et al 1969) of the input–output map $\boldsymbol{f}$.

Beginning with the state $\boldsymbol{x}$ at time $t = -q$ and applying an input sequence represented by

$$\boldsymbol{u} = \sum_{t=-q}^{0} \boldsymbol{u}(t) z^{-t},$$

it is not difficult to show from (6.21) that the state at one interval of time after the end of this sequence (i.e. at $t = 1$) is

$$\boldsymbol{x} \circ \boldsymbol{u} = \boldsymbol{A}^{q+1}\boldsymbol{x} + \sum_{t=-q}^{0} \boldsymbol{A}^{-t}\boldsymbol{B}\boldsymbol{u}(t).$$

But $\boldsymbol{B}\boldsymbol{u}(t) = \sum_{j=1}^{m} \boldsymbol{b}_j u_j(t)$, so that

$$\boldsymbol{x} \circ \boldsymbol{u} = \boldsymbol{A}^{q+1}\boldsymbol{x} + \sum_{j=1}^{m} \sum_{t=-q}^{0} \boldsymbol{A}^{-t}\boldsymbol{b}_j u_j(t).$$

Now $\boldsymbol{x} \mapsto \boldsymbol{A}\boldsymbol{x}$ in vector space notation is equivalent to $\boldsymbol{x} \mapsto z\boldsymbol{x}$ in module notation, Thus, replacing $\boldsymbol{A}$ by z gives

$$\begin{aligned}\boldsymbol{x} \circ \boldsymbol{u} &= z^{q+1}\boldsymbol{x} + \sum_{j=1}^{m} \boldsymbol{b}_j \left(\sum_{t=-q}^{0} u_j(t) z^{-t} \right) \\ &= z^{q+1}\boldsymbol{x} + \sum_{j=1}^{m} \boldsymbol{b}_j u_j,\end{aligned}$$

where $u_j = \sum_{t=-q}^{0} u_j(t) z^{-t}$. Starting with the zero state, the state reached at one interval of time after the end of the input sequence $\boldsymbol{u}$ is therefore given by

$$\boldsymbol{0} \circ \boldsymbol{u} = \sum_{j=1}^{m} u_j \boldsymbol{b}_j,$$

a linear combination of $\boldsymbol{b}_1, ..., \boldsymbol{b}_m$. Moreover,

$$\boldsymbol{0} \circ p(z)\boldsymbol{u} = \sum_{j=1}^{m} p(z) u_j \boldsymbol{b}_j$$

for any $p(z) \in \mathbb{R}[z]$. Thus, the map $\bar{B}_\Sigma : \Omega \to X_\Sigma$, given by $\boldsymbol{u} \mapsto \boldsymbol{0} \circ \boldsymbol{u}$, is an $\mathbb{R}[z]$-module homomorphism and its image is the submodule of X_Σ generated by $\{\boldsymbol{b}_1, ..., \boldsymbol{b}_m\}$. This is the submodule of reachable states and $\boldsymbol{b}_1, ..., \boldsymbol{b}_m$ are called the *accessible generators* of X_Σ.

There is also a module homomorphism from the state module X_Σ into the output module Γ. The output $\boldsymbol{y}$ arising from an initial state $\boldsymbol{x} = \boldsymbol{x}(1)$ corresponds to the formal inverse-power series

$$\bar{\boldsymbol{y}}(z^{-1}) = \sum_{t=1}^{\infty} (\boldsymbol{C}\boldsymbol{A}^{t-1}\boldsymbol{x}) z^{-t} = \bar{C}_\Sigma(\boldsymbol{x}),$$

where $\bar{C}_\Sigma : X_\Sigma \to \Gamma$ is a module homomorphism since

$$\begin{aligned}\bar{C}_\Sigma(z\boldsymbol{x}) = \bar{C}_\Sigma(\boldsymbol{A}\boldsymbol{x}) &= \sum_{t=1}^{\infty} (\boldsymbol{C}\boldsymbol{A}^{t}\boldsymbol{x}) z^{-t} \\ &= \boldsymbol{A} \sum_{t=1}^{\infty} (\boldsymbol{C}\boldsymbol{A}^{t-1}\boldsymbol{x}) z^{-t} = \boldsymbol{A}\bar{C}_\Sigma(x) = z\bar{C}_\Sigma(\boldsymbol{x}),\end{aligned}$$

proving the second equation of (6.16); the first one is easy to prove.

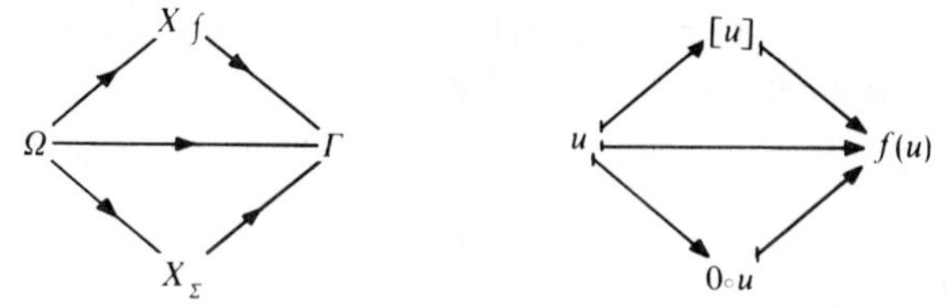

Fig. 6.11 Input–output map $f: \Omega \to \Gamma$

To sum up, the input–output map $f: \Omega \to \Gamma$ can be decomposed in two ways (Fig. 6.11):

(i) through the homomorphism $\Omega \to X_f$ into Nerode equivalence classes of inputs by the isomorphism $X_f \cong \boldsymbol{f}(\Omega)$,

(ii) through the state module via two homomorphisms $\Omega \to X_\Sigma \to \Gamma$. Since X_Σ is a finitely generated torsion module, so is its image $\boldsymbol{f}(\Omega)$ in Γ, and hence so is the isomorphic module X_f. As was pointed out earlier, X_f is a finitely generated $\mathbb{R}[z]$-module; however, it will not be a free module unless $\ker \boldsymbol{f} = \{0\}$. The kernel of $\boldsymbol{f}$ is a submodule of Ω and is free and of rank not greater than m.[21]

The final example of this chapter illustrates several of the concepts associated with an input–output map $\boldsymbol{f}: \Omega \to \Gamma$.

Example 6.22

Consider the system of Example 4.6, now written in terms of z:

$$zx_1 = x_2,$$

$$zx_2 = -x_1 - 2x_2 + u_1,$$

$$zx_3 = -2x_3 + u_2,$$

$$y = x_1 + x_3.$$

The transfer function (1.15) for this system is given by

$$\boldsymbol{G}(z) = \left[\frac{1}{(z+1)^2}, \frac{1}{z+2} \right].$$

Consider the equivalence class of inputs that give zero output, remembering that any output belonging to $\mathbb{R}[z]$ is considered a zero output since outputs are formal power series in the indeterminate z^{-1} with coefficients in $\mathbb{R}$. Then any input of the form

$$\bar{u}(z) = \begin{bmatrix} (z+1)^2 p(z) \\ (z+2) q(z) \end{bmatrix}, \qquad p, q \in \mathbb{R}[z],$$

gives an output $\bar{y}(z) = \boldsymbol{G}(z)\bar{\boldsymbol{u}}(z) = p(z) + q(z) \in \mathbb{R}[z]$ and so $\bar{y}(z) = 0$.

The kernel of f is then given by

$$\ker \boldsymbol{f} = \left\{ \begin{bmatrix} (z+1)^2 p(z) \\ (z+2)q(z) \end{bmatrix} \in \Omega : p, q \in \mathbb{R}[z] \right\}.$$

Notice that $\ker f$ is generated by the vectors $(z^2 + 2z + 1, 0)$ and $(0, z + 2)$.

The quotient module $\Omega/\ker \boldsymbol{f} = X_f$ consists of all the equivalence classes $\ker \boldsymbol{f} + \bar{\boldsymbol{u}}$ $(\bar{\boldsymbol{u}} \in \Omega)$. Suppose, for example, that

$$\bar{\boldsymbol{u}}(z) = \begin{bmatrix} z^3 + 2 \\ z^2 - 1 \end{bmatrix}.$$

Now $z^3 + 2$ divided by $(z + 1)^2$ yields $z - 2$ with remainder $3z + 4$, and so

$$z^3 + 2 \equiv 3z + 4 \pmod{(z+1)^2}.$$

Similarly $z^2 - 1$ divided by $z + 2$ yields $z - 2$ with remainder 3, so

$$z^2 - 1 \equiv 3 \pmod{(z+2)}.$$

Then

$$\begin{bmatrix} z^3 + 2 \\ z^2 - 1 \end{bmatrix} \equiv \begin{bmatrix} 3z + 4 \\ 3 \end{bmatrix} \left(\operatorname{mod} \begin{bmatrix} (z+1)^2 \\ z + 2 \end{bmatrix} \right).$$

It the set of all inputs congruent to $\begin{bmatrix} az + b \\ c \end{bmatrix}$, for given $a,b,c \in \mathbb{R}$, is denoted by $\begin{bmatrix} az + b \\ c \end{bmatrix}$, then

$$X_f = \left\{ \begin{bmatrix} az + b \\ c \end{bmatrix} : a,b,c \in \mathbb{R} \right\}.$$

As an example of the operations in the $\mathbb{R}[z]$-module X_f, consider

$$\begin{bmatrix} 3z + 4 \\ 3 \end{bmatrix} + \begin{bmatrix} z - 1 \\ 2 \end{bmatrix} = \begin{bmatrix} 4z + 3 \\ 5 \end{bmatrix} \in X_f,$$

$$z^2 \begin{bmatrix} 3z + 4 \\ 3 \end{bmatrix} = \begin{bmatrix} z + 2 \\ 12 \end{bmatrix}.$$

On the other hand, if $\bar{\boldsymbol{u}} = (z, 3)$, then

$$\begin{aligned} \bar{y} &= \left[\frac{1}{(z+1)^2}, \frac{1}{z+2} \right] \begin{bmatrix} z \\ 3 \end{bmatrix} = \frac{z}{(z+1)^2} + \frac{3}{z+2} \\ &= z \cdot z^{-2}(1 + z^{-1})^{-2} + 3z^{-1}(1 + 2z^{-1})^{-1} \\ &= z^{-1}(1 - 2z^{-1} + 3z^{-2} - \cdots) + 3z^{-1}(1 - 2z^{-1} + 4z^{-2} + \cdots) \\ &= 4z^{-1} - 8z^{-2} + 15z^{-3} - \cdots, \end{aligned}$$

which is a formal power series in z^{-1} and so is a nonzero output.

Exercises

1. Show that the set $C[0, 1]$ of all continuous functions $f: [0, 1] \to \mathbb{R}$ defined by $t \mapsto f(t)$ is a vector space over $\mathbb{R}$ by proving that $C[0, 1]$ is an additive abelian group (see Section 4.3) and that equations (6.3) are satisfied. Here addition and scalar multiplication are defined as:

$$(f+g)(t) = f(t) + g(t) \quad \forall\, f, g \in C[0, 1],$$

$$(\alpha f)(t) = \alpha f(t) \quad \forall\, \alpha \in \mathbb{R} \quad \forall\, f \in C[0, 1].$$

2. Show that the set of all solutions $\{x\{\bullet\}\}$ of the differential equation $\ddot{x} + x = 0$ is a vector space.
3. Show that the vectors $\boldsymbol{x}_1 = (1, 2, -3, 4)$, $\boldsymbol{x}_2 = (3, -1, 2, 1)$, and $\boldsymbol{x}_3 = (1, -5, 8, -7)$ are linearly dependent. Determine a subset of linearly independent vectors and express those remaining as a linear combination of that subset.
4. The vectors $\boldsymbol{x}_1 = (2, -1, 3)$, $\boldsymbol{x}_2 = (1, 2, -1)$, and $\boldsymbol{x}_3 = (1, -1, -1)$ are basis vectors in the vector space $\mathbb{R}^3$. If a vector $\boldsymbol{x}$ relative to this basis is given by (1, 2, 3) find the components of $\boldsymbol{x}$ relative to the basis $\{(1, 0, 0), (0, 1, 0), (0, 0, 1)\}$.
5. If $\boldsymbol{x}_Z$ and $\boldsymbol{x}_W$ represent a vector $\boldsymbol{x}$ with respect to two bases

$$Z\colon \quad \{(1, 0, 0), (1, 0, 1), (1, 1, 1)\},$$

$$W\colon \quad \{(1, 1, 2), (2, 2, 1), (1, 2, 2)\},$$

and if $\boldsymbol{y}_Z = \boldsymbol{A}\boldsymbol{x}_Z$ represents a linear transformation relative to the Z basis then find the same transformation $\boldsymbol{y}_W = \boldsymbol{B}\boldsymbol{x}_W$ relative to the W basis, where

$$\boldsymbol{A} = \begin{bmatrix} 1 & 1 & 2 \\ 2 & 2 & 1 \\ 3 & 1 & 2 \end{bmatrix}.$$

6. Check that End M is a ring under the definitions (6.15).
7. Check that the set of all $m \times n$ matrices with entries from $\mathbb{R}[z]$, denoted by $\mathbb{R}^{m \times n}[z]$, is an $\mathbb{R}[z]$-module and that, when $m = n$, the vector space $\mathbb{R}^{n \times n}[z]$ is actually a ring which is noncommutative when $n > 1$. Give another way of looking at the matrix

$$\begin{bmatrix} 1 + 3z & 2 + z^2 \\ z^2 & 1 + z^2 \end{bmatrix} \in \mathbb{R}^{2 \times 2}[z].$$

8. Check the module axioms for the $F[x]$-module via α and V in Example 6.16.
9. Show that the map $\varphi\colon \mathbb{C} \to \mathbb{C}$ defined by $z \mapsto \bar{z}$, where $\bar{z}$ is the complex conjugate of z, is a ring homomorphism but not a $\mathbb{C}$-module homomorphism.
10. With $\mathbb{Z}_4$ as a $\mathbb{Z}$-module find the order ideals of elements 2 and 3 in $\mathbb{Z}_4$.
11. Let T be the set of torsion elements in an R-module M, where R is an integral domain. Prove that T is a submodule of M and that M/T is torsion free.
12. Prove that $[\boldsymbol{u}] = (\boldsymbol{u})_f$ for all $\boldsymbol{u} \in \Omega$.

Notes for Chapter 6

1. Notice that, in $\alpha \boldsymbol{x}$, the scalar α precedes the vector $\boldsymbol{x}$ (i.e. $\boldsymbol{x}$ is premultiplied by α) whereas, in $\boldsymbol{b}u(t)$ of (6.1), the scalar $u(t)$ follows the vector $\boldsymbol{b}$ (i.e. $\boldsymbol{b}$ is postmultiplied by $u(t)$). Either form can be used because, if we define

$$\alpha \boldsymbol{x} \triangleq (\alpha x_1, ..., \alpha x_n), \qquad \boldsymbol{x}\alpha \triangleq (x_1 \alpha_1, ..., x_n \alpha),$$

for any $\alpha \in \mathbb{R}$ and any $\boldsymbol{x} = (x_1, ..., x_n) \in \mathbb{R}^n$, then these two vectors are equal since $\alpha x_i = x_i \alpha$ ($i = 1, ..., n$) by virtue of the commutative law for multiplication of real numbers. Similar remarks can be made for scalar multiplication of matrices:

$$\alpha \boldsymbol{A} = \boldsymbol{A}\alpha \quad \forall\, \boldsymbol{A} \in \mathbb{R}^{m \times n} \quad \forall\, \alpha \in \mathbb{R}.$$

2. Equations (6.5) are equivalent to the single statement

$$f(\alpha x + \beta y) = \alpha f(x) + \beta f(y) \quad \forall\, \alpha, \beta \in \mathbb{R} \quad \forall\, x, y \in X,$$

so that the image of a linear combination is the same as the linear combination of the images.

3. If 0_X is the null element of X and 0_Y that of Y, then choosing $\alpha = 0$ and $x = 0_X$ in (6.5) yields $f(00_X) = 0f(0_X)$. But $00_X = 0_X$ and $0f(0_X) = 0_Y$, by definition. Therefore $f(0_X) = 0_Y$, so that a linear map $f: X \to Y$ always maps the zero element of X into the zero element of Y.

4. Any nonzero vector (a, b) can be normalized, i.e. made into a unit vector with the same direction as (a, b) but with unit length. This is done by calculating the length $\sqrt{(a^2 + b^2)}$ of the vector (a, b) and multiplying that vector by the positive real number $1/\sqrt{(a^2 + b^2)}$. Thus, if $\boldsymbol{v} = (a, b)$ and $\boldsymbol{v}$ is normalized to the unit vector $\hat{\boldsymbol{v}}$, then

$$\hat{\boldsymbol{v}} = \frac{1}{\sqrt{(a^2 + b^2)}}(a, b) = \left(\frac{a}{\sqrt{(a^2 + b^2)}}, \frac{b}{\sqrt{(a^2 + b^2)}}\right).$$

5. In much of the literature a basis is an *ordered* set in which case it is called an *ordered basis*. In Example 6.7, the ordered basis ((2, 1), (1, 2)) is a different basis from ((1, 2), (2, 1)). This is important with regard to the matrix of a linear transformation with respect to given bases (Tropper, 1969).

6. Given three nonparallel vectors in $\mathbb{R}^2$, each must lie in the plane ($\mathbb{R}^2$) defined by the other two, so one is linearly dependent on (i.e. a linear combination of) the other two.

7. Note that any linear combination of vectors $\boldsymbol{u}_1, ..., \boldsymbol{u}_n$ such as

$$c_1 \boldsymbol{u}_1 + \cdots + c_n \boldsymbol{u}_n,$$

may be expressed in the form

$$[\boldsymbol{u}_1, ..., \boldsymbol{u}_n] \begin{bmatrix} c_1 \\ \vdots \\ c_n \end{bmatrix}.$$

Thus, (6.8) can be written as

$$[\boldsymbol{e}_1 ,..., \boldsymbol{e}_n]\begin{bmatrix} c_1 \\ \vdots \\ c_n \end{bmatrix} = \begin{bmatrix} 0 \\ \vdots \\ 0 \end{bmatrix}.$$

But the matrix $[\boldsymbol{e}_1 ,..., \boldsymbol{e}_n]$, with vectors $\boldsymbol{e}_1 ,..., \boldsymbol{e}_n$ as columns, is simply the unit matrix $\boldsymbol{I}_n$ of order n. Thus, (6.8) may be written as $\boldsymbol{I}_n\boldsymbol{c} = \boldsymbol{0}$, where $\boldsymbol{c} = (c_1 ,\cdots, c_n)$, and so again $\boldsymbol{c} = \boldsymbol{0}$.

8. Generally, in any vector space of finite dimensions, a basis is
(a) a minimal spanning set (remove one vector and the remainder do not span),
(b) a maximal linearly independent set (add another vector and the new set is linearly dependent).

9. The input discussed in Example 1.3 was a finite sequence. In (6.10) the same input has been represented as an infinite sequence, since this ties in better with the discussion on infinite sequences representing polynomials in Section 5.2. Note that, in (6.10), the semicolon is used to denote the end of the input sequence at $t = 0$.

10. A vector space V over the field $\mathbb{R}$ is therefore an $\mathbb{R}$-module.

11. In (6.13) both + signs in (ii) denote addition in the abelian group M whereas, in (iii), the + on the left-hand side is addition in R but + on the right-hand side is addition in M. Also, r_2m_1 on the left-hand side of (iv) denotes the module product whereas r_1r_2 on the right-hand side denotes the ring product.

12. In (6.14) the minus sign in $(-r_1)m_1$ denotes the additive inverse of r_1 in R, whereas the minus sign in $-(r_1m_1)$ denotes the additive inverse of r_1m_1 in M.

13. Note, however, that $\{\boldsymbol{e}_1 ,..., \boldsymbol{e}_m\}$ is not a basis of $\mathbb{R}^m[z]$ as an $\mathbb{R}$-vector space.

14. A discrete-time system is *causal* (or *nonanticipative*) if the implication

$$u(i) = 0 \quad \Rightarrow \quad y(i) = 0$$

holds for all $i < p$. For convenience this is usually strengthened to: $y(i) = 0$ for all $i \leqslant p$. Usually p is chosen to be zero, so that

$$..., u(-2) = 0, \quad u(-1) = 0 \quad \Rightarrow \quad ..., y(-2) = 0, \quad y(-1) = 0, \quad y(0) = 0.$$

15. There is no question of convergence here; each power series is treated purely algebraically as an element of $\mathbb{R}[\![z^{-1}]\!]$.

16. The zero of End M is the map $\varphi(0_R)$, which maps every element of M into 0_M. The identity of End M is the identity map

$$\imath = \mathrm{id}_{\mathrm{End}\, M} = \varphi(1_R)\colon M \to M, \text{ defined by } \imath(m) = m.$$

17. Different α's will yield different maps ψ, and hence different $F[x]$-modules.

18. Note that M and N are modules over the *same* ring R.

19. That is to say, φ can be extended to an R-homomorphism ψ:

$$m = \sum_i r_i m_i, \qquad \psi(m) = \sum_i r_i \psi(m_i) = \sum_i r_i \varphi(m_i).$$

The extending homomorphism ψ is unique. A free basis is a spanning set of M linearly independent over R.

20. The module X_Σ has been constructed for a given A. Different system matrices A will give rise to different modules X_Σ and so the $R[z]$-module is said to be constructed from X via A.
21. Here the rank of a free R-module is the number of elements in a basis of that module.

7

METRIC AND NORMED SPACES

The previous six chapters of this book have been mainly concerned with concepts from algebra, particularly those which have relevance to linear dynamical systems. For the study of nonlinear systems, a basic knowledge of some topics from analysis is also required. Much of the remaining five chapters is concerned with generalizations of concepts such as convergence of sequences, the extension of the differential calculus to general mappings, and the fundamental ideas of topology and differential geometry.

The reader will be familiar with the idea of distance (or difference) between two real numbers on the real line $\mathbb{R}$. Indeed, the distance between any two points in $\mathbb{R}^2$ or $\mathbb{R}^3$ is also well known from coordinate geometry (see Note 6.4 where the distance between the origin (0, 0) and the point (a, b) in $\mathbb{R}^2$ is given by $\sqrt{(a^2 + b^2)}$). It is natural to ask if 'distance' between two points in more general spaces such as $G\ell(n, \mathbb{R})$ and $C[0, 1]$ is a sensible concept. Of course it is, otherwise it would not have been mentioned here! For example, in the study of dynamical systems, the concept of nearness of graphs of two functions f and g is often of importance. If these two functions are elements of a particular set of functions, say $C[0, 1]$, then the 'distance' between f and g needs to be defined so that nearness can be interpreted in a sensible way. Such a distance measure is called a metric, and a certain subset of metrics arise from what are known as norms. A space on which is defined a metric or a norm is called a metric space or a normed space respectively. It is these concepts which are discussed in this chapter.

7.1 Metric spaces

The distance between any two real numbers a and b is given by the modulus $|a - b|$, which is defined as the value $a - b$ if $a \geqslant b$ or as $b - a$ if $a < b$. For example, $|3 - 1| = 2$ and $|4 - 7| = |-3| = 3$. Clearly, for $x \in \mathbb{R}$, we have $|x| \geqslant 0$ and furthermore $|x| > 0$ unless $x = 0$. Note that, by definition, $|0| = 0$ (Note 1.11). Results concerning the modulus include

$$|a + b| \leqslant |a| + |b|, \tag{7.1}$$

$$|a - b| \leqslant |a| + |b|,$$

$$|a - b| \geqslant \begin{cases} |a| - |b|, \\ |b| - |a|. \end{cases}$$

A more general result follows immediately for any finite set of real numbers a_i $(i = 1, \ldots, n)$, namely

$$|a_1 + a_2 + \cdots + a_n| \leqslant |a_1| + |a_2| + \cdots + |a_n|.$$

Another inequality of considerable importance is the *Schwartz inequality*. This states that, if $a_1, \ldots, a_n, b_1, \ldots, b_n$ are any $2n$ real numbers, then

$$(a_1 b_1 + \cdots + a_n b_n)^2 \leqslant (a_1^2 + \cdots + a_n^2)(b_1^2 + \cdots + b_n^2)$$

and hence

$$a_1 b_1 + \cdots + a_n b_n \leqslant \sqrt{(a_1^2 + \cdots + a_n^2)}\sqrt{(b_1^2 + \cdots + b_n^2)}, \tag{7.2}$$

with equality holding only when $a_1/b_1 = \cdots = a_n/b_n$.[1]

That is to say, equality holds in (7.2) only when there exists a real number λ such that $a_i = \lambda b_i$ for all $i = 1, \ldots, n$; in other words, when the vector $\boldsymbol{a} = (a_1, \ldots, a_n)$ is proportional to vector $\boldsymbol{b} = (b_1, \ldots, b_n)$, namely $\boldsymbol{a} = \lambda \boldsymbol{b}$.

It may seem most unnatural to talk about the distance between two matrices from the space $\mathbb{R}^{m \times n}$. Nevertheless, the generalization of distance from $\mathbb{R}$ to more general spaces, such as $\mathbb{R}^{m \times n}$, does prove to be extremely fruitful. The clue on how to proceed with such a generalization comes from noting that the modulus of the difference between any two real numbers a and b defines a function $f: \mathbb{R}^2 \to \mathbb{R}^+$ given by $(a, b) \mapsto |a - b|$ (cf. Example 3.4). The modulus also has the following properties for all $a,b,c \in \mathbb{R}$:[2]

(i) $|a - b| \geqslant 0$ and $|a - b| = 0$ iff $a = b$,

(ii) $|a - b| = |b - a|$,

(iii) $|a - b| \leqslant |a - c| + |c - b|$.

Guided by these properties of the modulus, suppose that A is a non-empty set with an associated function $\rho: A \times A \to \mathbb{R}^+$. Then ρ is called a *metric* on A iff

(i) $\rho(a, b) \geqslant 0$, and $\rho(a, b) = 0$ iff $a = b$,

(ii) $\rho(a, b) = \rho(b, a)$ (symmetric property), (7.3)

(iii) $\rho(a, b) \leqslant \rho(a, c) + \rho(c, b)$ (triangle inequality),

for all $a,b,c \in A$. Property (iii) is known as the triangle inequality because, when $A = \mathbb{R}^2$, it states that, for the triangle formed by points $A,B,C \in \mathbb{R}^2$, the length of vector $\overrightarrow{AB}$ is never greater than that of vector $\overrightarrow{AC} + \overrightarrow{CB}$ (Fig. 7.1). Note that the sum $\overrightarrow{AC} + \overrightarrow{CB}$ represents the usual vector sum obtained by the parallelogram law.

The real number $\rho(a, b)$ is called the distance between a and b. A non-empty set A on which is defined a metric ρ is called a *metric space* and denoted by (A, ρ). Thus, $\mathbb{R}$ together with the modulus metric $f: \mathbb{R}^2 \to \mathbb{R}^+$

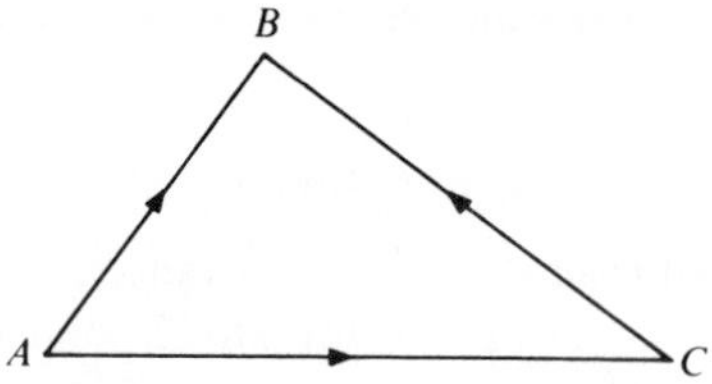

Fig. 7.1 The triangle inequality

defined above, form the metric space $(\mathbb{R}, f)$. When proving that a given function on a space A is a metric, it is usually simple to establish properties (i) and (ii) of (7.3). Property (iii) is often a little more difficult.

Example 7.1

$A = \mathbb{R}^2$. The function $\rho: \mathbb{R}^2 \times \mathbb{R}^2 \to \mathbb{R}^+$ defined by

$$\rho((x_1, x_2), (y_1, y_2)) = \sqrt{[(x_1 - y_1)^2 + (x_2 - y_2)^2]} \quad \text{(Fig. 7.2)}$$

is called the *Euclidean metric* or *usual metric on* $\mathbb{R}^2$. From the properties of the square-root function, $\rho((x_1, x_2), (y_1, y_2))$ is always non-negative and is zero only when $x_1 = y_1$ and $x_2 = y_2$, i.e. when the coordinates (x_1, x_2) and (y_1, y_2) represent the same point in $\mathbb{R}^2$. Property (i) in (7.3) is therefore satisfied. Property (ii) is also valid, because

$$\begin{aligned} \rho((y_1, y_2), (x_1, x_2)) &= \sqrt{[(y_1 - x_1)^2 + (y_2 - x_2)^2]} \\ &= \sqrt{[(x_1 - y_1)^2 + (x_2 - y_2)^2]} \\ &= \rho((x_1, x_2), (y_1, y_2)). \end{aligned}$$

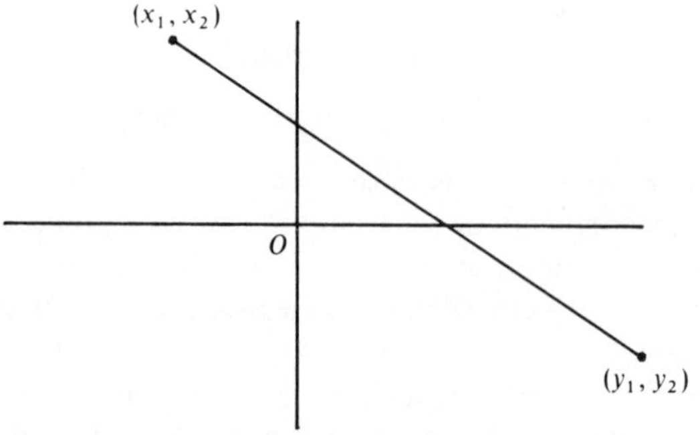

Fig. 7.2 The Euclidean metric

Finally, Property (iii) implies that, for all (x_1, x_2), (y_1, y_2), and (z_1, z_2) in $\mathbb{R}^2$, we have

$$\rho((x_1, x_2), (y_1, y_2)) \leqslant \rho((x_1, x_2), (z_1, z_2)) + \rho((z_1, z_2), (y_1, y_2)) \tag{7.4}$$

or equivalently,

$$\sqrt{[(x_1 - y_1)^2 + (x_2 - y_2)^2]} \leqslant \sqrt{[(x_1 - z_1)^2 + (x_2 - z_2)^2]} + \sqrt{[(z_1 - y_1)^2 + (z_2 - y_2)^2]}.$$

To show that this is true, consider the square of the right-hand side of (7.4). Writing $\boldsymbol{x} = (x_1, x_2)$, $\boldsymbol{y} = (y_1, y_2)$, and $\boldsymbol{z} = (z_1, z_2)$, we get

$$[\rho(\boldsymbol{x}, \boldsymbol{z}) + \rho(\boldsymbol{z}, \boldsymbol{y})]^2 = [\rho(\boldsymbol{x}, \boldsymbol{z})]^2 + [\rho(\boldsymbol{z}, \boldsymbol{y})]^2 + 2\rho(\boldsymbol{x}, \boldsymbol{z})\rho(\boldsymbol{z}, \boldsymbol{y}). \tag{7.5}$$

But the Schwartz inequality for any real numbers a_1, a_2, b_1, and b_2 yields

$$(a_1^2 + a_2^2)(b_1^2 + b_2^2) \geqslant (a_1 b_1 + a_2 b_2)^2. \tag{7.6}$$

So, with the substitutions $a_1 = x_1 - z_1$, $a_2 = x_2 - z_2$, $b_1 = z_1 - y_1$, and $b_2 = z_2 - y_2$, (7.6) becomes

$$[(x_1 - z_1)^2 + (x_2 - z_2)^2][(z_1 - y_1)^2 + (z_2 - y_2)^2] \geqslant [(x_1 - z_1)(z_1 - y_1) + (x_2 - z_2)(z_2 - y_2)]^2,$$

i.e. $[\rho(\boldsymbol{x}, \boldsymbol{z})]^2[\rho(\boldsymbol{z}, \boldsymbol{y})]^2 \geqslant [(x_1 - z_1)(z_1 - y_1) + (x_2 - z_2)(z_2 - y_2)]^2$. Thus,

$$\rho(\boldsymbol{x}, \boldsymbol{z})\rho(\boldsymbol{z}, \boldsymbol{y}) \geqslant (x_1 - z_1)(z_1 - y_1) + (x_2 - z_2)(z_2 - y_2).$$

Using this last inequality in (7.5) gives

$$\begin{aligned}[\rho(\boldsymbol{x}, \boldsymbol{z}) + \rho(\boldsymbol{z}, \boldsymbol{y})]^2 &\geqslant (x_1 - z_1)^2 + (x_2 - z_2)^2 + (z_1 - y_1)^2 + (z_2 - y_2)^2 \\ &\quad + 2[(x_1 - z_1)(z_1 - y_1) + (x_2 - z_2)(z_2 - y_2)] \\ &= [(x_1 - z_1) + (z_1 - y_1)]^2 + [(x_2 - z_2) + (z_2 - y_2)]^2 \\ &= (x_1 - y_1)^2 + (x_2 - y_2)^2 \\ &= [\rho(\boldsymbol{x}, \boldsymbol{y})]^2.\end{aligned}$$

Since $\rho(\boldsymbol{x}, \boldsymbol{z})$, $\rho(\boldsymbol{z}, \boldsymbol{y})$, and $\rho(\boldsymbol{x}, \boldsymbol{y})$ are all non-negative by definition, the square root of both sides of the above inequality gives

$$\rho(\boldsymbol{x}, \boldsymbol{z}) + \rho(\boldsymbol{z}, \boldsymbol{y}) \geqslant \rho(\boldsymbol{x}, \boldsymbol{y}),$$

which is Property (iii) of (7.3).

This example can be generalized to the case $A = \mathbb{R}^n$, for which $\rho\colon \mathbb{R}^n \times \mathbb{R}^n \to \mathbb{R}^+$ is defined by

$$\rho(\boldsymbol{x}, \boldsymbol{y}) = \sqrt{\left(\sum_{i=1}^{n} (x_i - y_i)^2\right)}.$$

Example 7.2

The Hilbert space ℓ_2 is the set of all infinite sequences $\boldsymbol{x} = (x_1, x_2, \ldots, x_i, \ldots)$ of real numbers such that the series

$$\sum_{i=1}^{\infty} |x_i|^2$$

is convergent (see Chapter 8). Here the metric ρ used in defining the metric space (ℓ_2, ρ) is understood to be given by

$$\rho(\boldsymbol{x}, \boldsymbol{y}) = \left(\sum_{i=1}^{\infty} (x_i - y_i)^2 \right)^{\frac{1}{2}}$$

for all $\boldsymbol{x}, \boldsymbol{y} \in \ell_2$ (cf. Example 7.1, where $i = 1, 2$).

Example 7.3

Consider the set $\mathbb{C}$ of all complex numbers $z = a + jb$. A suitable metric on $\mathbb{C}$, called the *usual metric on* $\mathbb{C}$, is given by

$$\rho(z_1, z_2) = |z_1 - z_2| = [(z_1 - z_2)(\overline{z_1 - z_2})]^{\frac{1}{2}},$$

where $\overline{z_1 - z_2}$ is the conjugate of the complex number $z_1 - z_2$. The reader should check that this choice of ρ satisfies the metric properties (7.3).

7.2 Normed spaces

In dealing with $\mathbb{R}^n$, it is convenient to generalize the distance between two points in $\mathbb{R}^2$ or $\mathbb{R}^3$. Thus, the length of a vector $\boldsymbol{x} = (x_1, \ldots, x_n) \in \mathbb{R}^n$ can be defined as

$$\|\boldsymbol{x}\| \triangleq \sqrt{(x_1^2 + \cdots + x_n^2)}$$

and $\|\boldsymbol{x}\|$ is called a *norm of* $\boldsymbol{x}$. If $n = 1, 2$, or 3, then $\|\boldsymbol{x}\|$ represents the usual expression for the distance of the point $\boldsymbol{x}$ from the origin in $\mathbb{R}$, $\mathbb{R}^2$, and $\mathbb{R}^3$ respectively. In particular, if $x = (x_1)$ then $\|x\|$ is simply the modulus $|x_1|$ of the real number x_1.

Suppose X is a vector space over $\mathbb{R}$ on which is defined a real-valued function $f: X \to \mathbb{R}^+$ given by $x \mapsto \|x\|$. This function f is called a *norm* and has the following properties:

(i) $\|x\| \geqslant 0$ and $\|x\| = 0$ iff $x = 0$,

(ii) $\|\lambda x\| = |\lambda| \, \|x\|$, [3] (7.7)

(iii) $\|x + y\| \leqslant \|x\| + \|y\|$,

for all $x,y \in X$ and all $\lambda \in \mathbb{R}$. A *normed vector space* over $\mathbb{R}$ is a vector space X over $\mathbb{R}$ on which is defined a norm. Some similarity is apparent between properties (7.7) and those of (7.3). In fact, all norms define metrics although not all metrics arise from norms.

It is easy to show that, if $\|\bullet\|$ is a norm on a vector space X, then a function $\rho: X \times X \to \mathbb{R}^+$ defined by $\rho(x, y) = \|x - y\|$ is a metric on X. This is because, for all $x,y \in X$, we have firstly $\rho(x, y) = \|x - y\| \geqslant 0$, by definition, and

$$\rho(x, y) = 0 \quad \Leftrightarrow \quad \|x - y\| = 0 \quad \Leftrightarrow \quad x - y = 0 \quad \Leftrightarrow \quad x = y;$$

secondly $\rho(y, x) = \|y - x\| = \|(-1)(x - y)\| = \|x - y\| = \rho(x, y)$, where property (ii) of (7.7) has been used; and finally, by (iii) of (7.7),

$$\begin{aligned}\rho(x, z) = \|x - z\| = \|(x - y) + (y - z)\| &\leqslant \|x - y\| + \|y - z\| \\ &= \rho(x, y) + \rho(y, z).\end{aligned}$$

Thus, all properties of (7.3) are satisfied by the function ρ which is called the *canonical metric induced by the given norm* on the normed vector space X. Thus, any normed vector space is a metric space in the natural way described above. Notice that the norm defined by [4]

$$\|\boldsymbol{x} - \boldsymbol{y}\| = \sqrt{[(x_1 - y_1)^2 + (x_2 - y_2)^2]}$$

is identical to the metric of Example 7.1. This norm is called the *Euclidean* norm of the element $\boldsymbol{x} - \boldsymbol{y}$ and can be generalized as indicated in Example 7.1.

Although it is of no practical significance as far as this book is concerned, the following fact may be of interest. A norm always gives rise to the *canonical* metric but not every metric can be obtained from a norm, as the next example illustrates. This example uses the *discrete metric* just to illustrate the above fact, but again it has no practical relevance here.

Example 7.4

The discrete metric ρ on $C[0, 1]$ is defined, for all $f,g \in C[0, 1]$, as

$$\rho(f, g) = \begin{cases} 0 & \text{if } f = g, \\ 1 & \text{if } f \neq g. \end{cases}$$

This metric on $C[0, 1]$ cannot be obtained from a norm because it could not satisfy (ii) of (7.7), viz

$$\rho(\lambda f, \lambda g) = \|\lambda f - \lambda g\| = |\lambda| \, \|f - g\| = |\lambda| \rho(f, g)$$

which must be true for all $\lambda \in \mathbb{R}$. If $f \neq g$, then $\lambda \neq 0$ implies that $\rho(\lambda f, \lambda g) = 1 = \rho(f, g)$. Hence, the only possible values of λ are ± 1. Since the equation

$\rho(\lambda f, \lambda g) = |\lambda| \rho(f, g)$ is not true for all $\lambda \in \mathbb{R}$, the discrete metric cannot be obtained from a norm.

The Euclidean norm is by no means the only norm suitable for $\mathbb{R}^n$. Some others are given in the next example.

Example 7.5

For $\boldsymbol{x} = (x_1, \dots, x_n) \in \mathbb{R}^n$, we may define

(i) $\|\boldsymbol{x}\| = |x_1| + \cdots + |x_n|$ (octahedral norm),

(ii) $\|\boldsymbol{x}\| = (|x_1|^p + \cdots + |x_n|^p)^{1/p}$

(p-norm, where $1 < p < \infty$),

(iii) $\|\boldsymbol{x}\| = \max\{|x_1|, \dots, |x_n|\}$ (cubic norm).

The Schwartz inequality (7.2) can be written in terms of the Euclidean norm. If $\boldsymbol{a} = (a_1, \dots, a_n)$ and $\boldsymbol{b} = (b_1, \dots, b_n)$ are any two column vectors in $\mathbb{R}^n$, then (7.2) can be written as

$$|\boldsymbol{a}^{\mathrm{T}}\boldsymbol{b}| \leqslant \|\boldsymbol{a}\| \, \|\boldsymbol{b}\|,$$

where

$$\boldsymbol{a}^{\mathrm{T}}\boldsymbol{b} = \sum_{i=1}^{n} a_i b_i, \tag{7.8}$$

with equality holding only when there exists $\lambda \in \mathbb{R}$ such that $\boldsymbol{a} = \lambda \boldsymbol{b}$. The term $\boldsymbol{a}^{\mathrm{T}}\boldsymbol{b}$ in (7.8) is called an *inner product* of the two vectors $\boldsymbol{a}$ and $\boldsymbol{b}$, a concept which is discussed in more detail later in this section.

The usefulness of the Schwartz inequality can be demonstrated by considering the optimization problem of minimizing a function of several variables numerically. Suppose that it is required to find a point $\boldsymbol{x}^* \in \mathbb{R}^n$ which minimizes the value $f(\boldsymbol{x})$ of a function $f: X \to \mathbb{R}$, with $X \subseteq \mathbb{R}^n$, where the derivative of f is a continuous function (see Chapter 10). A typical descent algorithm used for the numerical solution of this problem is to choose an initial point $\boldsymbol{x}_0 \in \mathbb{R}^n$ and an initial search direction $\boldsymbol{p} \in \mathbb{R}^n$. [5] Then for a step-length $\alpha > 0$, a Taylor expansion (see Chapter 10) gives a good approximation to the value of f at a neighbouring point $\boldsymbol{x}_1 = \boldsymbol{x}_0 + \alpha \boldsymbol{p}$, provided that α is sufficiently small. That is,

$$\begin{aligned} f(\boldsymbol{x}_1) &= f(\boldsymbol{x}_0 + \alpha \boldsymbol{p}) = f(\boldsymbol{x}_0) + \sum_{i=1}^{n} \left(\frac{\partial f}{\partial x_i} \right)_{\boldsymbol{x}_0} (\boldsymbol{x}_1 - \boldsymbol{x}_0)_i \\ &= f(\boldsymbol{x}_0) + \left(\frac{\partial f}{\partial \boldsymbol{x}} \right)_{\boldsymbol{x}_0} (\boldsymbol{x}_1 - \boldsymbol{x}_0) \\ &= f(\boldsymbol{x}_0) + \alpha \boldsymbol{g}^{\mathrm{T}}(\boldsymbol{x}_0) \boldsymbol{p} \quad (\text{since } \boldsymbol{x}_1 - \boldsymbol{x}_0 = \alpha \boldsymbol{p}), \end{aligned}$$

where $g(x_0) \in \mathbb{R}^n$ is the gradient vector, i.e. $g^T(x_0)$ is

$$\frac{\partial f}{\partial x} = \left[\frac{\partial f}{\partial x_1}, \ldots, \frac{\partial f}{\partial x_n}\right]$$

evaluated at the point x_0. Provided that $g^T(x_0)p < 0$, the step from x_0 to $x_0 + \alpha p$ yields a decrease in the value of the function f. By the Schwartz inequality, we have

$$|g^T p| \leqslant \|g\| \, \|p\|,$$

where equality holds only when $p = \lambda g$ for some $\lambda \in \mathbb{R}$. In this case, $|g^T p|$ assumes its maximum value. Thus, for a given step-length α, by choosing $p = -g \neq 0$ so that $g^T p = -g^T g < 0$ (since $g^T g$ is a sum of squares and so always non-negative), the greatest decrease in the value of f is obtained. The choice of the gradient vector g for the direction of search p is the basis of the steepest-descent method for determining the minimum value of a function by numerical methods (Walsh 1975; Wolfe 1978).

Multiplication of two vectors that gives another vector is not defined in a general vector space. For this, a vector space needs more structure, to give what is called a *linear algebra* (see Chapter 11). However, (7.8) defines a multiplication of two vectors $a, b \in \mathbb{R}^n$ giving a scalar quantity $a^T b$ which is called an inner product (or scalar product) of vectors a and b. As with norms, there are many different inner products of two vectors x and y in a vector space X. In the general case such an inner product is usually denoted by $\langle x, y \rangle$, and all inner products must satisfy certain properties. Given a vector space X over $\mathbb{R}$, an inner product is a function $X \times X \to \mathbb{R}$ defined by $(x, y) \mapsto \langle x, y \rangle$ and such that

(i) $\langle x, x \rangle \geqslant 0$, with $\langle x, x \rangle = 0$ iff $x = 0$,

(ii) $\langle x, y \rangle = \langle y, x \rangle$,

(iii) $\langle \alpha x, y \rangle = \langle x, \alpha y \rangle = \alpha \langle x, y \rangle$, [6]

for all $x, y \in X$ and all $\alpha \in \mathbb{R}$. Note that the Euclidean norm of a vector x can be expressed as $\sqrt{(x^T x)}$. A vector space X on which is defined an inner product is called an *inner-product space* (or *pre-Hilbert space*).

A given vector space will have many different norms (see Horn and Johnson 1985), as Example 7.5 illustrates for $\mathbb{R}^n$. For the vector space $\mathbb{R}^{n \times n}$ one possible norm of $A = [a_{ij}] \in \mathbb{R}^{n \times n}$ is obtained by considering A as an n^2-dimensional vector

$$(a_{11}, \ldots, a_{n1}; a_{12}, \ldots, a_{n2}; \ldots; a_{1n}, \ldots, a_{nn}).$$

Then, using the Euclidean norm, we get

$$\|A\| = \left(\sum_{i=1}^{n} \sum_{j=1}^{n} a_{ij}^2\right)^{\frac{1}{2}}.$$

Alternatively, an extension of the cubic norm (Example 7.5 (iii)) yields

$$\|A\| = \max_{i,j} |a_{ij}|.$$

For the function space $C[0, 1]$, a possible norm can be chosen for $f \in C[0, 1]$ as

$$\|f\| = \max_{t \in [0, 1]} |f(t)|. \tag{7.9}$$

Alternatively,

$$\|f\| = \left(\int_0^1 [f(t)]^2 \, dt \right)^{\frac{1}{2}}. \tag{7.10}$$

In practical problems, it is often very important to choose the right norm (or metric) for the job. An example of such a choice is discussed later in Section 8.5.

Although a given vector space X has many different norms, all of them define the same topological structure (see Section 9.5). More precisely, two norms $\|\bullet\|_1$ and $\|\bullet\|_2$ on X are *equivalent* if there exist two positive real numbers c_1 and c_2 such that

$$\|x\|_1 \leqslant c_1 \|x\|_2 \quad \text{and} \quad \|x\|_2 \leqslant c_2 \|x\|_1 \qquad \forall\, x \in X.$$

Exercises

1. Check that the following sets A and functions ρ satisfy the properties (7.3) of a metric space.

(a) $A = \mathbb{R}^2$, $\rho(\boldsymbol{x}, \boldsymbol{y}) = |x_1 - y_1| + |x_2 - y_2|$, where $\boldsymbol{x} = (x_1, x_2)$ and $\boldsymbol{y} = (y_1, y_2)$.

(b) Any set A, $\rho(a, b) = \left\{ \begin{matrix} 0 & \text{if } a = b \\ 1 & \text{if } a \neq b \end{matrix} \right\}$. The function ρ is called the discrete metric on A (see Example 7.4).

2. Let (X, ρ_X) and (Y, ρ_Y) be metric spaces and let $p = (x_1, y_1)$ and $q = (x_2, y_2)$ be points in $X \times Y$. Prove that ρ, defined by

$$\rho(p, q) = \max\,[\rho_X(x_1, x_2), \rho_Y(y_1, y_2)],$$

is a metric on $X \times Y$.

3. Let (A, ρ) be a metric space. Show that ρ', defined by

$$\rho'(x, y) = \frac{\rho(x, y)}{1 + \rho(x, y)},$$

is also a metric on A. This metric transformation has no practical significance in this book.

4. Given $\boldsymbol{x},\boldsymbol{y} \in \mathbb{R}^n$, prove that $\|\boldsymbol{x}+\boldsymbol{y}\| \leqslant \|\boldsymbol{x}\| + \|\boldsymbol{y}\|$ where $\|\bullet\|$ is the Euclidean norm.
5. Prove that the Euclidean norm defined by $\|\boldsymbol{x}\| = \sqrt{\sum_{i=1}^{n} x_i^2}$, for $\boldsymbol{x} = (x_1, \dots, x_n) \in \mathbb{R}^n$ satisfies the properties (7.7) of a norm.
6. In a normed vector space X, prove that

$$\|x\| - \|y\| \leqslant \|x - y\| \quad \forall\, x, y \in X.$$

Notes for Chapter 7

1. The proof is sketched briefly here for $n = 3$. Consider the difference

$$(a_1^2 + a_2^2 + a_3^2)(b_1^2 + b_2^2 + b_3^2) - (a_1 b_1 + a_2 b_2 + a_3 b_3)^2.$$

This may be expanded as

$$a_1^2 b_1^2 + \cdots + a_2^2 b_3^2 + a_3^2 b_2^2 + \cdots - (a_1^2 b_1^2 + 2a_2 b_2 a_3 b_3 + \dots),$$

which can be regrouped as

$$[a_2 b_3 - a_3 b_2]^2 + [a_3 b_1 - a_1 b_3]^2 + [a_1 b_2 - a_2 b_1]^2;$$

since this is a sum of squares of real numbers, it can never be negative. Moreover, the difference is always greater than zero unless the three squared brackets are all zero, i.e. unless

$$a_2 b_3 = a_3 b_2, \qquad a_3 b_1 = a_1 b_3, \qquad a_1 b_2 = a_2 b_1,$$

or, what is the same thing, unless $a_1/b_1 = a_2/b_2 = a_3/b_3$.
2. (i) and (ii) are easy to prove. Proof of (iii) is not too difficult since

$$|a - b| = |(a - c) + (c - b)| \leqslant |a - c| + |c - b|,$$

by (7.1).
3. Property (ii) of (7.7) is a property of linearity.
4. The element $\boldsymbol{x} - \boldsymbol{y}$ is defined as the ordered pair $(x_1 - y_1, x_2 - y_2)$ and so can be considered as an element of $\mathbb{R}^2$.
5. Note the different interpretations: $\boldsymbol{x}_0$ is considered a point in $\mathbb{R}^n$, whereas $\boldsymbol{p}$ is considered a vector in $\mathbb{R}^n$ parallel to vector $\boldsymbol{p}$ radiating from the origin of coordinates in $\mathbb{R}^n$.
6. Property (iii) is a property of linearity.
7. If $a,b \in \mathbb{R}^+$ are such that $0 \leqslant a \leqslant b$, then

$$a/(1+a) \leqslant b/(1+b),$$

because $a/(1+a) \leqslant 1 - 1/(1+a)$ and

$$1 + a \leqslant 1 + b \quad \Rightarrow \quad 1/(1+a) \geqslant 1/(1+b).$$

Therefore $1 - 1/(1+a) \leqslant 1 - 1/(1+b)$ and the result follows. Now choose $\alpha = \rho(x, z)$ and $b = \rho(x, y) + \rho(y, z)$.

8

LIMITS, CONVERGENCE, AND BOUNDEDNESS

The reader will no doubt have met the idea of series in the form of binomial, Maclaurin, and Taylor series expansions of elementary functions. Such series generate associated sequences, and the limits of these sequences are intimately linked with the sum of the corresponding series. Much of this chapter is taken up describing the basic concepts of sequences and series of real numbers, since nonmathematicians are not always exposed to this kind of mathematical analysis. However, once the background knowledge has been acquired, the same basic ideas of limits, convergence, and boundedness are carried over into spaces other than $\mathbb{R}$. It is not too surprising to learn that the study of limits and convergence of series involving elements from general spaces requires the concepts of metric and norm discussed in the last chapter.

8.1 Convergence of sequences and series

Consider the case $n = 1$ and $\boldsymbol{u} = \mathbf{0}$ in (1.28), so that the mathematical model of a dynamical system takes the form of a first-order scalar nonautonomous differential equation

$$\dot{x} = f(x, t) \tag{8.1}$$

with initial condition

$$x(t_0) = x_0 \in \mathbb{R}. \tag{8.2}$$

The problem is to find the $x(\cdot)$ that not only satisfies the differential equation but also has an initial value x_0. Integrating (8.1) over the interval $[t_0, t]$, subject to (8.2), yields

$$x(t) - x_0 = \int_{t_0}^{t} f\big(x(\tau), \tau\big)\,\mathrm{d}\tau. \tag{8.3}$$

The integration in (8.3) cannot be carried out, because $x(\tau)$ is not known except when $\tau = 0$. However, if $x(\tau)$ in the integrand $f\big(x(\tau), \tau\big)$ of (8.3) is replaced by the initial value x_0, then a first approximation to $x(t)$ is given by

$$x^{(1)}(t) = x_0 + \int_{t_0}^{t} f(x_0, \tau)\,\mathrm{d}\tau. \tag{8.4}$$

Since x_0 is known, it may be possible to evaluate the integral in (8.4) and obtain $x^{(1)}(t)$ in terms of x_0 and t. A second approximation $x^{(2)}(t)$ can then be obtained by replacing $x(\tau)$ in (8.3) by $x^{(1)}(\tau)$ and again evaluating the integral. This procedure can be repeated over and over again to produce a sequence x_0, $x^{(1)}(t)$, $x^{(2)}(t)$,..., which is usually abbreviated to $(x^{(k)}(t))$ in which $k = 0, 1, \ldots$, and $x^{(0)}(t) = x_0$ for all t. An example will illustrate the method.

Example 8.1

$$\dot{x} = 2tx, \qquad x(0) = 1. \tag{8.5}$$

Integrating over the interval $[0, t]$ gives the solution to (8.5) as that of the integral equation

$$x(t) = 1 + 2\int_0^t \tau x(\tau)\,\mathrm{d}\tau.$$

Approximating $x(\tau)$ by $x(0)$, which is unity, we obtain

$$x^{(1)}(t) = 1 + 2\int_0^t \tau\,\mathrm{d}\tau = 1 + t^2.$$

Repeating the process gives

$$x^{(2)}(t) = 1 + 2\int_0^t \tau(1 + \tau^2)\,\mathrm{d}\tau = 1 + t^2 + \tfrac{1}{2}t^4,$$

$$\begin{aligned} x^{(3)}(t) &= 1 + 2\int_0^t \tau(1 + \tau^2 + \tfrac{1}{2}\tau^4)\,\mathrm{d}\tau \\ &= 1 + t^2 + \frac{1}{2!}t^4 + \frac{1}{3!}t^6. \end{aligned}$$

Continuing in the same way indefinitely produces an infinite sequence $x^{(0)}(t)$, $x^{(1)}(t)$, $x^{(2)}(t)$, ...

The $(k + 1)$th term of the sequence obtained in Example 8.1 is

$$x^{(k)}(t) = \sum_{i=0}^{k} \frac{1}{i!}(t^2)^i \tag{8.6}$$

and is called the kth partial sum. The zeroth partial sum $(k = 0)$ [1] is the real number 1, but the values of the subsequent partial sums depend on the value of t. Therefore, associated with the infinite sequence obtained from successive approximations to the solution of (8.5) is an infinite sequence of functions $x^{(0)}, x^{(1)}, x^{(2)}, \ldots$, each function $x^{(k)}: [0, \infty) \to \mathbb{R}$ defined by (8.6). The sequence of graphs for these functions is shown in Fig. 8.1.

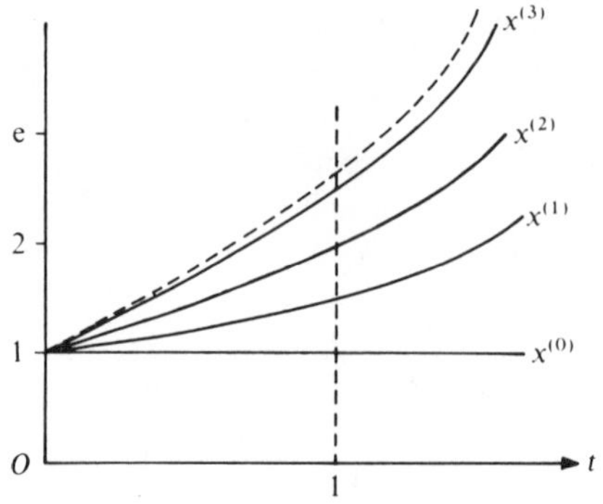

Fig. 8.1 Sequence of function graphs

Two questions immediately arise concerning the above sequence. First, does the sequence *converge*? That is to say, does the sequence of graphs for the functions $x^{(k)}$ in Fig. 8.1 approach some limiting graph (such as the one shown dotted) as k increases without limit? Put another way, does the sequence of values $x^{(k)}(t)$ approach some function value $x(t)$ for all $t \geqslant 0$ as k is chosen larger and larger? Secondly, if the sequence $(x^{(k)}(t))$ does converge to some $x(t)$ for each value of t, is the function x the solution of the differential equation and initial condition (8.5)?

To simplify the discussion, choose the value of t as unity (i.e. $t = 1$) in the sequence $x^{(0)}(t), x^{(1)}(t), x^{(2)}(t), \ldots$. Then from (8.6) the kth partial sum $x^{(k)}(1)$ is given by

$$x^{(k)}(1) = \sum_{i=0}^{k} \frac{1}{i!}. \tag{8.7}$$

The sequence of partial sums in this case is $1, 1 + 1, 1 + 1 + \frac{1}{2}, 1 + 1 + \frac{1}{2} + \frac{1}{6}$, ... , i.e. the sequence $1, 2, 2\frac{1}{2}, 2\frac{2}{3}, \ldots$. Does the kth partial sum $x^{(k)}(1)$ approach a limit, a real number l, say, as k increases indefinitely? No doubt the reader will have recognized that $x^{(k)}(1)$ given by (8.7) is the sum of the first $k + 1$ terms of the infinite series for the irrational number e so that the sequence $(x^{(k)}(1))$ does converge to the limit e as k tends to infinity (written as $k \to \infty$). The convergence of sequence $(x^{(k)}(1))$ to limit e is denoted by either

$$\lim_{k \to \infty} x^{(k)}(1) = \mathrm{e} \quad \text{or} \quad x^{(k)}(1) \to \mathrm{e} \quad \text{as } k \to \infty.$$

Notice that every term in the sequence is the sum of a finite series (8.7), and the limiting value e can be written as the sum of an infinite series

$$\sum_{i=0}^{\infty} \frac{1}{i!},$$

called a convergent series. [2]

Returning to the finite series in (8.6), perhaps the reader will now recognize $x^{(k)}(t)$ as the kth partial sum of the infinite series for e^{t^2}. Now e^{t^2} is a solution of (8.5), because

$$\frac{d}{dt}(e^{t^2}) = 2t(e^{t^2}) \quad \text{and} \quad [e^{t^2}]_{t=0} = e^0 = 1.$$

Hence [3] $x(t) = e^{t^2}$ is the solution of (8.5).

Many algorithms written for computer calculation involve iterative procedures which produce sequences of real numbers such as $(x^{(k)}(1))$ in Example 8.1. The method of steepest descent mentioned in Section 7.2 is used to find a minimum value of a function $f: \mathbb{R}^n \to \mathbb{R}$ given by $\boldsymbol{x} = (x_1, ..., x_n) \mapsto f(\boldsymbol{x}) = f(x_1, ..., x_n)$. This method is based upon the iterative procedure

$$\boldsymbol{x}^{(k+1)} = \boldsymbol{x}^{(k)} - \alpha \boldsymbol{g}^{(k)},$$

in which α is the step-length chosen sufficiently small to ensure a decrease in the function value when moving from the point $\boldsymbol{x}^{(k)} \in \mathbb{R}^n$ to $\boldsymbol{x}^{(k+1)} \in \mathbb{R}^n$. Also $\boldsymbol{g}^{(k)\mathrm{T}}$ is the gradient row vector $[\partial f/\partial x_1, ..., \partial f/\partial x_n]$ evaluated at $\boldsymbol{x}^{(k)}$. Two sequences arise from this algorithm: $(\boldsymbol{x}^{(k)})$ is the sequence of points in $\mathbb{R}^n$ and $(f(\boldsymbol{x}^{(k)}))$ the sequence of function values in $\mathbb{R}$ corresponding to points $\boldsymbol{x}^{(k)}$. In the output-feedback problem discussed in Section 2.2, a matrix $\boldsymbol{F} \in \mathbb{R}^{m \times l}$ is required which minimizes a quadratic form $\boldsymbol{x}_0^{\mathrm{T}} \boldsymbol{K}(\boldsymbol{F}) \boldsymbol{x}_0$. An algorithm based upon an order relationship for positive definite matrices would yield a sequence of matrices $(\boldsymbol{F}^{(k)})$ for which

$$\boldsymbol{x}_0^{\mathrm{T}} \boldsymbol{K}(\boldsymbol{F}^{(j)}) \boldsymbol{x}_0 < \boldsymbol{x}_0^{\mathrm{T}} \boldsymbol{K}(\boldsymbol{F}^{(i)}) \boldsymbol{x}_0 \quad \forall j > i,$$

with $\boldsymbol{F}^{(k)} \in \mathbb{R}^{m \times l}$.

These and other algorithms are useful only if such sequences converge to a limit in the appropriate space, e.g. only if $(\boldsymbol{x}^{(k)})$ converges to some $\boldsymbol{x}^* \in \mathbb{R}^n$, with $(f(\boldsymbol{x}^{(k)}))$ converging to $f(\boldsymbol{x}^*)$ and $(\boldsymbol{F}^{(k)})$ converging to some $\boldsymbol{F}^* \in \mathbb{R}^{m \times l}$, and so on. From the discussion thus far, it should be clear that sequences of elements in a particular space are of considerable importance. However, before discussing the general theory of convergence and associated ideas, attention will be focused first on the theory of infinite sequences of real numbers, i.e. sequences in $\mathbb{R}$. Most of the analysis can then be generalized in a fairly obvious way to sequences in more general spaces.

If an infinite sequence of real numbers (a_k) tends to a finite limit l as k tends to infinity, then any open interval in $\mathbb{R}$ containing l, no matter how small, will contain all but a finite number of the elements a_k. The definition of the limit l of an infinite sequence (a_k) may be expressed formally in the following way. Given any positive number ε, no matter how small, there exists some number k_0 depending on ε (written $k_0(\varepsilon)$) such that $|a_k - l| < \varepsilon$ whenever $k > k_0$. The

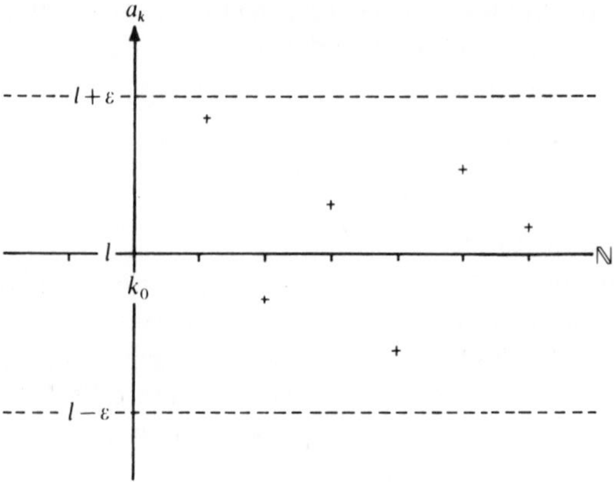

Fig. 8.2 A convergent sequence

sequence (a_k) is said to be *convergent* or to converge to the limit l. Diagrammatically, this means that, for $k > k_0$, all terms a_k have values lying between $l - \varepsilon$ and $l + \varepsilon$ (Fig. 8.2).

Example 8.2

The sequence $(1, \frac{1}{2}, \frac{1}{3}, \ldots, (1/k), \ldots)$ converges to the limit 0. Here $a_k = 1/k$ $(k = 1, 2, \ldots)$ and $l = 0$. Choosing any positive number ε, no matter how small, there always exists a number k_0 such that $|1/k - 0| < \varepsilon$ whenever $k > k_0$. This is easily seen to be true, because

$$|1/k - 0| = |1/k| = 1/k < \varepsilon \quad \Leftrightarrow \quad k > 1/\varepsilon.$$

Thus, by choosing any $k_0 > 1/\varepsilon$, it is always true that $|1/k - 0| < \varepsilon$ whenever $k > k_0$.[4]

If a finite limit l does not exist for a sequence, then that sequence is said to be *divergent*. For example, the sequence (a_k) with $a_k = k \in \mathbb{N}$ diverges because $a_k \to \infty$ as $k \to \infty$. Also, the sequence (a_k) with $a_k = (-1)^k$ yields the oscillating sequence $(-1, 1, -1, 1, \ldots)$ and is divergent because it does not converge to a finite limit l according to the above definition of such a limit. However, unlike the sequence (k), the sequence $((-1)^k)$ does not tend to ∞ (or to $-\infty$). It actually oscillates between the values 1 and -1 as $k \to \infty$.

8.2 Supremum and infimum

The sequence (s_k) in which

$$s_k = \sum_{i=0}^{k} \frac{1}{i!} \qquad \text{(eqn (8.7))}$$

is obtained from Example 8.1 and converges to the limit e. This irrational number e is approximately 2·718 and is certainly less than 3. The real number 3 is called an *upper bound* of the sequence and that sequence is said to be *bounded above*. Any number greater than or equal to e is also an upper bound for the sequence. The sequence $(1/k)$ of Example 8.2 is *bounded below*, with zero or any negative real number being a *lower bound*. A sequence (a_k) of real numbers which is both bounded above and below is said to be *bounded*. That is to say, there exists a fixed number K, independent of k, such that

$$|a_k| < K \quad (\text{i.e. } -K < a_k < K) \quad \text{for all } k.$$

In fact, both sequences (s_k) and $(1/k)$ of examples 8.1 and 8.2 are bounded (e.g. K could be chosen as 3 and $\frac{3}{2}$ respectively).

In order to illustrate how results can be obtained from the definitions already given, it will now be proved that all convergent sequences are bounded. Suppose that (a_k) is a convergent sequence so that $a_k \to l$ as $k \to \infty$ for some finite l. Choose $\varepsilon = 1$ so that $|a_k - l| < 1$ whenever $k > k_0$. Then

$$|a_k| = |(a_k - l) + l| \leqslant |a_k - l| + |l| < 1 + |l|$$

whenever $k > k_0$. Let [5] $h = \max\{|a_1|, \ldots, |a_{k_0}|\}$, so that

$$|a_k| \begin{cases} \leqslant h & \text{if } k \leqslant k_0, \\ < 1 + |l| & \text{if } k > k_0. \end{cases}$$

Then, choosing $K = h + 1 + |l|$, it follows that $|a_k| < K$ for all k, and hence (a_k) is a bounded sequence.

Note that the sequence $((-1)^k)$ is a bounded sequence because, for example, $|a_k| < 2$ for all k. However, this sequence is not convergent, as noted earlier. Thus, although all convergent sequences are bounded, the converse is not true. That is, not all bounded sequences are convergent.

A result which is now quoted without proof is that an increasing sequence (a_k) in $\mathbb{R}$ (i.e. $a_k \leqslant a_{k+1}$ for all $k \in \mathbb{N}$) that is bounded above converges to a real number. [6] This enables one to deduce that any decimal expansion converges to a real number.

Example 8.3

The real number with decimal expansion 2·47863 ⋯ can be written as the series

$$2 + \frac{4}{10} + \frac{7}{10^2} + \frac{8}{10^3} + \frac{6}{10^4} + \frac{3}{10^5} + \cdots.$$

Clearly the sequence of partial sums for this series is increasing and bounded above (e.g. 3 is an upper bound by inspection). Hence, the partial sums converge to a real number, which is the same thing as saying that the decimal expansion 2.47863 ⋯ converges to a real number. Clearly this analysis can be repeated for the decimal expansion of any real number. In particular, the irrational number π represented by the nonterminating expansion 3.14159⋯ is the limit of the associated series

$$3 + \frac{1}{10} + \frac{4}{10^2} + \frac{1}{10^3} + \frac{5}{10^4} + \frac{9}{10^5} + \cdots$$

and is therefore a real number.

Clearly there are an infinite number of upper bounds and lower bounds for a bounded sequence (a_k). In the set of all upper bounds, there is a least upper bound (l.u.b.) called the *supremum of* (a_k) and denoted by $\sup(a_k)$. Thus, for every upper bound M of (a_k), we have

$$\sup(a_k) \leqslant M.$$

The existence of a supremum depends on the definition of a real number and the interested reader should consult Parzynski and Zipse (1987). The relevant theorem here is that every non-empty set $S \subset \mathbb{R}$ that is bounded above has a least upper bound.

Similarly, there exists a greatest lower bound (g.l.b.) called the *infimum of* (a_k) and denoted by $\inf(a_k)$. For every lower bound m of (a_k), $\inf(a_k) \geqslant m$. Note that a sequence bounded below may not contain its g.l.b. as illustrated next.

Example 8.4

The sequence $S = (1, \frac{1}{2}, \frac{1}{3}, \frac{1}{4}, \ldots, 1/k, \ldots) = (1/k)$ is certainly bounded below, and $\inf S = 0$. However, the real number 0 is not an element of S; a number $1/k$ can be found as close to zero as desired by taking k sufficiently large, but no k can be found which gives the actual value zero (again avoiding such an idea as $k = \infty$). However, for the particular sequence S, $\sup S = 1$ which does belong to S.

Similar remarks can be made concerning sequences bounded above; they may or may not contain their l.u.b. To sum up, a bounded sequence may not always have a greatest (maximum) or least (minimum) element, but the supremum and infimum will always exist.

An important result, which is now quoted without proof, concerns the boundedness of continuous functions (see Chapter 10). If $f: [a, b] \to \mathbb{R}$ is a continuous function on the closed interval $[a, b] \subset \mathbb{R}$, then there exists $K \in \mathbb{R}$ such that $|f(x)| \leqslant K$ for all $x \in [a, b]$, and f is said to be bounded. [7]

A series in which all terms are positive must converge if the kth partial sum is bounded above. The only alternative is for the partial sums to increase without limit as $k \to \infty$. Some series have both positive and negative (alternating) terms such as the geometric progression with common ration $-\frac{1}{2}$:

$$1 - \tfrac{1}{2} + \tfrac{1}{4} - \tfrac{1}{8} + \cdots .$$

This series converges to 2/3. If the modulus of each term is taken to form a new series

$$1 + \tfrac{1}{2} + \tfrac{1}{4} + \tfrac{1}{8} + \cdots ,$$

this also is a geometric progression with common ratio $\frac{1}{2}$. This series converges to 2. Thus, by changing the negative signs in the first series to positive signs the resulting series still converges. When such is the case, the original series is said to be *absolutely convergent*. On the other hand, if a series containing both positive and negative terms is convergent but the series made up of the moduli of the original terms is divergent, then the original series is called *conditionally convergent*.

It is always possible to form a *subsequence* from any sequence by selecting arbitrary elements from that sequence. However, in this process the order of the terms must be preserved; that is to say, no interchange of terms is permissible.

Example 8.5

Sequence $(a_k) = (1, \frac{1}{2}, \frac{1}{3}, \frac{1}{4}, \frac{1}{5}, \frac{1}{6}, \frac{1}{7}, \frac{1}{8}, \dots)$. A subsequence of (a_k) is $(1, \frac{1}{4}, \frac{1}{5}, \frac{1}{7}, \dots)$ but $(1, \frac{1}{5}, \frac{1}{4}, \frac{1}{7}, \dots)$ is not a subsequence of (a_k).

Formally, given a sequence of real numbers (a_k) and a sequence of positive integers (i_k) such that $i_1 < i_2 < \dots$, then the sequence (a_{i_k}) is a subsequence of (a_k).

8.3 Cauchy sequences and completeness

In the above discussion on convergence of sequences, the limit l (if it exists) plays an important role. However, if there is no information about the limit l

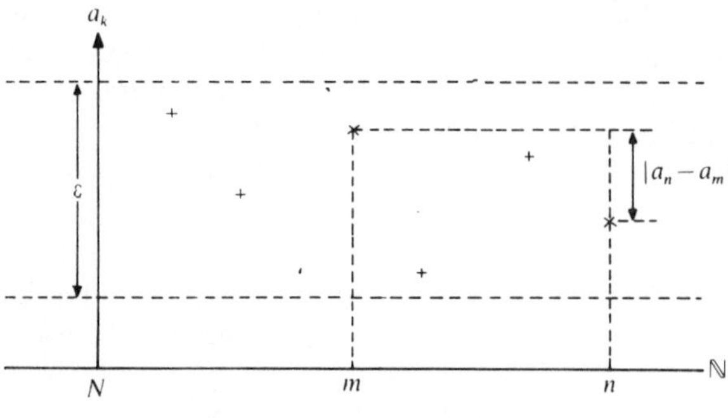

Fig. 8.3 A Cauchy sequence

(even whether or not it exists) *a priori*, then the *Cauchy convergence test* can be employed. This states that a sequence of real numbers (a_k) is convergent if, given $\varepsilon > 0$, there exists an integer $N(\varepsilon)$ such that $|a_n - a_m| < \varepsilon$ whenever $n,m > N$ (Fig. 8.3). A sequence (a_k) satisfying this criterion is called a *Cauchy sequence*. In other words, the terms of a Cauchy sequence (a_k) become arbitrarily close to each other as $k \to \infty$. It is important to note that every Cauchy sequence in $\mathbb{R}$ is convergent. However, when a Cauchy sequence is defined in a general metric space X (see Section 8.5 below) this is no longer true.

Suppose that (a_k) is an infinite Cauchy sequence in $\mathbb{R}$ and that (a_{i_k}) is an infinite subsequence of (a_k) which converges to a limit a. Then (a_k) converges to a. The proof of this result will now be given, because it illustrates a method which is typical for this kind of analysis. Given $\varepsilon > 0$, there exists k_0 such that $|a_k - a_{k_1}| < \frac{1}{2}\varepsilon$ whenever $k,k_1 > k_0$, because (a_k) is a Cauchy sequence. Furthermore, k_1 can be chosen so that a_{k_1} is an element of the subsequence (a_{i_k}) and also sufficiently large to ensure that $|a_{k_1} - a| < \frac{1}{2}\varepsilon$, since $a_{i_k} \to a$ as $k \to \infty$. Thus,

$$\begin{aligned} |a_k - a| = |(a_k - a_{k_1}) + (a_{k_1} - a)| &\leqslant |a_k - a_{k_1}| + |a_{k_1} - a| \\ &< \tfrac{1}{2}\varepsilon + \tfrac{1}{2}\varepsilon = \varepsilon. \end{aligned}$$

Hence, $k > k_0 \Rightarrow |a_k - a| < \varepsilon$, which means that (a_k) converges to a.

A set S of real numbers is said to be *complete* if every Cauchy sequence (a_k), with $a_k \in S$ for all k, converges to an element of S. Note that, since every convergent sequence of real numbers converges to a real number, the set $\mathbb{R}$ is complete.

Example 8.6

The set of positive integers $\mathbb{N}$ is complete. To see this, suppose that (a_k) is any Cauchy sequence in $\mathbb{N}$ and choose $\varepsilon = \frac{1}{2}$. Then, whenever k and k_1 are greater than some number K, we have

$$|a_k - a_{k_1}| < \tfrac{1}{2}$$

which implies that $a_k = a_{k_1} = a$, where $a \in \mathbb{N}$. Thus, the Cauchy sequence is of the form $(a_1, ..., a_K, a, a, ...)$, and clearly this sequence converges to $a \in \mathbb{N}$. Therefore $\mathbb{N}$ is complete.

On the other hand, the set of rationals $\mathbb{Q}$ is not complete, since the sequence of decimal approximations to $\sqrt{2}$, namely (1, 1.4, 1.41, 1.414, ...), of terms in $\mathbb{Q}$ converges to $\sqrt{2}$ which is not in $\mathbb{Q}$.

8.4 Uniform convergence

The sequence $(x^{(k)}(t)) = (x^{(0)}(t), x^{(1)}(t), ...)$ with functions $x^{(k)}: [0, 1] \to \mathbb{R}$ $(k = 0, 1, ...)$ defined by (8.6) converges to a limit $x(t) = e^{t^2}$ which defines the function $x: [0, 1] \to \mathbb{R}$. By this is meant that $\lim_{k \to \infty} x^{(k)}(t) = x(t)$ for each value of t in the domain $[0, 1]$. Applying the formal definition of convergence stated earlier in this chapter, we obtain: given $\varepsilon > 0$, there exists $k_0(\varepsilon)$ such that, for all $k > k_0$ and for all $t \in [0, 1]$ (Fig. 8.4),

$$|x^{(k)}(t) - e^{t^2}| < \varepsilon, \tag{8.8}$$

i.e. $e^{t^2} - \varepsilon < x^{(k)}(t) < e^{t^2} + \varepsilon$ (cf. Fig. 8.2). The important point here is that, for each $t \in [0, 1]$, the sequence $(x^{(k)}(t))$ converges to $x(t)$. In the proof of convergence, the same number $k_0(\varepsilon)$ can be chosen independently of the choice of t in $[0, 1]$ and this is what makes the convergence *uniform* over $[0, 1]$. This is certainly so for the sequence $(x^{(k)}(t))$ defined in (8.6) and the

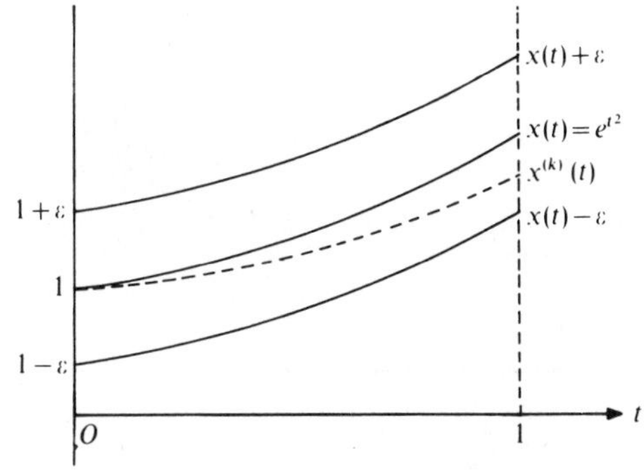

Fig. 8.4 Uniform convergence

sequence is said to *converge uniformly* over $[0, 1]$. The next example gives a convergent sequence of functions for which the convergence is *non-uniform*.

Example 8.7

Consider the infinite sequence of functions $x^{(k)}$ $(k = 1, 2, \ldots)$, where $x^{(k)}: [0, 1] \to \mathbb{R}$ is defined by $x^{(k)}(t) = t^k$. The corresponding sequence of function values is $(t, t^2, t^3, \ldots)$, and the limit of this sequence depends on the value of t. The graphs of the functions in the sequence are shown in Fig. 8.5. For $t \in [0, 1)$, the sequence has the limit zero, e.g. $t = \frac{1}{2}$ gives the sequence $(\frac{1}{2}, \frac{1}{4}, \frac{1}{8}, \ldots, 1/2^k, \ldots) \to 0$ as $k \to \infty$. However, for $t = 1$, the sequence is $(1, 1, 1, \ldots)$ which has the limit 1. Thus, given $\varepsilon > 0$ no matter how small and certainly less than unity, a real number k_0 is required such that

$$|t^k - x(t)| < \varepsilon \quad \forall\, t \in [0, 1]$$

whenever $k > k_0$, with $x(t) = \lim_{k \to \infty} t^k$. But, for any $t \in [0, 1)$, we have

$$|t^k - x(t)| = |t^k - 0| = t^k;$$

now, by choosing t sufficiently close to unity, we can ensure that $t^k > \varepsilon$ no matter how large k_0 was chosen to be. In fact, $t^k > \varepsilon$ implies that $t > \sqrt[k]{\varepsilon}$, so any $t \in (\sqrt[k]{\varepsilon}, 1)$ will do nicely. The reason for this behaviour can be seen from Fig. 8.5. As $k \to \infty$, more of the graph of t^k stays close to the t axis. However, every graph has to eventually leave the neighbourhood of the t axis in order to reach the value 1 at $t = 1$. This means that, no matter how large k is chosen, some t can be found sufficiently close to $t = 1$ such that $t^k > \varepsilon$.

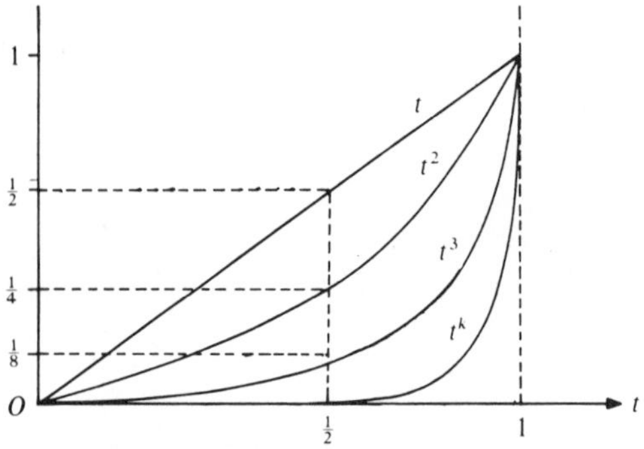

Fig. 8.5 Graphs of sequence (t^k)

The next example illustrates uniform convergence arising in a controlled dynamical system.

Example 8.8

A first-order control system, with some given control function $u: [0, 1] \to [-1, 1]$ and with continuous state $x: [0, 1] \to \mathbb{R}$, is governed by

$$\dot{x} = u, \qquad x(0) = 0. \tag{8.9}$$

Subdivide $[0, 1]$ into $2k$ equal subintervals, $k \in \mathbb{N}$. Let

$$I_i = [i/2k, (i+1)/2k] \quad (i = 0, \ldots, 2k-1)$$

(Fig. 8.6 (a)). Define

$$u^{(k)}(t) = \begin{cases} 1 & \text{if } t \in I_i \quad (i \text{ odd}), \\ -1 & \text{if } t \in I_i \quad (i \text{ even}) \end{cases}$$

(Fig. 8.6 (b, c)), and choose $u(t) = u^{(k)}(t)$, for $t \in [0, 1]$.

If $t \in I_{2j}$ $(j = 0, \ldots, k-1)$, integration of (8.9) with respect to time from 0 to t gives

$$\begin{aligned} x(t, k) &= \int_0^t u(\tau)\,\mathrm{d}\tau \\ &= \int_0^{1/2k} (-1)\,\mathrm{d}\tau + \int_{1/2k}^{2/2k} (1)\,\mathrm{d}\tau + \cdots + \int_{2j/2k}^{t} (-1)\,\mathrm{d}\tau \\ &= -\frac{1}{2k} + \left(\frac{2}{2k} - \frac{1}{2k}\right) + \cdots + \left(-t + \frac{2j}{2k}\right) = -t + \frac{2j}{2k}, \end{aligned}$$

and so

$$x\left(\frac{2j+1}{2k}, k\right) = -\frac{2j+1}{2k} + \frac{2j}{2k} = -\frac{1}{2k}$$

(Fig. 8.6 (d)). Similarly, if $t \in I_{2j+1}$, then

$$\begin{aligned} x(t, k) &= \int_0^{1/2k} (-1)\,\mathrm{d}\tau + \int_{1/2k}^{2/2k} (1)\,\mathrm{d}\tau \\ &\quad + \cdots + \int_{2j/2k}^{(2j+1)/2k} (-1)\,\mathrm{d}\tau + \int_{(2j+1)/2k}^{t} (1)\,\mathrm{d}\tau \\ &= -\frac{1}{2k} + \left(\frac{2}{2k} - \frac{1}{2k}\right) + \cdots + \left(t - \frac{2j+1}{2k}\right) \\ &= -\frac{1}{2k} + t - \frac{2j+1}{2k} = t - \frac{j+1}{k}, \end{aligned}$$

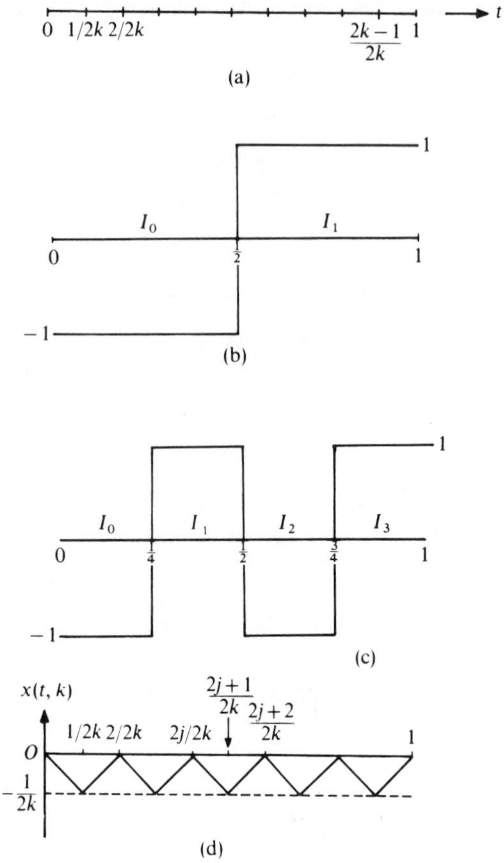

Fig. 8.6 Control functions of Example 8.8

and so $x\big((j+1)/k, k\big) = 0$ (Fig. 8.6(d)). It is also easy to see that $x(1, k) = 0$. The graph of $x(t, k)$ is therefore in the form of a saw-tooth wave (Fig. 8.6(d)).

Clearly, for any fixed t, we have $x(t, k) \to 0$ as $k \to \infty$. Also, given $\varepsilon > 0$, there exists $k_0 = 1/2\varepsilon$ such that $k > k_0$ implies

$$|x(t, k) - 0| = -x(t, k) \leqslant 1/2k < \varepsilon \quad \forall\, t \in [0, 1],$$

(because $k > 1/2\varepsilon \Rightarrow 1/2k < \varepsilon$). Since k_0 depends only on ε and not on t, the convergence $x \to 0$ is uniform on $[0, 1]$, where 0 is the zero function defined for all $t \in [0, 1]$ by $x(t) = 0$. [8]

It should be noted that the infinite power series for e^{t^2} which arises from Example 8.1, namely

$$1 + t^2 + \tfrac{1}{2}t^4 + \cdots + \frac{1}{k!}t^{2k} + \cdots$$

actually converges for all values of t. This is by no means true for all power series since, for example,

$$\ln(1+t) = t - \tfrac{1}{2}t^2 + \tfrac{1}{3}t^3 - \cdots + \frac{(-1)^k}{k}t^k + \cdots \tag{8.10}$$

converges for $-1 < t < 1$ (i.e. $|t| < 1$) but diverges for $|t| > 1$. The open interval $(-1, 1)$ is called the *interval of convergence* for the power series in (8.10). Generally, nothing can be said about the end values of an interval of convergence; they have to be considered separately. For example, the power series in (8.10) actually converges for the value $t = 1$ but diverges for $t = -1$.

The full story on power series is that, given any power series $S = \sum_k a_k t^k$, with $a_k \in \mathbb{R}$, one of the following holds:

either (i) S converges for all values of $t \in \mathbb{R}$
or (ii) S converges only for $t = 0 \in \mathbb{R}$
or (iii) There is some $R \in \mathbb{R}$, with $R > 0$, such that
S converges for all t in the interval $|t| < R$,
S diverges for all t in the region $|t| > R$,

S converges uniformly on $[0, \beta]$ for all $\beta \in (0, R)$. In (i), S converges uniformly on $[0, \beta]$ for all $\beta > 0$.

8.5 Generalizations

In the definition of convergence for a sequence in $\mathbb{R}$, the modulus function $f: \mathbb{R}^2 \to \mathbb{R}$ given by $f(a, b) = |a - b|$ plays an important part. It is the obvious distance measure to use in $\mathbb{R}$; but it will come as no surprise that different metrics or norms are required when the idea of convergence is applied to sequences in spaces other than $\mathbb{R}$, because the elements of such sequences may be matrices, mappings, polynomials, or whatever.

Given a metric space (X, ρ), a sequence $(x^{(k)})$ in X is said to converge if

$$\lim_{k\to\infty} \rho(x^{(k)}, x) = 0$$

for some $x \in X$. When it exists, the element x is called the limit of $(x^{(k)})$, and the fact that the sequence $(x^{(k)})$ converges to x is written as

$$\lim_{k\to\infty} x^{(k)} = x \quad \text{or} \quad x^{(k)} \to x \text{ as } k \to \infty.$$

Alternatively, for every $\varepsilon > 0$, no matter how small, there is some N such that

$$\rho(x^{(k)}, x) < \varepsilon \quad \text{whenever } k > N.$$

As mentioned previously in Section 7.2, for a given space, there are many metrics and norms which can be chosen. For example, consider the space $C[a, b]$ of all continuous real functions $x:[a, b] \to \mathbb{R}$, where $[a, b] \subset \mathbb{R}$. One metric for this space, defined by

$$\rho(x, y) = \sup_{t \in [a, b]} |x(t) - y(t)| \tag{8.11}$$

for any $x, y \in C[a, b]$, is called the *uniform* metric because any convergent sequence $(x^{(k)})$ in $C[a, b]$ is uniformly convergent, i.e. $(x^{(k)})$ converges uniformly on $[a, b]$. Note that the elements $x^{(k)}$ of the sequence are all elements of the space $C[a, b]$, and that these elements are themselves functions defined on the interval $[a, b]$. Spaces such as $C[a, b]$ in which the elements are functions or mappings are called *function spaces.*

The use of the supremum in (8.11) rather than the maximum is because the modulus $|x(t) - y(t)|$ could attain its maximum value at one of the end points of the interval $[a, b]$, either at a or at b (or even possibly at both). In this case, the use of the supremum enables the function spaces $C[a, b]$, $C[a, b)$, $C(a, b]$, and $C(a, b)$ to be treated together since $\sup_{[a, b]} |x(t) - y(t)|$, $\sup_{[a, b)} |x(t) - y(t)|$, $\sup_{(a, b]} |x(t) - y(t)|$, and $\sup_{(a, b)} |x(t) - y(t)|$ all represent the same value.

Another possible metric for the space $C[a, b]$ is

$$\rho(x, y) = \int_0^1 |x(t) - y(t)| \, dt. \tag{8.12}$$

The choice of a particular metric for a given space will depend upon the problem being investigated. For example, if $x(t)$ lies in a small neighbourhood of $y(t)$ for t throughout the closed interval $[a, b]$, then metric (8.11) would be a suitable choice because the small value of $\rho(x, y)$ would reflect the close proximity of the two functions. If, on the other hand, $x(t)$ differs greatly from $y(t)$ but only over a very small time interval $(\alpha - \varepsilon, \alpha + \varepsilon)$, as shown in Fig. 8.7, then using (8.11) would give a relatively large value for $\rho(x, y)$, suggesting that the functions x and y differ by a significant amount. This would not represent the situation at all well, since $x(t)$ approximates $y(t)$ satisfactorily except in a small neighbourhood of $t = \alpha$. In this case, (8.12) would be a more suitable metric since the contribution to the integral from the interval $(\alpha - \varepsilon, \alpha + \varepsilon)$ would not distort the value of $\rho(x, y)$ too much. Such a situation arises when a dynamical system is subjected to persistent disturbances. These may be generally small in magnitude; however, large

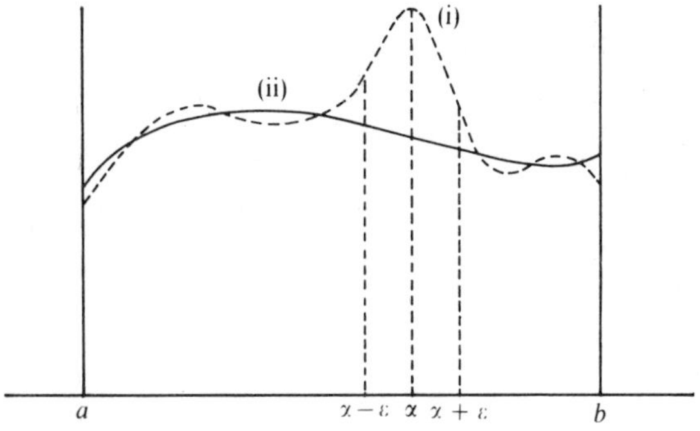

Fig. 8.7 Functions in C$[a, b]$; (i) $x(\cdot)$, (ii) $y(\cdot)$

disturbances may occasionally act on the system but only for very short intervals of time (Krasovskii 1963).

The definition of a bounded sequence given in Section 8.2 can easily be generalized by defining a bounded set. A non-empty subset $A \subset X$ is a *bounded* set if its diameter

$$d(A) = \sup_{x, y \in A} \rho(x, y)$$

is finite. A sequence $(x^{(k)})$ in X is then a bounded sequence if the corresponding point set (i.e. the set $\{x^{(k)} : k \in \mathbb{N}\}$) is a bounded subset of X.

If $A \subseteq \mathbb{R}$, then a function $f: A \to \mathbb{R}$ is said to be bounded on A if there exist real numbers k and K such that $k \leqslant f(x) \leqslant K$ for all $x \in A$. In other words, a bounded function is one whose range is a bounded set.

Example 8.9

$f: (1, 2) \to \mathbb{R}$ is given by $x \mapsto 1/x$. This function f is bounded on $(1, 2)$. Here

$$\inf\{1/x: 1 < x < 2\} = \tfrac{1}{2}, \qquad \sup\{1/x: 1 < x < 2\} = 1.$$

Example 8.10

$f: [2, \infty) \to \mathbb{R}$ is given by $x \mapsto 1/x$. This function f is bounded on $[2, \infty)$. Here

$$\inf\{1/x: x \geqslant 2\} = 0, \qquad \sup\{1/x: x \geqslant 2\} = \tfrac{1}{2}.$$

Note that in Example 8.10 the infimum 0 is not attained, since the set $\{1/x : x \geqslant 2\}$ does not contain the real number zero. This set therefore has no minimum value. However, the supremum $\frac{1}{2}$ is attained and is therefore equal to the maximum value of the set. With regard to the bounded functions of

Examples 8.9 and 8.10, it is worth remarking that the function $f:(0, 1) \to \mathbb{R}$ given by $x \mapsto 1/x$ is not bounded on $(0, 1)$.

The usual definition of a bounded general mapping is different from that given above for a function $f: A \to \mathbb{R}$, with $A \subseteq \mathbb{R}$. Suppose that $(X, \|\bullet\|_X)$ and $(Y, \|\bullet\|_Y)$ are two normed vector spaces and that f is a linear mapping $A \to Y$ where $A \subset X$. Then the norms on spaces X and Y are used in the definitions of boundedness of f. The mapping f is said to be bounded if there exists $K \in \mathbb{R}$ such that

$$\| f(x) \|_Y \leqslant K \| x \|_X \quad \forall\, x \in A.$$

The least possible value of K is obtained from

$$\frac{\| f(x) \|_Y}{\| x \|_X} \leqslant K.$$

The infimum of K is called the *norm of* f and denoted by $\| f \|$. Clearly

$$\| f \| = \sup_{\substack{x \in A \\ x \neq 0}} \left\{ \frac{\| f(x) \|}{\| x \|} \right\}, \tag{8.13}$$

where the spaces X and Y have been dropped from the norms for convenience. It is easy to show that $\|f\|$ satisfies the conditions of a norm (Exercise 4).

The space over which the supremum is taken in (8.13) is $A \backslash \{0\}$, the origin being excluded because, from Section 6.1, $f(0_X) = 0_Y$. Thus,

$$\frac{\|f(x)\|}{\| x \|} \leqslant \|f\|,$$

or alternatively,

$$\|f(x)\| \leqslant \|f\| \; \| x \|. \tag{8.14}$$

Cauchy sequences and completeness are easily defined following the ideas generated for $\mathbb{R}$ in Section 8.3. Suppose that (X, ρ) is a metric space and $(x^{(k)})$ a sequence of elements in X. Then $(x^{(k)})$ is said to be a Cauchy sequence if, for every $\varepsilon > 0$ no matter how small, there is some $K(\varepsilon)$ such that

$$\rho(x^{(m)}, x^{(n)}) < \varepsilon$$

whenever $m,n > K$.

Every convergent sequence in a metric space is a Cauchy sequence. The proof of this is now given because it demonstrates the way in which the metric chosen for the space plays the role of the modulus in the earlier discussions on sequences in $\mathbb{R}$. If $x^{(k)} \to x$ as $k \to \infty$ then, for every $\varepsilon > 0$, there exists $K(\varepsilon)$ such that

$$\rho(x^{(k)}, x) < \tfrac{1}{2}\varepsilon \quad \forall\, k > K.$$

Hence, by the triangle inequality ((iii) of (7.3)), for $m,n > K$, we have

$$\rho(x^{(m)}, x^{(n)}) \leqslant \rho(x^{(m)}, x) + \rho(x, x^{(n)})$$
$$< \tfrac{1}{2}\varepsilon + \tfrac{1}{2}\varepsilon = \varepsilon.$$

This shows that $(x^{(k)})$ is a Cauchy sequence.

As mentioned earlier in Section 8.3, the converse of the above result is not true, i.e. not all Cauchy sequences are convergent. However, a space X is complete if every Cauchy sequence in X converges, i.e. every Cauchy sequence has a limit that is an element of X. For example, it can be shown (Exercise 5) that, for $[a, b] \subset \mathbb{R}$, the function space $C[a, b]$ is complete, by taking $\rho(x^{(m)}, x^{(n)}) = \max_{t \in [a, b]} |x^{(m)}(t) - x^{(n)}(t)|$. An example of a metric space which is not complete is the space of all polynomials considered as functions of t on $[a, b]$ and with metric

$$\rho(x, y) = \max_{t \in [a, b]} |x(t) - y(t)|.$$

To see this, consider a Cauchy sequence of polynomials that converges uniformly on $[a, b]$ to a continuous function that is not a polynomial. Then the limit of the Cauchy sequence does not belong to the space of all polynomials defined above.

A space that is not complete, such as $\mathbb{Q}$, may be embedded in a complete space, $\mathbb{R}$ in the case of $\mathbb{Q}$. These ideas lead to extension fields, an important concept in Galois theory (Stewart 1973). At the time of writing, differential Galois theory (Pommaret 1983) looks as though it could be relevant to nonlinear dynamical system theory (Fliess 1986).

A complete normed vector space is called a *Banach space.* A complete inner-product space (Section 7.2) is called a *Hilbert space* (see Example 7.2). Problems in system theory are often formulated in a Banach or Hilbert space. For example, in an optimization problem, a numerical algorithm may generate a Cauchy sequence of vectors, the limit of which is the optimal vector desired. In a Banach or Hilbert space, the existence of this limit is guaranteed.

The concepts of infimum and supremum can be generalized to partially ordered sets. Suppose A is a partially ordered set and $S \subset A$. Then $a \in A$ is a lower bound of S iff $a \preccurlyeq S$ for all $s \in S$. As in the case of $A = \mathbb{R}$ (although $\mathbb{R}$ is a totally ordered set) discussed earlier, there will be many lower bounds for S but if there is a lower bound l such that $a \preccurlyeq l$ for all lower bounds a, then l is the g.l.b. and is called the infimum of S, denoted by inf S. Similarly, $b \in A$ is an upper bound of S iff $s \preccurlyeq b$ for all $s \in S$. The l.u.b. is an element L such that $L \preccurlyeq b$ for all upper bounds b of S, and L is called the supremum of S, denoted by sup S.

Example 8.11

Let $A = \{a, b, c, d, e, f, g\}$ be a partially ordered set in which $a \preccurlyeq c$, $b \preccurlyeq c$, $b \preccurlyeq d$, $c \preccurlyeq e$, $d \preccurlyeq e$, $e \preccurlyeq f$, and $e \preccurlyeq g$. Suppose that $S = \{c, d, e\}$, as shown in Fig. 8.8. Then e, f, g are upper bounds of S, and b is the only lower bound of S since $a \not\preccurlyeq d$. Note that $\sup S = e$, which belongs to S, but $\inf S = b$, which does not; i.e. $\sup S \in S$ but $\inf S \notin S$.

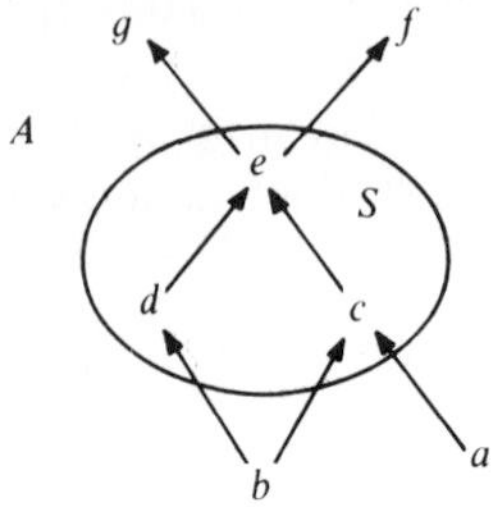

Fig. 8.8 Partially ordered set A

Sometimes $\sup S$ and $\inf S$ do not exist, even for a finite set S. For example, if in Example 8.11 it is also given that $a \preccurlyeq d$, then both a and b are lower bounds of S, but $\inf S$ does not exist because nothing is known concerning the relation between a and b.

8.6 Other topics, including Zorn's lemma and the contraction-mapping theorem

Remember that a totally ordered set is a partially ordered set in which every two elements are comparable. A totally ordered set is sometimes called a *chain*. If X is a partially ordered set and every chain $A \subset X$ has an upper bound in X, [9] then X has at least one maximal element (i.e. an element $m \in X$ such that $m \preccurlyeq x$ holds only for $x = m$. This result is known as *Zorn's lemma*. It is equivalent to the axiom of choice (Halmos 1974).

Suppose that (X, ρ) is a metric space. Then a mapping $f: X \to X$ is called a *contraction* on X if there exists a real number λ, with $0 \leqslant \lambda < 1$, such that

$$\rho(f(x_1), f(x_2)) \leqslant \lambda \rho(x_1, x_2) \quad \forall\ x_1, x_2 \in X.$$

Thus, in a contraction, the distance between the images of two elements is less than the distance between the elements.

If (X, ρ) and (Y, d) are two metric spaces, then a mapping $f: X \to Y$ is said to be *isometric* or an *isometry*, if,

$$d(f(x_1), f(x_2)) = \rho(x_1, x_2) \quad \forall\ x_1, x_2 \in X.$$

That is, the distance between any two elements in X is the same as the distance between their images in Y. So an isometry preserves distances, whereas a contraction reduces distance.

Suppose that (X, ρ) is a complete metric space and that $f: X \to X$ a contraction on X. Then f has a unique fixed point $a \in X$ defined by $f(a) = a$. This result is known as the *contraction-mapping theorem* or the *Banach fixed-point theorem*. A way to prove this theorem is to construct a sequence (x_n) and show that it is a Cauchy sequence, so that it converges in the complete space X. The limit a of this sequence is then shown to be a fixed point of f, i.e. $f(a) = a$. Finally, a is shown to be unique by assuming that a and $\bar{a}$ are two fixed points of f, and then showing that $\bar{a} = a$. [10]

Exercises

1. Prove that $(1 + 1/k)^k$ tends to a real number as $k \to \infty$. The limit is the irrational number e.

2. Show that the sequence $(x^{(k)}(t))$ exhibits nonuniform convergence on $\mathbb{R}$, where $x^{(k)}: \mathbb{R} \to \mathbb{R}$ is defined by $x^{(k)}(t) = 1/(1 + t^{2k})$ $(k = 0, 1, \ldots)$. The result proved in Note 10.3 is required for this exercise.

3. Consider the space $\mathscr{B}(\mathbb{R}^2)$ of all bounded real-valued functions defined on $\mathbb{R}^2$, i.e. $\mathscr{B}(\mathbb{R}^2) = \{f: \mathbb{R}^2 \to \mathbb{R} \mid f \text{ is bounded on } \mathbb{R}^2\}$, and let $S = \mathscr{B}(\mathbb{R}^2) \times \mathscr{B}(\mathbb{R}^2)$. Verify that the function $\rho: S \times S \to \mathbb{R}$ defined by

$$\begin{aligned}\rho(f, g) &= \rho\big(\big(f_1(x), f_2(x)\big), \big(g_1(x), g_2(x)\big)\big) \\ &= \sup_{x \in \mathbb{R}^2} \{[f_1(x) - g_1(x)]^2 + [f_2(x) - g_2(x)]^2\}^{\frac{1}{2}}\end{aligned}$$

is a metric on S.

4. Show that the norm $\|f\|$ (8.13) of a mapping $f: A \to Y$, where A and Y are two normed vector spaces, satisfies the conditions of a norm. Prove that $\|f\|$ as given in (8.13) can be written equivalently as

$$\sup_{\substack{y \in A \\ \|y\| = 1}} \|f(y)\| \qquad \text{(cf. (7.9))}.$$

5. C[0, 1] is the set of all real-valued continuous functions defined on the interval $[0, 1] \subset \mathbb{R}$. Prove that

$$\|x\| = \max_{t \in [0, 1]} |x(t)|$$

is a norm on C[0, 1]. Show also that C[0, 1] with this norm is a complete normed space.

6. A metric for C[0, 1] is given by

$$\rho(x, y) = \int_0^1 |x(t) - y(t)| \, dt.$$

Show that the metric space (C[0, 1], ρ) is not complete.

Notes for Chapter 8

1. Remember that $0! \triangleq 1$ and that $(t^2)^0 \triangleq 1$ for all $t \in \mathbb{R}$.
2. It is very important that the reader should not be misled by the notation: $x^{(k)}(1) \to \mathrm{e}$ means that the sequence $(x^{(k)}(1))$ converges to a real number e. However, $k \to \infty$ means that k increases without limit, not that k converges to a number ∞. Infinity is kept at arms length, avoiding such unacceptable statements as $k = \infty$.

Similarly, $\sum_{i=0}^{k} \frac{1}{i!}$ is a sum of $k + 1$ terms. On the other hand, $\sum_{i=0}^{\infty} \frac{1}{i!}$ is not the sum of an infinite number of terms, but is the limit of the sequence of kth partial sums as $k \to \infty$.
3. A solution of a differential equation $\dot{x} = f(x, t)$ that satisfies a given initial condition $x(t_0) = x_0$ is the only solution, provided that certain conditions on the function f are satisfied (see Section 10.1).
4. It is assumed in Example 8.2 that the elements of the sequence $(1/k)$ lie in the space $\mathbb{R}$ or perhaps $[0, 1]$. If the space under consideration is $(0, 1]$, then the sequence $(1/k)$ does not converge in that space, since the limit 0 does not belong to $(0, 1]$. Thus, the concept of convergence depends not only on the sequence but also on the space in which the sequence elements lie.
5. $h = \max\{|a_1|, \ldots, |a_{k_0}|\}$ means that h is chosen to be equal to the maximum value in the set of values $\{|a_1|, \ldots, |a_{k_0}|\}$.
6. This statement is sometimes called the *axiom of completeness* for the real numbers. The result could be used as a definition of a real number.
7. Note that it is essential for f to have a closed domain. The function $f: [0, \frac{1}{2}\pi) \to \mathbb{R}$ defined by $f(x) = \tan x$ is certainly defined and continuous—but not bounded—on $[0, \frac{1}{2}\pi)$. This is possible because $[0, \frac{1}{2}\pi)$ is not a closed interval.
8. There is some danger of confusion in Example 8.8. The differential equation $\dot{x} = u$ is to be solved but x is not differentiable everywhere since u is discontinuous (see Chapter 10). The limit of $u^{(k)}(t)$ as $k \to \infty$ is incomprehensible and, in the limit, $\dot{x} \neq u$! More will be said about this in Chapter 9.
9. The upper bound must be in X because otherwise, for example, the set $X = \{x \in \mathbb{R} : x > 0, x^2 < 2\}$ has upper bounds in $\mathbb{R}$ but has no maximal element in X.
10. A proof of the Banach fixed-point theorem is as follows. Apply f repeatedly to obtain $\rho\big(f^n(x), f^n(y)\big) \leqslant \lambda^n \rho(x, y)$. Choose any $x_0 \in X$ and construct the sequence (x_n) given by x_0, $x_1 = f(x_0)$, $x_2 = f(x_1) = f^2(x_0), \ldots, x_n = f^n(x_0), \ldots$. To show that this sequence is Cauchy, note that

$$\begin{aligned}
\rho(x_m, x_{m+1}) &= \rho\big(f(x_{m-1}), f(x_m)\big) \\
&\leqslant \lambda \rho(x_{m-1}, x_m) \\
&= \lambda \rho\big(f(x_{m-2}), f(x_{m-1})\big) \\
&\leqslant \lambda^2 \rho(x_{m-2}, x_{m-1}) \\
&\;\;\vdots \\
&\leqslant \lambda^m \rho(x_0, x_1).
\end{aligned}$$

Thus, assuming $n > m$, we have

$$\begin{aligned}\rho(x_m, x_n) &\leqslant \rho(x_m, x_{m+1}) + \rho(x_{m+1}, x_n)\\ &\leqslant \rho(x_m, x_{m+1}) + \rho(x_{m+1}, x_{m+2}) + \rho(x_{m+2}, x_n)\\ &\vdots \qquad\qquad\qquad\qquad \vdots\\ &\leqslant \rho(x_m, x_{m+1}) + \rho(x_{m+1}, x_{m+2}) + \cdots + \rho(x_{n-1}, x_n).\end{aligned}$$

Hence, from the above result,

$$\begin{aligned}\rho(x_m, x_n) &\leqslant (\lambda^m + \lambda^{m+1} + \cdots + \lambda^{n-1})\rho(x_0, x_1)\\ &= \lambda^m(1 + \lambda + \cdots + \lambda^{n-m-1})\rho(x_0, x_1)\\ &= \lambda^m \frac{1 - \lambda^{n-m}}{1 - \lambda}\rho(x_0, x_1).\end{aligned}$$

Since $0 \leqslant \lambda < 1$, this last expression can be made as small as required by making m sufficiently large (and consequently n since $n > m$). Hence, (x_n) is a Cauchy sequence and has a limit $a \in X$ since X is complete.

The triangle inequality ensures that

$$\rho\big(a, f(a)\big) \leqslant \rho(a, x_m) + \rho\big(x_m, f(a)\big).$$

But

$$\rho\big(x_m, f(a)\big) = \rho\big(f(x_{m-1}), f(a)\big) \leqslant \lambda\rho(x_{m-1}, a),$$

so that

$$\rho\big(a, f(a)\big) \leqslant \rho(a, x_m) + \lambda\rho(x_{m-1}, a).$$

By making m sufficiently large, the right-hand side of the last inequality can be made as small as required, because $x_{m-1}, x_m \to a$ as $m \to \infty$. Thus,

$$\rho\big(a, f(a)\big) = 0,$$

which implies $f(a) = a$, and so a is a fixed point.

To prove that a is unique, suppose that $\bar{a}$ is any other fixed point of the mapping f. Then a is unique if $\bar{a} = a$. To prove this, note that

$$f(a) = a, \qquad f(\bar{a}) = \bar{a},$$

and $\rho(a, \bar{a}) = \rho(f(a), f(\bar{a})) \leqslant \lambda\rho(a, \bar{a})$, i.e.

$$(1 - \lambda)\rho(a, \bar{a}) \leqslant 0.$$

But $0 \leqslant \lambda < 1$ and $\rho(a, \bar{a}) \geqslant 0$, so that $\rho(a, \bar{a}) = 0$ and $\bar{a} = a$.

9

SETS, CONVEXITY, AND TOPOLOGY

In Chapter 2, it was shown that the fundamental ideas of set theory are extremely useful for the description of dynamical system behaviour. In the present chapter, attention is directed towards special sets with properties that are particularly attractive for obtaining specific results in mathematical analysis. For example, a convex set can ensure the existence of a minimum value in an optimization problem. Another example is a compact set which can guarantee the existence of the solution of a differential equation. Sets of open sets lead to the concept of a topology in which, for example, the ideas of convergence of sequences can be discussed without reference to a particular metric or norm. The scene is then set for a discussion in Chapter 10 of continuity and differentiability in general spaces, and provides a foundation for the study of differential topology (Chillingworth 1976), which is essential for the understanding of nonlinear system theory. However, despite the attraction of assuming sets with special properties in order to obtain mathematical results, if such assumptions give a mathematical model which is divorced from the practical system, then those assumptions must be discarded. In other words, the practical problem must dictate which assumptions are acceptable and which are not.

9.1 Open and closed sets

In many practical problems, some control variables (and sometimes state variables as well) can have values only within some given finite interval. For example, an aircraft engine is designed to give thrusts up to a maximum value $T_{\max}$, say. Of course, the minimum thrust magnitude is zero, attained when the engine is switched off. In practice some very small thrusts may not be feasible, but for a mathematical model of the engine it is usually assumed that the pilot of the aircraft can choose any thrust T constrained by the inequalities $0 \leqslant T \leqslant T_{\max}$. In other words, T can be any real number in the closed interval $[0, T_{\max}]$. On the other hand, some constraints involve strict inequalities ($<$ or $>$), because the end value of an interval may be a critical value in some sense and must be avoided. An example arises in damped harmonic motion of a system with natural frequency ω_n and damping factor ζ. Such a system is governed by the differential equation

$$\ddot{x} + 2\zeta\omega_n\dot{x} + \omega_n^2 x = 0, \qquad x(0) \neq 0.$$

If damped oscillatory motion is required, then ζ must be chosen from the open interval $(0, 1)$. For some damping to be present in the system, ζ must be greater than zero, and it must be less than unity because a complete oscillation is not obtained when $\zeta = 1$. The critical case $\zeta = 1$ divides the underdamped sinusoidal solutions from the overdamped exponential time behaviour of the system.

When a dynamical system involves several control variables $u_1, \ldots, u_m$, all of which are unconstrained in value, then the control vector $\boldsymbol{u}(t) = (u_1(t), \ldots, u_m(t))$ can be chosen as any element in $\mathbb{R}^m$. In practice, the values of the control variables are usually restricted to lie between given finite limits. For example, with $m = 3$, the values $u_1(t)$, $u_2(t)$, and $u_3(t)$ might all have to lie in the closed interval $[0, 1]$, so that $0 \leqslant u_1(t) \leqslant 1$, $0 \leqslant u_2(t) \leqslant 1$, and $0 \leqslant u_3(t) \leqslant 1$ for all time t. Then, for all t, the control vector $\boldsymbol{u}(\mathrm{t}) = (u_1(t), u_2(t), u_3(t))$ would be constrained to lie either inside or on the unit cube in $\mathbb{R}^3$ shown in Fig. 9.1. Generally a control vector $\boldsymbol{u}(t)$ is assumed to lie in some subset U of $\mathbb{R}^m$.

The concepts of open and closed intervals on the real line $\mathbb{R}$ can be extended to more general sets. $\mathbb{R}^n \backslash \{\mathbf{0}\}$ is an open subset of $\mathbb{R}^n$ in much the same way as $(0, 1)$ is an open subset of $[0, 1]$. The set $\mathrm{G}\ell(n, \mathbb{R})$ of all real nonsingular $n \times n$ matrices is an open set in the sense that there are $n \times n$ matrices with determinants as small as desired but still nonzero and therefore elements of $\mathrm{G}\ell(n, \mathbb{R})$. Again, this is exactly the same concept as an open interval $(a, b) \subset \mathbb{R}$ in which there are real numbers arbitrarily close to a or b and yet still are in (a, b) even though a and b are not. The complement of the open set $\mathrm{G}\ell(n, \mathbb{R})$ is the set of all real singular $n \times n$ matrices:

$$\{A \in \mathrm{M}_n(\mathbb{R}) : \det A = 0\}$$

(cf. Fig. 2.1) and is a closed set. In general, the complement of an open set is closed and vice versa.

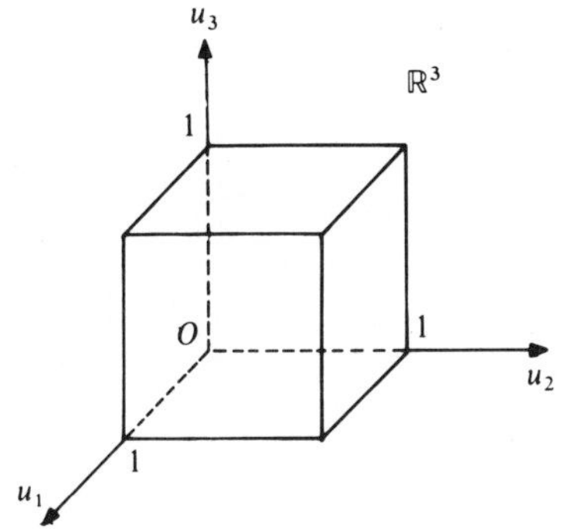

Fig. 9.1 Bounded controls

A reason why it is sometimes important to assume that a set is open is illustrated by an example from ordinary max–min theory. Suppose that it is required to find a point $x^* \in (0, 1)$ at which a function $f:(0, 1) \to \mathbb{R}$, defined by $x \mapsto f(x)$, attains a minimum value, and let the graph of f be as shown in Fig. 9.2. There is no global minimum of $f(x)$ in $(0, 1)$, since smaller and smaller values are obtained as x is chosen closer and closer to 1, but $f(1)$ is not defined, because 1 does not belong to the domain of f. However, there is a local minimum at x^*. Because $x^* \in (0, 1)$, there exists an ε such that $x^* + \varepsilon \in (0, 1)$ provided that $|\varepsilon|$ is sufficiently small. Then, by Taylor's theorem (see Chapter 10),

$$f(x^* + \varepsilon) = f(x^*) + \varepsilon f'(x^*) + o(\varepsilon^2), \tag{9.1}$$

where $f'(x^*)$ denotes the value of $\mathrm{d}f/\mathrm{d}x$ at x^*. A necessary condition for $f(x^*)$ to be a local minimum of f is that the difference

$$\Delta = f(x^* + \varepsilon) - f(x^*) \tag{9.2}$$

must be positive for all sufficiently small $|\varepsilon|$, irrespective of the sign of ε. From (9.1) and (9.2),

$$\Delta = \varepsilon f'(x^*) + o(\varepsilon^2)$$

and, for sufficiently small ε, the term $\varepsilon f'(x^*)$ determines the sign of Δ. But, since x^* belongs to the open interval $(0, 1)$, the quantity ε can be chosen either as a positive real number or as a negative one, provided that the magnitude $|\varepsilon|$ is sufficiently small to ensure that $x^* + \varepsilon \in (0, 1)$. Since $f'(x^*)$ is a known value at x^*, it follows that, if $f'(x^*) \neq 0$, then the sign of $\varepsilon f'(x^*)$ (and hence that of Δ) can be changed simply by changing the sign of ε. Such a possibility would violate the above necessary condition for a minimum value of f. Hence $f'(x^*) = 0$.

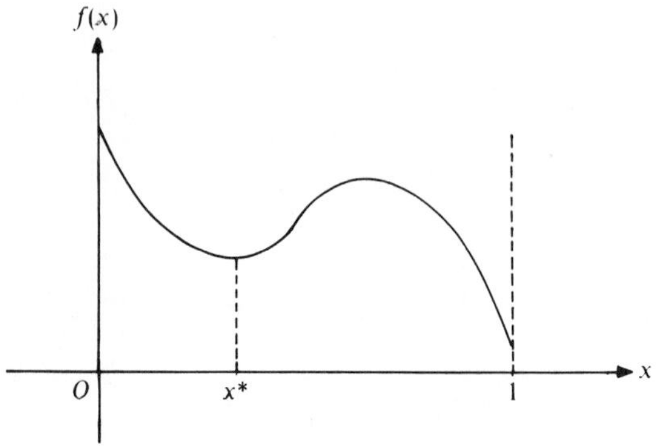

Fig. 9.2 Minimization of function f

If the domain of f in the above example had been the closed interval $[0, 1]$ with the graph of Fig. 9.2, then it would not have been possible to assume that ε could be chosen either positive or negative. For the possibility would then exist of x^* being equal to either end value in the interval $[0, 1]$. In the search for x^*, suppose that the case $x^* = 1$ is considered. Then, as before, $f'(x^*)$ will be a known value, and the expression $\varepsilon f'(x^*)$ must be positive for sufficiently small ε. However, this time, ε can only be negative since $x^* + \varepsilon = 1 + \varepsilon$, with $\varepsilon > 0$, would be a real number outside of $[0, 1]$ and so would not be an admissible value for x. Thus, since ε must be negative, in order for $\varepsilon f'(x^*) > 0$ (and hence $\Delta > 0$), it is necessary that

$$f'(x^*) < 0.$$

A similar argument for the case $x^* = 0$ would lead to $f'(x^*) > 0$. These conditions on $f'(x^*)$ arise only for the closed interval $[0, 1]$ and not when the variable x is restricted to the open interval $(0, 1)$.

A similar situation arises in optimal control theory. Given the problem described in Section 1.4, (1.20) and (1.21), necessary conditions for optimality are obtained by assuming that a control profile $u^*(\bullet)$ is an optimal control. A family of control variations $u^*(\bullet) + \delta u(\bullet)$ (Fig. 9.3) is then used to determine the consequent change in the performance index (1.21), just as $x^* + \varepsilon$ was used to change the value of $f(x^*)$ above. If the control u belongs to some open set Ω of control functions and $u^* \in \Omega$, then it is known that there exist such control variations $u^* + \delta u$ which also belong to Ω, provided δu is sufficiently small so that $u^* + \delta u$ is near to u^*.

As discussed in Chapters 7 and 8, the idea of elements of a space being near to some given element of that space can be expressed by using a metric (or norm) for the space. The concepts of open and closed sets can also be applied to a general metric (or normed) space (X, ρ). Given $a \in X$ and $\varepsilon > 0$, the set

$$\mathrm{B}(a, \varepsilon) = \{x \in X : \rho(a, x) < \varepsilon\}$$

is called the *open ball*, centre a and radius ε. The word 'open' is used because

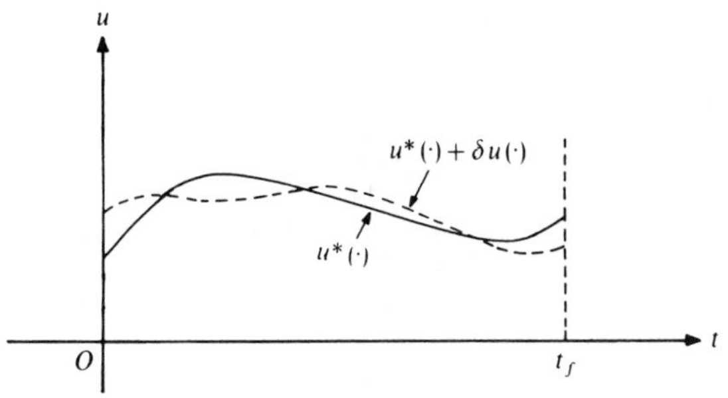

Fig. 9.3 Control variation

of the strict inequality $\rho(a, x) < \varepsilon$. Such an open ball is sometimes called a *spherical neighbourhood* or *ε-neighbourhood* of a. A *neighbourhood* of a point $x \in X$ is any subset of X which contains an ε-neighbourhood of x.

Example 9.1

Select the Euclidean norm in $\mathbb{R}^2$; if $\boldsymbol{x}, \boldsymbol{y} \in \mathbb{R}^2$, with $\boldsymbol{x} = (x_1, x_2)$ and $\boldsymbol{y} = (y_1, y_2)$, then

$$\rho(\boldsymbol{x}, \boldsymbol{y}) = \|\boldsymbol{x} - \boldsymbol{y}\| = \sqrt{[(x_1 - y_1)^2 + (x_2 - y_2)^2]}.$$

An ε-neighbourhood of the point $\boldsymbol{a} = (a_1, a_2) \in \mathbb{R}^2$ is the set $B(\boldsymbol{a}, \varepsilon)$ which in this case consists of all points $\boldsymbol{x} = (x_1, x_2)$ *inside* the circle

$$(x_1 - a_1)^2 + (x_2 - a_2)^2 = \varepsilon^2,$$

with centre at (a_1, a_2) and radius ε (Fig. 9.4).

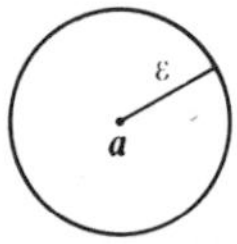

Fig. 9.4 The open ball $B(\boldsymbol{a}, \varepsilon)$ of Example 9.1

The reader should beware of thinking that a spherical neighbourhood is necessarily 'spherical'; it all depends on the metric or norm chosen. For example, in the metric space $(\mathbb{R}^2, \rho)$ in which

$$\rho(\boldsymbol{x}, \boldsymbol{y}) = \max\{|x_1 - y_1|, |x_2 - y_2|\} \quad \forall\, \boldsymbol{x}, \boldsymbol{y} \in \mathbb{R}^2,$$

the open ball $B(\boldsymbol{0}, 1)$, for which the point y is the origin (0, 0), is the interior of the square

$$\max\{|x_1|, |x_2|\} < 1 \qquad \text{(Fig. 9.5).}$$

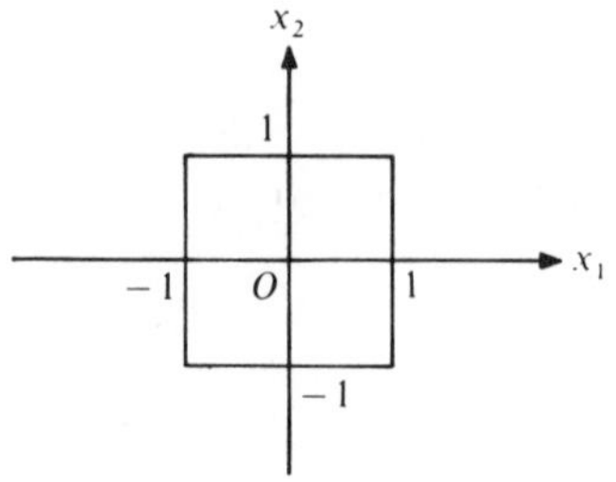

Fig 9.5 $\max\{|x_1|, |x_2|\} < 1$

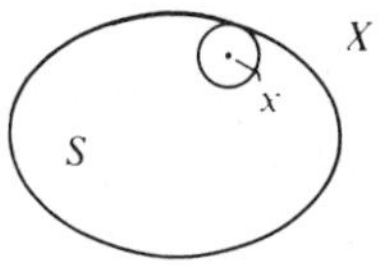

Fig. 9.6 Interior point x of set S

Suppose that X is a metric space and $S \subset X$. A point $x \in S$ is called an *interior point* of S if there is some open ball, with centre x, which is contained in S. Put another way, there is always an ε-neighbourhood of an interior point $x \in S$ such that all points of this neighbourhood lie inside S, none lying on what can be called the boundary of S (Fig. 9.6). The set of all interior points of S is called the *interior of* S and denoted by int S. Thus,

$$x \in \operatorname{int} S \quad \text{iff} \quad \mathrm{B}(x, \varepsilon) \subseteq S \text{ for some } \varepsilon > 0.$$

The interior of a set S is used to define an open set in the following way. If $\operatorname{int} S = S$, then S does not contain its boundary points and S is said to be open. [1] In other words, if S is a subset of a metric space X, then S is an open set if, given any $x \in S$, there is some $\varepsilon > 0$ such that $\mathrm{B}(x, \varepsilon) \subseteq S$. For example, $\{(x, y) \in \mathbb{R}^2 : x^2 + y^2 < 1\}$ is an open set (disc) in $\mathbb{R}^2$. Any union of open sets is open, and any finite intersection of open sets is open (Exercises 4 and 5).

The fact of whether a subset S of a metric space X is open or not depends on the space X, as the next example shows.

Example 9.2

$S = (0, 1)$ is an open subset of the metric space $X = \mathbb{R}$; but it is not open if $X = \mathbb{C}$, because any ε-neighbourhood of a point in S contains complex numbers that are not contained in S (Fig. 9.7). See Example 7.3 for the usual metric on $\mathbb{C}$.

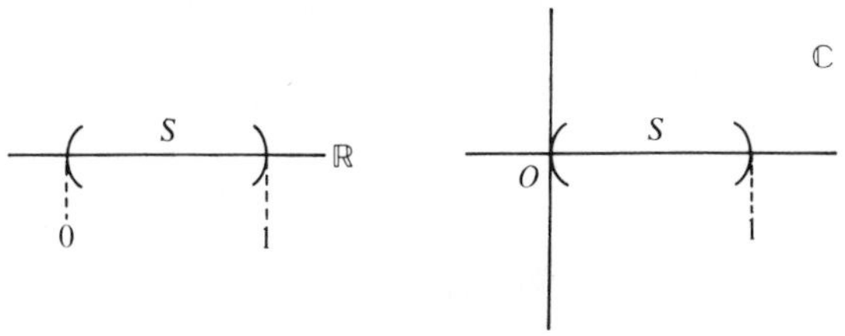

Fig. 9.7 Subset (0 , 1) in $\mathbb{R}$ and $\mathbb{C}$

An *exterior point* of S is an interior point of S', the complement of S. The set of all exterior points of S is called the *exterior of* S and denoted by ext S. The *boundary of S* is the set of points x such that every open set containing x contains points from both S and S'. The boundary of S is usually denoted by ∂S. [2] Note that $\operatorname{int} S$, $\operatorname{ext} S$, and ∂S are disjoint sets (provided none is the empty set), and X is the union of these disjoint sets. The null set $\varnothing$ and the universal set X are always open sets in X, and hence both are also closed in X.

If $S \subset X$, then a point $x \in X$ is called a *limit* (or *accumulation*) *point of S* when every open ball centred at x contains a point of S different from x. Notice that x does not necessarily belong to S. The definition implies that every ε-neighbourhood of a limit point contains an infinite number of elements of S. [3]

Example 9.3

Every point $a \in [0, 1]$ is a limit point of the set of real numbers $S = (0, 1]$, since every open ball (i.e. open interval on $\mathbb{R}$ in this case) containing a will contain points of $(0, 1]$ besides a. Notice that the accumulation point 0 is not an element of the set S.

Example 9.4 (Hermes and LaSalle 1969)

A system is governed by the equations

$$\dot{x}_1 = (1 - x_2^2), \qquad x_1(0, k) = 0, \tag{9.3}$$

$$\dot{x}_2 = u^{(k)}, \qquad x_2(0, k) = 0, \tag{9.4}$$

where the control $u^{(k)}$ is defined in Example 8.8. Let $X(1)$ be the set of all points $\big(x_1(1), x_2(1)\big)$ which can be attained by using all admissible controls $u^{(k)}(\cdot)$ in unit time. It will now be shown that the point $(1, 0)$ is an accumulation point of $X(1)$, but that $(1, 0) \notin X(1)$. In Example 8.8, it is shown that $x_2(t, k)$ converges uniformly to zero on the interval $[0, 1]$ as $k \to \infty$, and that

$$x_2(1, k) = 0 \quad \forall\, k. \tag{9.5}$$

Integrating (9.3) over the interval $[0, 1]$, we have

$$\begin{aligned} x_1(1, k) &= \int_0^1 (1 - x_2^2)\, dt \\ &= 1 - \int_0^1 x_2^2(t, k)\, dt \\ &< 1. \end{aligned} \tag{9.6}$$

Since $x_2(t, k)$ converges uniformly to zero as $k \to \infty$, it follows that

$x_1(1, k) \to 1$ as $k \to \infty$. Furthermore, by (9.5),

$$\big(x_1(1, k), x_2(1, k)\big) \to (1, 0) \quad \text{as } k \to \infty.$$

Thus, (1, 0) is an accumulation point of $X(1)$.

Now assume that $(1, 0) \in X(1)$. Then $x_1(1, k) = 1$ and $x_2(1, k) = 0$ for some k, so from (9.6) it follows that $x_2(t, k) = 0$ for all $t \in [0\,,1]$ except at points of discontinuity of $u^{(k)}$. But, by (9.4), this implies that $u^{(k)} \equiv 0$ which contradicts the definition of $u^{(k)}$. Thus, $(1, 0) \notin X(1)$.

Not every infinite set has an accumulation point. The set $\mathbb{N}$ as a subset of $\mathbb{R}$ has no such points since, for any $a \in \mathbb{R}$, we can find an $\varepsilon > 0$ sufficiently small that the open interval $(a - \varepsilon\,, a + \varepsilon)$ contains no point of $\mathbb{N}$ (unless $a \in \mathbb{N}$). However, there is an important result known as the *Bolzano–Weierstrass Theorem.* This states that every bounded infinite set has at least one accumulation point. This theorem can be used, for example, to prove that a continuous real-valued function on a closed interval $[a\,,b]$ is bounded, as discussed in Section 8.2 (Chillingworth 1976).

The concept of an accumulation or limit point provides a further definition for a closed set. A set is said to be *closed* iff it contains all its limit points. A set $S \subset X$ together with all its limit points is called the *closure* of S, denoted by $\bar{S}$. That is,

$$\bar{S} = S \cup \{s : s \text{ is a limit point of } S\}.$$

In Example 9.4, the point (1, 0) is a limit point of $X(1)$. However, $(1, 0) \notin X(1)$ so $X(1)$ is not closed. In some optimal control problems, the existence of an optimal control depends upon whether an attainable set such as $X(1)$ is closed or not.

Example 9.5

From Example 9.3 every point $a \in [0\,,1]$ is a limit point of the set $S = (0\,,1]$. Since 0 is the only limit point that does not belong to $(0\,,1]$, the closure of S is given by

$$\bar{S} = (0\,,1] \cup \{0\} = [0\,,1].$$

The closure of an open ball $\mathrm{B}(a, \varepsilon)$ defined earlier is a closed ball with centre $a \in X$ and radius $\varepsilon > 0$, given by

$$\bar{\mathrm{B}}(a, \varepsilon) = \{x \in X : \rho(a, x) \leqslant \varepsilon\}.$$

A necessary and sufficient condition for a set $S \subset X$ to be closed is that $\bar{S} = S$. The closure $\bar{S}$ of a set S is the intersection of all closed sets A that contain S, i.e. [4]

$$\bar{S} = \bigcap \{A : A \text{ is closed, } S \subseteq A\}.$$

and $\bar{S}$ is the smallest closed set containing S. The interior of a set $S \subset X$ is the largest open set in S.[5,6] Again, S is closed if its complement S' is open. It follows that S is closed iff, for each $x \notin S$, there is a neighbourhood U of x that is disjoint from S.

An open set $X \subset \mathbb{R}^n$ is said to be *connected* if it is impossible to express X as the union of two non-empty disjoint open subsets. X is said to be *arcwise connected* if any two points in X can be joined by a continuous curve that lies in X. This latter definition is obviously of importance in dynamical system theory, since normally the state vector $\boldsymbol{x}$, with values $\boldsymbol{x}(t) \in \mathbb{R}^n$, is continuous in time. So, in order to control the state from any point in $X \subset \mathbb{R}^n$ to any other point in X, the set X must be arcwise connected.

It can be shown that if X is connected in the sense of the first definition then it is also arcwise connected.

9.2 Convex sets

As mentioned earlier, in optimization problems arising from dynamical systems, the state and/or control variables are often constrained to take values only from some given bounded region of the state space or control space. The following two examples illustrate the type of region that such constraints may specify.

Example 9.6

$$0 \leqslant x_1 \leqslant 1,$$

$$x_2 \geqslant 0, \qquad x_2 \leqslant 5(2x_1 + 1), \qquad x_2 \leqslant -14x_1 + 17.$$

The feasible region in $\mathbb{R}^2$ defined by these inequalities from which the ordered pairs (x_1, x_2) may be chosen is shown in Fig. 9.8.

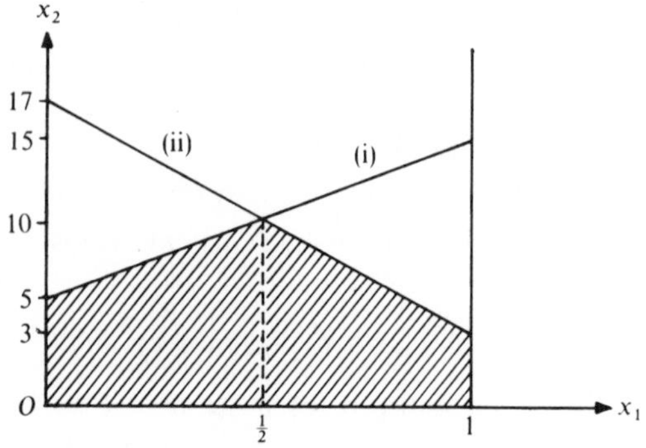

Fig. 9.8 Feasible region of Example 9.6: (i) $x_2 = 5(2x_1 + 1)$, (ii) $x_2 = -14x_1 + 17$

Example 9.7

$$0 \leqslant x_1 \leqslant 1, \qquad x_2 \geqslant 0,$$

$$x_2 \leqslant 20x_1(1 - x_1), \qquad x_2 \geqslant 12x_1(1 - x_1) + 1.$$

The feasible region in this case is shown in Fig. 9.9.

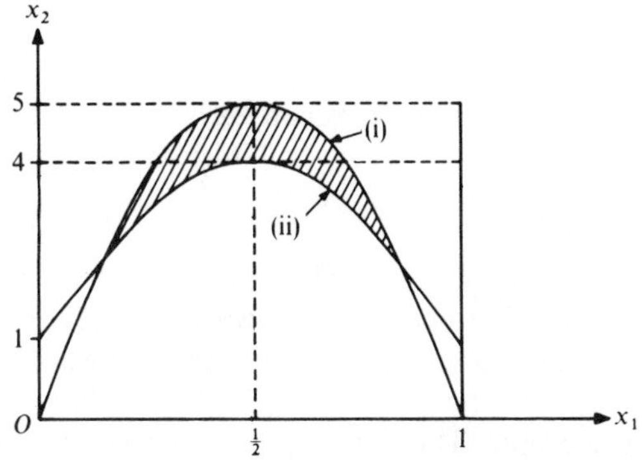

Fig. 9.9 Feasible region of Example 9.7: (i) $x_2 = 20x_1(1 - x_1)$, (ii) $x_2 = 12x_1(1 - x_1) + 1$

A fundamental difference between the two feasible subsets of $\mathbb{R}^2$ shown in Figures 9.8 and 9.9 is that a straight line joining any two points of the feasible set in Fig. 9.8 lies wholly within that set whereas this is untrue for the set of Fig. 9.9. The set of Fig. 9.8 is called a *convex set* and that of Fig. 9.9 a *nonconvex set*. Generalizing this concept of convexity to any vector space X over the field $\mathbb{R}$, the line segment joining any two points $x, y \in X$ is the set of all points of the form $\lambda x + (1 - \lambda)y$ with $\lambda \in [0, 1] \subset \mathbb{R}$. A set $S \subseteq X$ is said to be *convex* if, given $x, y \in S$, the line segment joining x and y lies wholly in S. In other words, if x and y are any two points in a convex set S, then all points of the form $\lambda x + (1 - \lambda)y$ with $0 \leqslant \lambda \leqslant 1$ are also in S. This is illustrated in Fig. 9.10.

Linear combinations of elements in any subspace of X lie in that subspace. Suppose that S is a subspace of X, let x and y be any elements of S, and let $\lambda \in [0, 1]$ be arbitrary. Then, since $\lambda x + (1 - \lambda)y$ is a linear combination of x and y, it belongs to S, and so the subspace S is convex. Thus, all subspaces (and linear varieties; see Luenberger 1969) are convex. The empty set is also always taken to be convex. The intersection of an arbitrary number of convex sets is convex. [7] To see why this is so, suppose $\mathscr{C}$ is an arbitrary collection of convex sets and let $C = \bigcap_{A \in \mathscr{C}} A$. If $C = \varnothing$ then C is convex as already stated. Now assume that $C \neq \varnothing$ and that $x_1, x_2 \in C$, and select $\lambda \in [0, 1]$. Then

Fig. 9.10 Convex and nonconvex sets

$x_1, x_2 \in A$ for any $A \in \mathscr{C}$, and so again $\lambda x_1 + (1 - \lambda)x_2 \in A$ since A is convex. Thus, $\lambda x_1 + (1 - \lambda)x_2 \in C$, and so C is convex.

A function $f: X \to \mathbb{R}$ is said to be convex if its domain X is a convex set and

$$f(\lambda x_1 + (1 - \lambda)x_2) \leqslant \lambda f(x_1) + (1 - \lambda)f(x_2)$$

for every $x_1, x_2 \in X$ and all $\lambda \in [0, 1]$. The situation is illustrated in Fig. 9.11, where the graph of f is said to be convex downwards. [8] From Fig. 9.11 it can be seen that, if the graph of a function f is convex downwards, then the chord joining any two points on the graph never lies below that part of the graph lying between those points. The significance of a convex function for optimization should now be clear to the reader. A convex function will have a global minimum, so that the existence of a solution to finding the minimum value of the function is guaranteed.

Of course, not all subsets of a vector space are convex. However, because convex sets make the analysis so much easier (although sometimes to the detriment of practical considerations!), it is often convenient, when we are dealing with a nonconvex set S, to work with the smallest convex set that contains S. Suppose that S is an arbitrary subset of a vector space X, and consider the collection $\mathscr{C}$ of all convex sets containing S. Since X is convex, it is a member of $\mathscr{C}$, and so $\mathscr{C}$ is not empty. The intersection of all members of $\mathscr{C}$

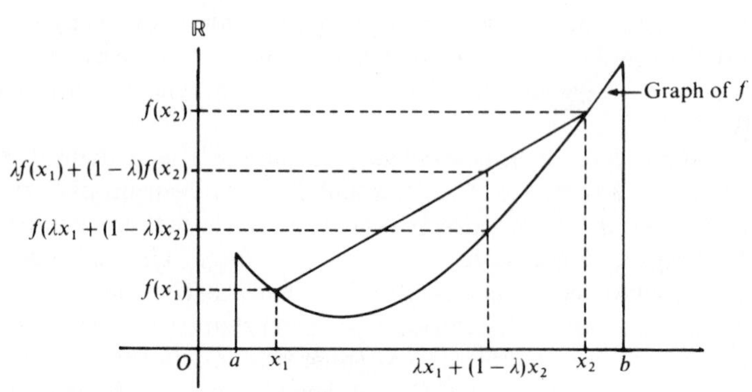

Fig. 9.11 A convex function

is clearly the smallest convex set containing S and is called the *convex hull of S, denoted by* co S. The existence of co S is guaranteed by the fact, proved above, that the intersection of an arbitrary number of convex sets is convex. Furthermore,

$$\text{co}\, S = \{x : x = \sum_i \lambda_i x_i,\ x_i \in S,\ \lambda_i \geqslant 0,\ \sum_i \lambda_i = 1\}.$$

Typical convex hulls are shown in Fig. 9.12.

There is a weaker property than convexity for a subset S of a vector space X. A set S is said to be *starlike with respect to* $x \in S$ when, for any $s \in S$, the line segment joining x and s lies wholly in S (Boothby 1975). This is illustrated in Fig. 9.13.

A set S in a vector space X is said to be a *cone with vertex at the origin* if $s \in S$ and $\lambda \geqslant 0$ imply that $\lambda s \in S$. A *cone with vertex* x is defined as a translation $x + S$ of a cone S with vertex at the origin. A *convex cone* is defined as a set that is both convex and a cone (Fig. 9.14). Convex cones arise in particular when attention is focused on vectors in certain subsets of a vector space. For example, Fig. 9.14(b) represents the convex cone in $\mathbb{R}^2$ given by the positive quadrant

$$S = \{\boldsymbol{x} : \boldsymbol{x} = (x_1, x_2),\ x_1 \geqslant 0,\ x_2 \geqslant 0\}.$$

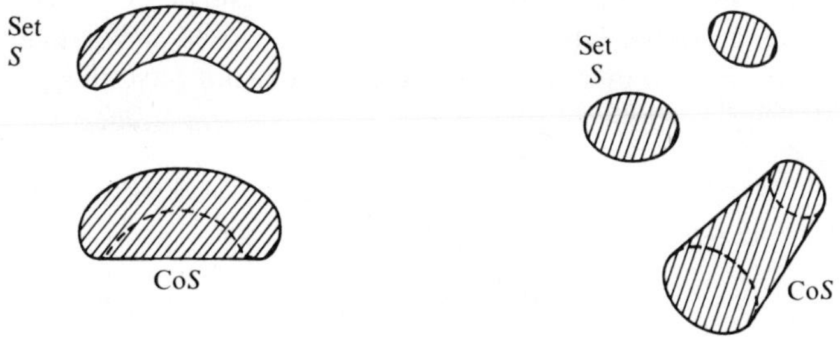

Fig. 9.12 Convex hulls

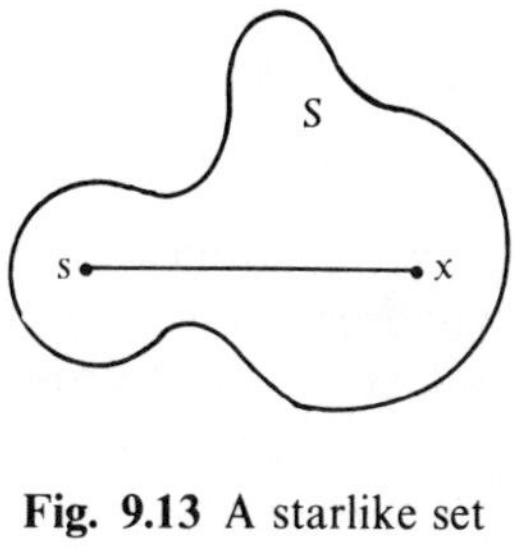

Fig. 9.13 A starlike set

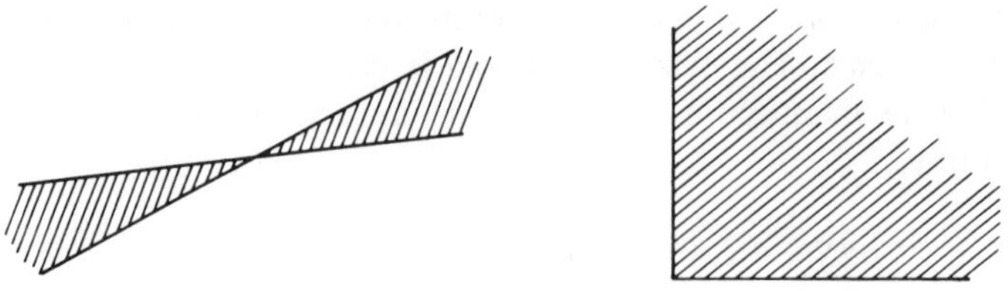

(a) (b)

Fig. 9.14 Convex cones

Convex sets in the plane illustrate the importance of 'corners'. The convex hull of three noncollinear points in $\mathbb{R}^2$ is a triangle, and the convex hull of four points in $\mathbb{R}^2$, no three of which are collinear, is a quadrilateral. Furthermore, the corners of these hulls are their only points that do not lie on any open line segment contained wholly within the hull. A generalization of this notion is the following definition. A point x in a convex set S is an *extreme* point of S if there is no open line segment in S that contains x. In other words, $x \in S$ is an extreme point of S if, for any $\lambda \in [0, 1]$ and any $y,z \in S$, the equation

$$x = \lambda y + (1 - \lambda)z$$

holds only when $\lambda = 0$ (i.e. $x = z$) or $\lambda = 1$ (i.e. $x = y$). In $\mathbb{R}^2$, a triangle and rectangle both have the property of being the convex hull of their corresponding set of extreme points. On the other hand, the closed upper half plane, although convex, has no extreme points and so does not have this property.

Example 9.8

Consider the linear-programming problem:

$$\text{maximize} \quad 7x + 11y$$

subject to the constraints

$$0 \leqslant x \leqslant 15, \qquad y \geqslant 0, \qquad x - y + 10 \geqslant 0, \qquad x + y - 20 \leqslant 0.$$

The constraints are represented by straight lines shown in Fig. 9.15. Writing the linear objective function, given above, as

$$7x + 11y = p,$$

this can also be represented by a straight line for any given value of p (Fig. 9.15). The problem is to find the point (x, y) that yields the largest value of p inside or on the boundary of the feasible region $OABCD$. Notice that this region is convex. The solution to the problem is to move the line $7x + 11y = p$, parallel to itself, further and further away from the origin O.

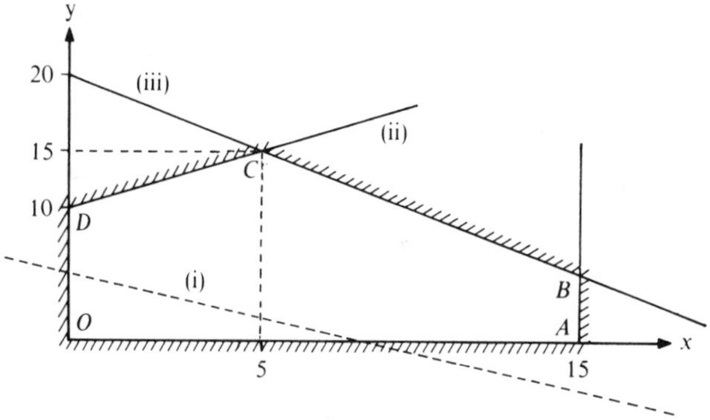

Fig. 9.15 Feasible region of Example 9.8: (i) $7x+11y=p$, (ii) $x-y+10=0$, (iii) $x+y-20=0$

The point at which this line leaves the convex feasible region is C, since the gradient of the line $x + y - 20 = 0$ is -1 whereas the gradient of $7x + 11y - p = 0$ is only $-7/11$. Thus, the maximum value of p is obtained at the point C where the lines $x - y = -10$ and $x + y = 20$ intersect. At this point, $x = 5$ and $y = 15$, so the maximum value of p is 200. The points O, A, B, C, D are extreme points of the convex region, and it is easy to see that generally the solution to a linear-programming problem such as the one above will be at an extreme point of the feasible region.

9.3 Compact sets

The solution of the differential equations governing a dynamical system should be well behaved in some practical sense. In particular, the value of any variable must not become indefinitely large as some finite value of time is approached. That such an explosive situation can arise from very simple differential equations is illustrated by the following example.

Example 9.9

An uncontrolled dynamical system is governed by the differential equation

$$\dot{x}(t) = x^2(t),\ x(0) = 1.$$

By separating the variables and integrating with respect to time, we obtain

$$\int_1^x \frac{dz}{z^2} = \int_0^t d\tau$$

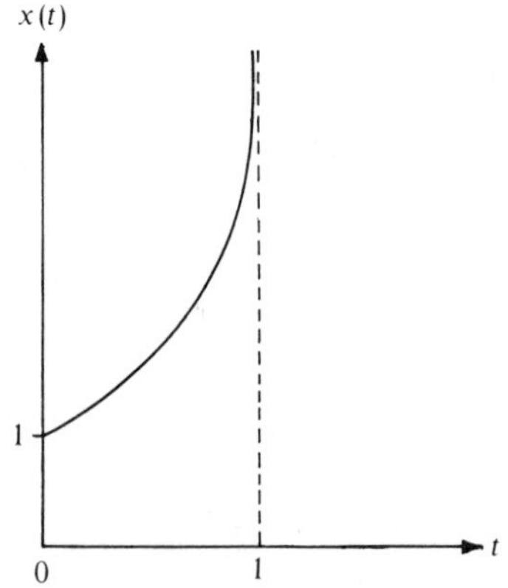

Fig. 9.16 Finite escape time

which yields $[-1/z]_1^x = [\tau]_0^t$, i.e. $-1/x + 1 = t$, whence

$$x(t) = \frac{1}{1-t}.$$

The graph of x against t is shown in Fig. 9.16. Clearly $x(t) \to \infty$ as $t \to 1$ and the dynamical system is said to *blow up* or to have a *finite escape time.*

Such behaviour as that shown in the last example is to be avoided when discussing the general theory of dynamical systems. In Example 9.9, if the domain of the function x is restricted to $[0, 1-\varepsilon]$, where ε is a small but positive real number, then x is bounded on this interval. The graph of x shown in Fig. 9.16 is then restricted to the rectangle $\{(t, x):0 \leqslant t \leqslant 1-\varepsilon, 1 \leqslant x(t) \leqslant x(1-\varepsilon)\}$, and the solution to the differential equation remains finite within this closed and bounded region. Although in what follows the general concept of a compact set is discussed, in much of the dynamical-systems literature a compact set simply means a closed and bounded set.

Suppose X is a metric space and $S \subset X$. Consider a (possibly infinite) set of subsets $\{A_i : i \in \mathscr{I}\}$ such that

$$S \subseteq \bigcup_{i \in \mathscr{I}} A_i.$$

Then the set $[A_i : i \in \mathscr{I}\}$ is called a *cover* (or an *open cover*) of S if each A_i is an open set (Fig. 9.17). In other words, if every element of S belongs to at least one of a collection of open sets, then that collection is said to be an open cover of S. Should there exist a finite set $\{A_1, \dots, A_n\}$ that is also a cover of S, then this set is called a *finite cover* of S. This latter case is particularly important: if

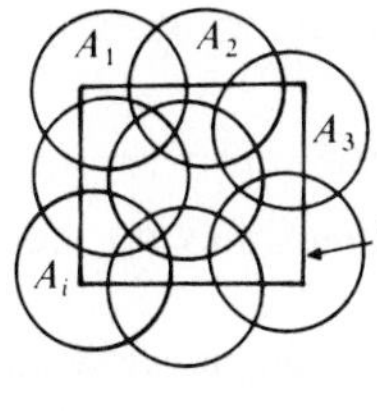

Fig. 9.17 A cover of set S

every open cover of S contains a finite subcollection of open sets that also covers S, then the set S is said to be *compact*. Obviously a set (or space) with only a finite number of elements is always compact.

Example 9.10

The infinite set $A = \{0, 1, \frac{1}{2}, \frac{1}{3}, \ldots, 1/n, \ldots\}$ is compact. To see this, let $C = \{C_i : i \in \mathscr{I}\}$ be any set of open sets (intervals) C_i that covers A. Then the element 0 belongs to some open set $C_k \in C$. There are only a finite number of points of A which are not in C_k (Fig. 9.18) and each of these remaining points require at most one more open set.

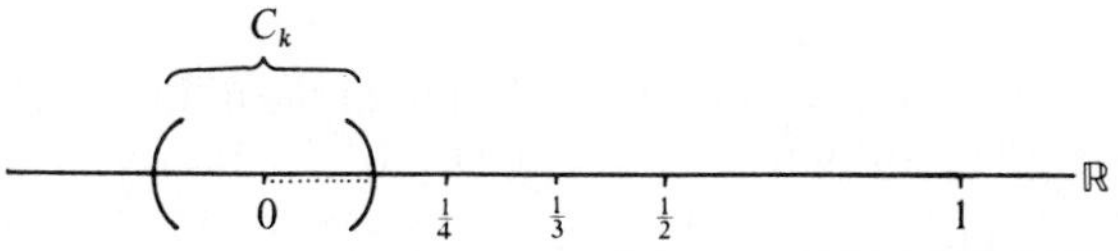

Fig. 9.18 A finite cover for a compact set

Example 9.11

The infinite set $A = \{\frac{1}{2}, \frac{1}{3}, \ldots, 1/n, \ldots\}$ is not compact. This may be seen by considering the infinite collection of open intervals

$$C = \{(\tfrac{1}{3}, \tfrac{2}{3}), (\tfrac{1}{4}, \tfrac{3}{4}), (\tfrac{1}{5}, \tfrac{4}{5}), \ldots\} = \{(\tfrac{1}{n}, 1 - \tfrac{1}{n}) : n \geq 3, n \in \mathbb{N}\}.$$

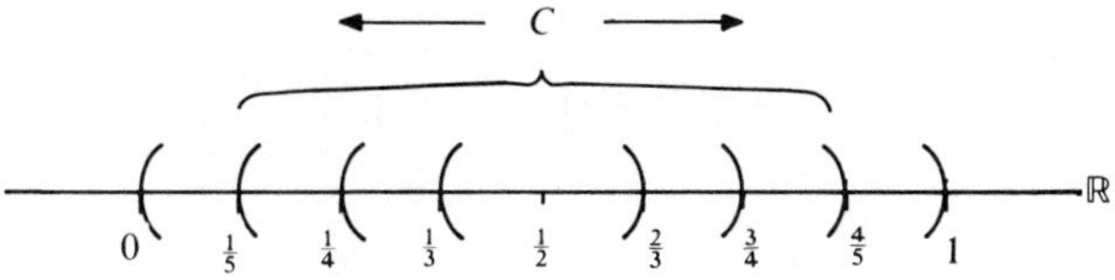

Fig. 9.19 A noncompact set

This is an open cover of A, but there is no finite collection of sets in C that covers A (Fig. 9.19).

A space X is said to be *locally* compact if every $x \in X$ has a neighbourhood whose closure is compact.

In general, it is not a trivial task to determine whether any particular set is compact. However, help is at hand for sets in $\mathbb{R}^n$ through the *Heine–Borel Theorem.* This states that any closed bounded subset $U \subset \mathbb{R}^n$ is compact. In particular, if $U = [a, b]$ is a closed interval of $\mathbb{R}$, then $[a, b]$ is compact. Furthermore, a subset S of any finite-dimensional normed space X is compact iff S is closed and bounded. If X is just a metric space and $S \subset X$, all that can be said is that, if S is compact, then S is closed and bounded. The converse is not true in general, i.e. not all closed and bounded subsets of a metric space are compact.

An alternative way of expressing compactness is as follows. Suppose $(\mathbb{R}^n, \|\cdot\|)$ is a normed linear space. A set $U \subset \mathbb{R}^n$ is compact iff every sequence $(x^{(k)})$ in S contains a subsequence $(x^{(k_i)})$ that converges to a limit in S. This is important in the proof of convergence of a sequence $(x^{(k)})$ in $\mathbb{R}^n$ arising, say, from a steepest-descent algorithm (Section 8.1). If the sequence $(x^{(k)})$ lies in a compact subset $S \subset \mathbb{R}^n$, and x^* is the unique limit point of $(x^{(k)})$, then that sequence converges to x^* as $k \to \infty$ (Wolfe 1978).

Returning to Example 9.9, the semi-infinite rectangle (Fig. 9.16) $\{(t, x): 0 \leqslant t < 1, x \geqslant 1\}$, in which the solution of the differential equation lies, is not compact. The infinite set $\{(0, 2), (1, 3), (2, 4), \ldots\}$ of open intervals covers the interval $[1, \infty)$ but there does not exist a finite cover. So, in order to avoid uncivilized behaviour (such as a finite escape) when discussing a general dynamical system

$$\dot{\boldsymbol{x}} = \boldsymbol{f}(\boldsymbol{x}, \boldsymbol{u}), \ \boldsymbol{x}(0) = \boldsymbol{x}_0 \in \mathbb{R}^n, \tag{9.7}$$

some constraint must be placed upon the state $\boldsymbol{x}$ such as (Hermes and LaSalle 1969)

$$\boldsymbol{x}^{\mathrm{T}} \boldsymbol{f}(\boldsymbol{x}, \boldsymbol{u}) \leqslant K(1 + \|\boldsymbol{x}\|^2) \tag{9.8}$$

for some constant K and for all admissible $\boldsymbol{x}$ and $\boldsymbol{u}$. It follows from this constraint that

$$\boldsymbol{x}^{\mathrm{T}} \dot{\boldsymbol{x}} \leqslant K(1 + \|\boldsymbol{x}\|^2),$$

i.e.

$$\frac{\mathrm{d}}{\mathrm{d}t} \|\boldsymbol{x}\|^2 \leqslant 2K(1 + \|\boldsymbol{x}\|^2). \tag{9.9}$$

Now write $y = \|\boldsymbol{x}\|^2$ and $y_0 = \|\boldsymbol{x}_0\|^2$; then (9.9) becomes

$$\frac{\mathrm{d}y}{\mathrm{d}t} \leqslant 2K(1 + y),$$

so that, by separation of variables,

$$\int_{y_0}^{y} \frac{\mathrm{d}z}{1 + z} \leqslant 2K \int_0^t \mathrm{d}\tau.$$

Integration yields

$$\ln\left(\frac{1 + y}{1 + y_0}\right) \leqslant 2Kt,$$

whence $\|\boldsymbol{x}(t)\|^2 \leqslant (1 + \|\boldsymbol{x}_0\|^2)\mathrm{e}^{2Kt} - 1$. Therefore,

$$\|\boldsymbol{x}(t)\|^2 \leqslant (1 + \|\boldsymbol{x}_0\|^2)\,\mathrm{e}^{2Kt}.$$

By choosing a real number $T \geqslant 0$, one need only consider the compact set

$$S = \{(t, x){:}0 \leqslant t \leqslant T,\ \|\boldsymbol{x}\|^2 < (1 + \|\boldsymbol{x}_0\|^2)\,\mathrm{e}^{2Kt}\}$$

of the $(n + 1)$-dimensional $(t, \boldsymbol{x})$-space. Any other constraint on the state $\boldsymbol{x}$ instead of (9.8) would be suitable provided that it restricts t and $\boldsymbol{x}$ to a compact set of the $(t, \boldsymbol{x})$-space. The semi-infinite rectangle $\{(t, x){:}0 \leqslant t < 1, x \geqslant 1\}$ of Example 9.9 is not a compact set of the (t, x)-space and this is why a finite escape time is possible.

In controlled dynamical systems, the control vector $\boldsymbol{u}(t) = (u_1(t), \ldots, u_m(t))$ is frequently constrained to lie in some chosen compact set U of $\mathbb{R}^m$. The unit m-cube

$$\{\boldsymbol{u}(t) \in \mathbb{R}^m{:}|u_j(t)| \leqslant 1\ (j = 1, \ldots, m) \quad \forall\, t \geqslant 0\}$$

is often taken as the compact set U, i.e. $\boldsymbol{u}(t) \in U \subset \mathbb{R}^m$. Since some or all of the control variables $u_1(t), \ldots, u_m(t)$ may be discontinuous with respect to t, there could be a finite or infinite number of times t at which the function $\boldsymbol{f}$ and hence the derivative $\dot{\boldsymbol{x}}$ in (9.7) are discontinuous. In order to deal formally with such behaviour, the ideas of null sets and measure must be mentioned.

9.4 Null sets

For simplicity, first consider a scalar control variable $u(t) \in \mathbb{R}$, and suppose that $u(\cdot)$ is discontinuous at a sequence of points t_i $(i \in \mathbb{N})$. A subset $A \subset \mathbb{R}$ is a *null set* (or has *measure zero*) if, for every arbitrarily small $\varepsilon > 0$, there exists a cover $\{A_1, A_2, \ldots\}$ of A by open (or closed) intervals A_i such that the sum of the lengths $\ell(A_i)$ of the intervals is less than ε, i.e.

$$\sum_{i=1}^{\infty} \ell(A_i) < \varepsilon.$$

Example 9.12

The infinite set of points $t = t_i$ ($i \in \mathbb{N}$) at which a control function $t \mapsto u(t)$ is discontinuous is a null set. To see this, let $\ell(A_i) = \varepsilon/2^{i+1}$ where A_i is the interval which includes t_i. Then

$$\sum_{i=1}^{\infty} \ell(A_i) = \varepsilon \sum_{i=1}^{\infty} 1/2^{i+1} = \varepsilon(\tfrac{1}{4} + \tfrac{1}{8} + \cdots) = \tfrac{1}{2}\varepsilon < \varepsilon.$$

Any countable set (see Section 3.3) on the real line $\mathbb{R}$ is null. When the function $\boldsymbol{f}$ in (9.7) is discontinuous at points where $\boldsymbol{u}$ is discontinuous, then certain equalities arising in the analysis may not be valid at those points. However, provided that $\boldsymbol{u}$ is discontinuous only on a null set, these equalities are said to hold *almost everywhere* (a.e.), i.e. everywhere except on a null set. In the theory of integration, one may virtually ignore the null sets because, if a function is integrable, then its values can be altered arbitrarily at all points of a null set without altering the value of the integral.

Example 9.13

The control variable $u^{(1)}$ in Fig. 8.6(b) has a discontinuity at $t = \frac{1}{2}$ in the interval $[0, 1]$. If the value of $u^{(1)}$ at $t = \frac{1}{2}$ is defined to be $u^{(1)}(\frac{1}{2}) = 0$, then the value of the integral

$$\int_0^1 u^{(1)}(t)\,\mathrm{d}t$$

is zero and remains at that value even when the value of $u^{(1)}$ at $t = \frac{1}{2}$ is changed to any other value, say $u^{(1)}(\frac{1}{2}) = 1$.

The above definition of a null set on $\mathbb{R}$ can be generalized to $\mathbb{R}^m$ by replacing the intervals A_i by m-dimensional cubes of arbitrarily small volumes (m-cubes) enclosing each of the elements of the null set. A subset $A \subset \mathbb{R}^m$ is a null set (or has measure zero) if, for every $\varepsilon > 0$, there exists a cover $\{A_1, A_2, \ldots\}$ of A by open (or closed) m-cubes A_i such that the sum of their volumes $v(A_i)$ is less than ε, i.e.

$$\sum_{i=1}^{\infty} v(A_i) < \varepsilon.$$

Any finite or countably infinite set of points in $\mathbb{R}^m$ has measure zero.

Two functions f and g defined on $A \subset \mathbb{R}^m$ that differ in value only on a null set are said to be equal almost everywhere (a.e.) on A, written as

$$f = g \quad \text{a.e. on } A.$$

Again, the function $f: \mathbb{R} \to \mathbb{R}$ defined by $f(t) = |t|$ has a derivative a.e., in fact everywhere except on the null set $\{0\}$.

The measurable sets on $\mathbb{R}^m$ are defined as the members of the smallest family of sets of $\mathbb{R}^m$ that contains all open sets, all closed sets, all null sets of $\mathbb{R}^m$, and also every difference, countable union, and countable intersection of its members. For a real interval $[a, b] \subset \mathbb{R}$, a real-valued function $f:[a, b] \to \mathbb{R}$ is called a *measurable function* when, for all $\alpha, \beta \in \mathbb{R}$, the set

$$\{t : t \in [a, b], \alpha < f(t) < \beta\}$$

is measurable in $\mathbb{R}$ (Barra 1981). The reader need not worry too much about this, because state and control vectors are usually assumed to be the values of measurable vector-valued functions. The assumption of measurable controls is particularly important in the proof of Pontryagin's maximum principle in optimal control (Pontryagin et al 1962).

Example 9.14

The Fuller problem (Fuller 1963) is the optimal control problem (Section 1.4):

$$\text{minimize} \quad \tfrac{1}{2} \int_0^{t_f} x_1^2 \, dt$$

subject to

$$\dot{x}_1 = x_2, \quad x_1(0) = x_1^0 \neq 0, \qquad \dot{x}_2 = u, \quad x_2(0) = x_2^0, \quad |u| \leqslant 1.$$

The solution to this problem for sufficiently large t_f is a control that switches repeatedly between the values 1 and -1 to drive the state to the origin in a finite time t'. An infinite number of switches occur in a neighbourhood of t'. This optimal control is measurable, and its existence justifies the assumption of measurable controls in the proof of Pontryagin's principle (McDanell and Powers 1971).

9.5 Topological spaces

In Chapter 7, the concepts of metric and norm were used to define the convergence of an infinite sequence of elements in a space. In the present chapter, the same concepts have been used to define open sets. It is possible and often convenient to define open sets without reference to metrics or norms. Convergence of sequences, continuity of mappings, and so on can actually be described by using general collections of these open sets, called *topologies*. The branch of mathematics in which this approach is used is called topology. In Chapter 11, the concept of a manifold will be introduced. A manifold is a special example of a general type of space that will now be defined: a topological space.

Given a non-empty set X, a collection $\mathcal{T}$ of subsets of X, called open sets, is called a *topology on* X iff

(i) $\varnothing \in \mathcal{T}$ and $X \in \mathcal{T}$,
(ii) the union of any collection (even if infinite and uncountable) of sets $T_i (i \in \mathcal{I})$ in $\mathcal{T}$ is also in $\mathcal{T}$, i.e. $\bigcup_{i \in \mathcal{I}} T_i \in \mathcal{T}$ (general unions). Here $\mathcal{I}$ is an index set, and to each i there exists a T_i, although the T_i may not be distinct.
(iii) $T_1 \in \mathcal{T}$ and $T_2 \in \mathcal{T}$ together imply that $T_1 \cap T_2 \in \mathcal{T}$ (finite intersections). [9]

The pair $(X, \mathcal{T})$ is called a *topological space*. If the collection $\mathcal{T}$ is well understood, then just the set X is often referred to as the topological space.

Example 9.15

$X = \mathbb{R}$ and $\mathcal{T}$ is the collection of open sets defined to be $\varnothing$, $\mathbb{R}$, and all sets that are unions of open intervals of $\mathbb{R}$. This collection $\mathcal{T}$ is called the *usual topology* on $\mathbb{R}$. This example illustrates why infinite intersections are not allowed in (iii) above, e.g.

$$\bigcap_{n \in \mathbb{N}} (-1/n\,,\, 1/n) = \{0\}$$

and $\{0\}$ is not an open interval.

Example 9.16

The class of all open sets in the plane $\mathbb{R}^2$ is a topology, and is called the *usual topology on* $\mathbb{R}^2$. The union of any number of open subsets in $\mathbb{R}^2$ is open, and the intersection of any finite number of open subsets in $\mathbb{R}^2$ is open, satisfying conditions (ii) and (iii) for a topology (Exercises 4 and 5).

The class of all subsets of any non-empty set X is called the *discrete topology*. If A is any non-empty subset of a topological space $(X, \mathcal{T})$, then the set of all intersections of A defined as

$$\mathcal{T}_A \triangleq \{T_1 \cap A : T_1 \in \mathcal{T}\}$$

is a topology on A called the *relative topology on* A.

The most important topological spaces which occur in practice are Hausdorff spaces. A topological space $(X, \mathscr{T})$ is called a *Hausdorff space* if, for any two distinct elements $x_1, x_2 \in X$, there are elements $T_1, T_2 \in \mathscr{T}$ such that $x_1 \in T_1$, $x_2 \in T_2$, and $T_1 \cap T_2 = \varnothing$. In other words, there are neighbourhoods T_1 of x_1 and T_2 of x_2 that are disjoint, giving rise to the mnemonic that any two points can be 'housed off' by two disjoint open sets! All metric spaces are Hausdorff spaces.

Just as all elements of a vector space can be represented as a linear combination of basis vectors, so it is useful to express all elements of a topology as the union of some special set of open sets. This idea leads to the concept of a *base* for a topology. A class $\mathscr{B}$ of open sets of X is a *base* or *basis* for a topology $\mathscr{T}$ iff every open set of $\mathscr{T}$ is the union of members of $\mathscr{B}$. Alternatively, $\mathscr{B}$ is a base iff for each $T_1 \in \mathscr{T}$ and any $x \in T_1$, there is some $E \subset \mathscr{B}$ such that $x \in E \subset T_1$.

Example 9.17

The open intervals $A = (1\,,2)$ and $B = (0\,,2)$ of $\mathbb{R}$ give rise to the cartesian product $A \times B$ which is represented by an open rectangle in $\mathbb{R}^2$ (Fig. 2.6). The set of all such rectangles forms a base $\mathscr{B}$ for the usual topology on $\mathbb{R}^2$ (Example 9.16) in the following way. Let T_1 be any open set in $\mathbb{R}^2$ and $x \in T_1$. Then there exists an open disc $\mathrm{B}(x, \varepsilon) \subset T_1$. Any rectangle $E \in \mathscr{B}$ that is inscribed in $\mathrm{B}(x, \varepsilon)$ satisfies $x \in E \subset \mathrm{B}(x, \varepsilon) \subset T_1$, i.e. $x \in E \subset T_1$ (Fig. 9.20) and so $\mathscr{B}$ is a base for the usual topology on $\mathbb{R}^2$.

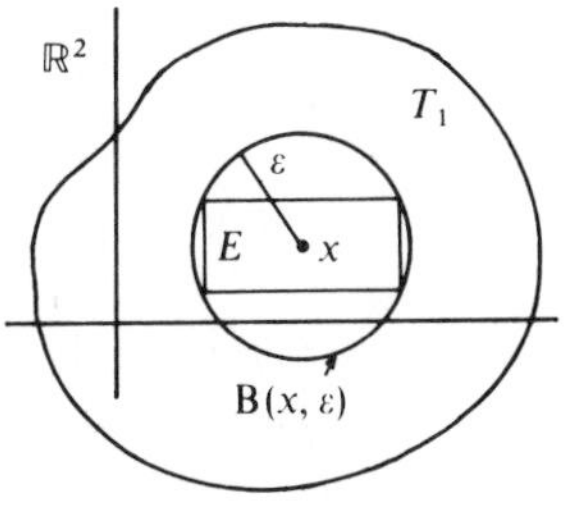

Fig. 9.20 A base for the usual topology on $\mathbb{R}^2$

A subset A of X is called *dense* in X iff $\bar{A} = X$. Here the closure of the topological space $A \subset X$ with topology $\mathscr{T}$ is defined as

$$\bar{A} = \bigcap \{Y : A \subseteq Y \subseteq X \text{ and } X \backslash Y \in \mathscr{T}\}.$$

A is said to be *nowhere dense* iff $\bar{A}'$ is dense in X, i.e. A is nowhere dense iff int $\bar{A} = \varnothing$. Alternatively,

$$\begin{aligned} A \text{ is nowhere dense} &\Leftrightarrow \bar{A} \text{ contains no non-empty open set} \\ &\Leftrightarrow \text{every non-empty open set contains an open ball disjoint from } \bar{A}. \end{aligned}$$

Example 9.18

The set of rational numbers $\mathbb{Q}$ is dense in $\mathbb{R}$. In the usual topology for $\mathbb{R}$, every real number $a \in \mathbb{R}$ is a limit point of $\mathbb{Q}$. Thus, $\bar{\mathbb{Q}} = \mathbb{R}$ and $\mathbb{Q}$ is dense in $\mathbb{R}$.[10]

Example 9.19

The set $G\ell(n, \mathbb{R})$ of all real nonsingular $n \times n$ matrices is dense in $M_n(\mathbb{R})$. Every matrix $A \in M_n(\mathbb{R})$ is a limit point of $G\ell(n, \mathbb{R})$. Thus, $\overline{G\ell(n, \mathbb{R})} = M_n(\mathbb{R})$ and so $G\ell(n, \mathbb{R})$ is dense in $M_n(\mathbb{R})$.

A topological space X is called *first countable* if, for each $x \in X$, there is a countable collection $\{T_i : i \in \mathbb{N}\}$ of open sets containing x such that for any neighbourhood A of x there exists an integer n such that $T_n \subseteq A$ (a local property). A topological space X is called *second countable* if it has a countable base (a global property).

Example 9.20

With the usual topology on $\mathbb{R}$, the set

$$\{(r - 1/n, r + 1/n) : r \in \mathbb{Q}, n \in \mathbb{N}\}$$

is a base for that topology which is countable. Hence, $\mathbb{R}$ is second countable.

A set X is said to be *separable* if it has a countable subset dense in X.

Example 9.21

The real line $\mathbb{R}$ is separable because the set $\mathbb{Q}$ of rational numbers in $\mathbb{R}$ is countable and $\mathbb{Q}$ is also dense in $\mathbb{R}$ (Example 9.18).

A covering $\{U_i : i \in \mathscr{I}\}$ of a topological space S is called *locally finite* if each point $u \in S$ has a neighbourhood U such that U intersects only a finite number of U_i's. Another covering $\{V_j : j \in \mathscr{I}\}$ of S is called a *refinement* of $\{U_i : i \in \mathscr{I}\}$ if, for each V_j, there exists a U_i such that $V_j \subset U_i$.

A Hausdorff space M is said to be *paracompact* if every covering $\{U_i : i \in \mathscr{I}\}$ of M has a locally finite refinement. Manifolds discussed in Chapter 11 are paracompact.

Most of the concepts discussed in Section 9.1 for normed vector spaces and metric spaces can be carried over to topological spaces without any difficulty. The main difference is that, instead of open balls which involve a metric or a norm, the elements of the topology, i.e. the open sets, are used. In topology, the "three c's" are of the greatest importance, namely *compactness*, *connectedness*, and *continuity*. A topological space S is compact iff every cover of S by open sets has a finite subcover. S is connected if it is impossible to express S as the union of two non-empty disjoint open subsets U and V, i.e. $S \neq U \cup V$. Continuity is discussed in Chapter 10.

Analogues of the Heine–Borel and Bolzano–Weierstrass theorems are available for topological spaces. For example, suppose that S is a topological space and $\{U_i : i \in \mathscr{I}\}$ is a family of open sets of S that covers S, i.e. $S \subset \bigcup_{i \in \mathscr{I}} U_i$. Then the Heine–Borel analogue states that there is a finite subfamily $\{U_1, \ldots, U_m\}$ that also covers S,

$$\text{i.e. } S \subset U_1 \cup \cdots \cup U_m.$$

In the Heine–Borel Theorem (Section 9.3), the closed bounded subset U of $\mathbb{R}^n$ is replaced by a topological space S, and the open sets that cover U are the open sets of the topology. Topological spaces that satisfy the analogue of the Heine–Borel Theorem are called compact spaces. The Bolzano–Weierstrass Theorem states that, if S is a first-countable compact space, then every sequence has a convergent subsequence.

Suppose X and Y are two topological spaces, and consider the cartesian product $X \times Y$. To construct a topology on $X \times Y$ from the topologies of X and Y, one might suggest defining a set in $X \times Y$ as open if it is of the form $U \times V$, where U is an open set in X and V an open set in Y. Unfortunately, this is not possible, as the next example shows.

Example 9.22

$X = \mathbb{R}$ and $Y = \mathbb{R}$ with the usual topology on $\mathbb{R}$ (Example 9.15). If U_1 and U_2 are given open intervals in X, and V_1 and V_2 given open intervals in Y, then $U_1 \times V_1$ and $U_2 \times V_2$ are rectangles in $X \times Y$ (Fig. 9.21). However, the

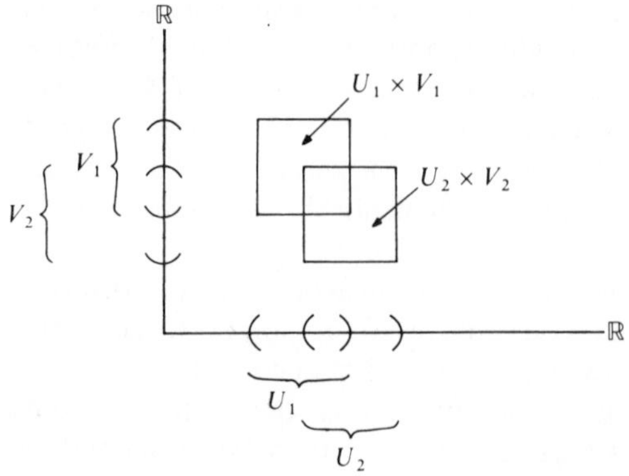

Fig. 9.21 Open sets in $\mathbb{R} \times \mathbb{R}$ of Example 9.22

union $(U_1 \times V_1) \cup (U_2 \times V_2)$ is generally not a product $U \times V$ with U open in X and V open in Y. Thus, Property (ii) of a topology is violated.

The way out of this difficulty is to define a subset of $X \times Y$ to be open if it is the union of sets of the form $U \times V$. The set of all such subsets is called the *product topology* on $X \times Y$.

Example 9.23

$X = \mathbb{R}$ and $Y = \mathbb{R}$. Any open subset of $\mathbb{R} \times \mathbb{R}$ is the union of open rectangles (see Example 9.17). These open rectangles form a basis for the product topology on $\mathbb{R} \times \mathbb{R}$.

Let X be a topological space and $\sim$ an equivalence relation on X. As usual, the equivalence classes decompose X into disjoint subsets. If the set of equivalence classes is denoted by $X/\sim$, then the natural map $\psi : X \to X/\sim$ is

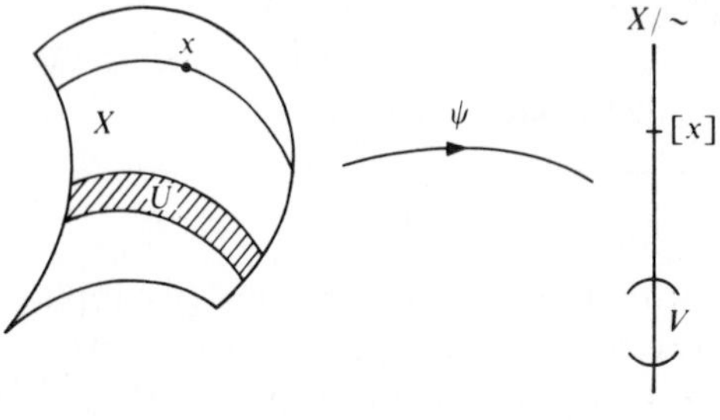

Fig. 9.22 The quotient topology

defined by $x \mapsto [x]$. The set $X/\sim$ inherits a topology from that of X as follows. Given any set $V \subset X/\sim$, let U be the subset of X consisting of the union of all equivalence classes $[x]$ that belong to V (Fig. 9.22). Define V to be open if $\psi^{-1}(V)$ is open in X. This yields a topology on $X/\sim$, called the *quotient topology.* [11]

Exercises

1. Find the open ball $\mathrm{B}(\cdot, \cdot)$ for the following metric spaces (X, ρ):

(i) $\mathrm{B}(0, \frac{1}{4})$, $X = \mathbb{R}$, $\rho(x, y) = |x - y|$;

(ii) $\mathrm{B}(0, \frac{1}{4})$, $X = [0, 1] \subset \mathbb{R}$, $\rho(x, y) = |x - y|$:

(iii) $\mathrm{B}(\mathbf{0}, 1)$, $X = \mathbb{R}^2$, $\boldsymbol{x} = (x_1, x_2)$, $\boldsymbol{y} = (y_1, y_2)$,

$$\rho = |x_1 - y_1| + |x_2 - y_2|.$$

2. Show that the open balls $\mathrm{B}(0, \frac{1}{4})$ and $\mathrm{B}(\frac{1}{16}, \frac{3}{16})$ in $[0, 1] \subset \mathbb{R}$, with metric $\rho(x, y) = |x - y|$, both represent the same set of real numbers.
3. Is the set $S = [0, \frac{1}{2})$ open in $[0, 1]$? Is it open in $\mathbb{R}$?
4. If $\{S_i : i \in \mathscr{I}\}$ is a given (possibly infinite) collection of open subsets of a metric space X, prove that $S = \bigcup_{i \in \mathscr{I}} S_i$ is an open set.
5. Prove that any finite intersection $\bigcap_{i=1}^{n} S_i$ of open subsets S_i of a metric space X is open. Explain why the result may not be true for an infinite intersection.
6. Show that the open intervals form a base for the usual topology on $\mathbb{R}$.
7. Given that c is a convex function on a convex set $X \supseteq S$, where $S = \{x \in X : c(x) \leqslant 0\}$, prove that S is convex.
8. Show that the set $\mathbb{Q}$ of rational numbers form a null set.
9. Write down the usual topology for $\mathbb{R}^2$.
10. Prove that the set $\mathbb{C}$ of complex numbers is separable.

Notes for Chapter 9

1. The empty set $\varnothing$ and the metric space X are both open subsets of X.
2. Regarding $\mathbb{Q}$ as a subset of $\mathbb{R}$, we can see that $\mathbb{Q}$ has no interior points or exterior points since any neighbourhood in $\mathbb{R}$ contains both rational and irrational points. Hence, $\operatorname{int}\mathbb{Q} = \varnothing$, $\operatorname{ext}\mathbb{Q} = \varnothing$, and $\partial\mathbb{Q} = \mathbb{R}$.
3. It is important to distinguish between the limit point of a set and the limit of a sequence. These notions may not coincide if a sequence is confused with its set of points. For example, the sequence $(-\frac{1}{2}, \frac{3}{4}, -\frac{5}{6}, \frac{7}{8}, \dots, (-1)^n(\frac{2n-1}{2n}), \dots)$ has no limit. However, if S is the set of elements of this sequence, (no ordering is implied in a set) then both 1 and -1 are limit points of S.

4. A set S containing only those elements common to the sets $S_1, ..., S_n$ can be written as $S = S_1 \cap \cdots \cap S_n$ or as $\bigcap \{S_i : 1 \leqslant i \leqslant n\}$ or simply as

$$S = \bigcap_{i=1}^{n} S_i.$$

5. The half-open interval $[0, 1)$ in $\mathbb{R}$ is neither open nor closed. In a discrete metric space (see Exercise 7.1(b)), every subset is both open and closed because each individual point is open.
6. Two useful results are (i) $a \in \bar{S}$ iff $\mathrm{B}(a, \varepsilon) \cap S \neq \varnothing$ for any $\varepsilon > 0$, (ii) S is closed iff $X \backslash S$ is open. Proof of (i): if $a \in \bar{S}$ then $a \in S$. Since $a \in \mathrm{B}(a, \varepsilon)$, it follows that $a \in \mathrm{B}(a, \varepsilon) \cap S$ and so $\mathrm{B}(a, \varepsilon) \cap S \neq \varnothing$. On the other hand, if $\mathrm{B}(a, \varepsilon) \cap S \neq \varnothing$, then there is an x such that $x \in \mathrm{B}(a, \varepsilon)$ and $x \in S$ for any ε no matter how small. This implies that a belongs to the set of all limit points of S (see Fig. 9.23) so that $a \in \bar{S}$. Proof of (ii): S is closed iff for any point $x \in X \backslash S$, there exists $\varepsilon > 0$ such that $\mathrm{B}(x, \varepsilon) \cap S = \varnothing$. This holds iff, for any $x \in X \backslash S$, there is an $\varepsilon > 0$ such that $\mathrm{B}(x, \varepsilon) \subset X \backslash S$; and this is true iff $X \backslash S$ is open.

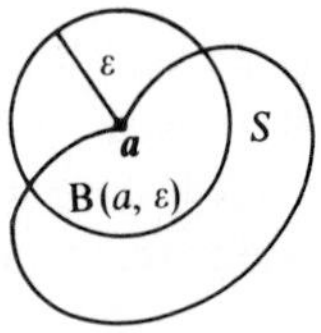

Fig. 9.23

7. Warning: the union of convex sets may not be convex.
8. Functions which are convex upwards (such as those which have graphs shown in Fig. 9.9) can be transformed to functions which are convex downwards simply by multiplying by -1.
9. A mnemonic for (ii) and (iii) is GUFI (pronounced Goofy!) standing for 'general unions, finite intersections'.
10. Exercise 9.8 shows that the rational numbers are also 'sparse' in $\mathbb{R}$!
11. Even if the topology on X comes from a metric, the quotient topology may not come from a metric on $X/\sim$. Indeed, such a metric will not generally exist and hence this is another reason for studying topological spaces rather than metric ones.
12. The open disc plays a role in the topology of the plane $\mathbb{R}^2$ which is analogous to the role of the open interval in the topology of $\mathbb{R}$. For example, the open discs form a base for the usual topology in $\mathbb{R}^2$.

10

CONTINUITY AND DIFFERENTIABILITY

The intuitive understanding of a continuous function $u:\mathbb{R}\to\mathbb{R}$ and the interpretation of the time derivative $\dot{x}$ of a function $x:\mathbb{R}\to\mathbb{R}$ has already been assumed of the reader in earlier chapters. In this chapter, things are put on a more formal footing and generalizations are made to cover mappings $f: X\to Y$

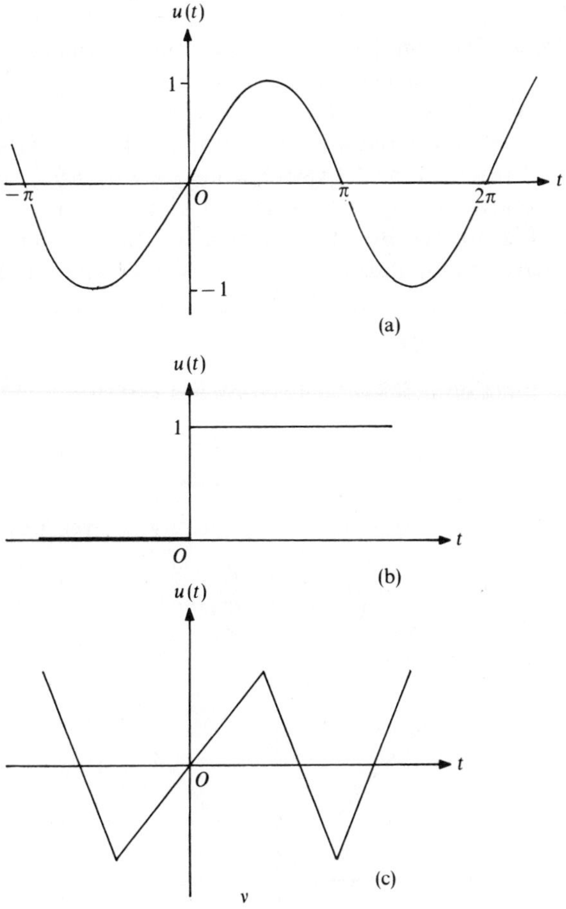

Fig. 10.1 Input functions

from one set X to another set Y. In particular, this leads to the development of a calculus on normed vector spaces. Naturally, the concepts of metric and norm play a leading role in these generalizations. In fact, much of the material in Chapters 7, 8, and 9 are brought together in this present chapter which, together with Chapter 11, lays the foundations for the study of nonlinear dynamical systems.

10.1 Continuity

The concept of a continuous curve or graph of a function will be familiar to the reader as one with no breaks or jumps. Indeed, this intuitive picture has been hinted at on a number of occasions already in this book. A typical continuous input signal u to a dynamical system is given by $u(t)=\sin t$ (which is the sort of smooth graph that intuition suggests for continuity), whereas an example of a discontinuous input is the unit step function (Fig. 10.1(a) and (b)). [1] On the other hand, many continuous functions have graphs which are anything but smooth (e.g. the saw-tooth graph of Fig. 10.1(c)).

If $u(t)=\sin t$, then the time derivative of u is given by $\dot{u}(t)=\cos t$, and so the function $\dot{u}$ is also a continuous function of time having a graph not too dissimilar to Fig. 10.1(a). But derivatives of continuous functions are not always themselves continuous functions, as the following example shows.

Example 10.1

The modulus function $u:\mathbb{R}\to\mathbb{R}$ defined by $u(t)=|t|$ (Fig. 10.2). Here

$$u(t)=\begin{cases} -t & \text{if } t\leqslant 0,\\ t & \text{if } t\geqslant 0.\end{cases}$$

The function u is continuous, but its derivative $\dot{u}$ is given by

$$\dot{u}(t)=\begin{cases} -1 & \text{if } t<0,\\ 1 & \text{if } t>0,\end{cases}$$

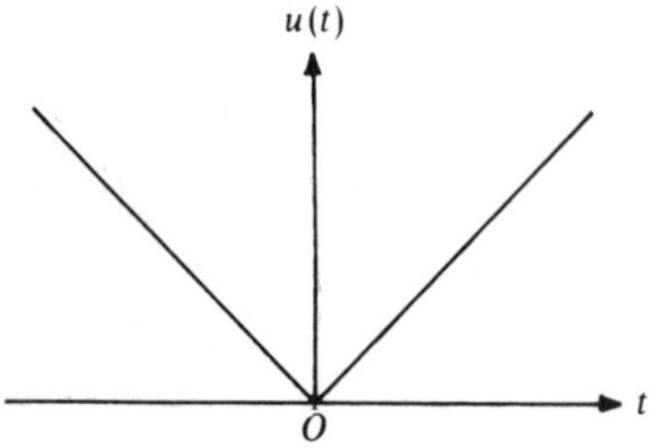

Fig. 10.2 The modulus function

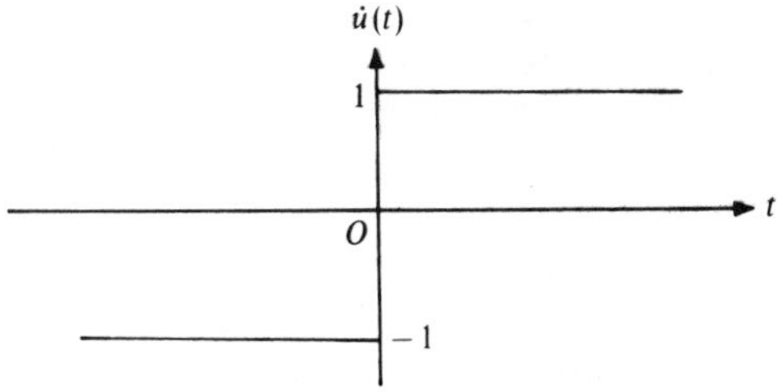

Fig. 10.3 The derivative of the modulus function

and $\dot{u}(t)$ is not defined at $t=0$ where the 'corner' occurs in the graph (see Fig. 10.2). The derivative $\dot{u}$ is represented by the graph shown in Fig. 10.3, and so clearly $\dot{u}$ is a discontinuous function even though u is continuous (see also Example 10.13 later).

Notice that the discontinuity of $\dot{u}$ at $t=0$ in Fig. 10.3 is a *finite* discontinuity, i.e. the instantaneous change in the value of $\dot{u}$ at the point $t=0$ is finite (total change of 2 from -1 to 1). Examples of *infinite* discontinuities are exhibited by the tangent function (Fig. 10.4) at points such as $t=-\frac{1}{2}\pi$ and $t=\frac{1}{2}\pi$.

The question of continuity of derivatives of functions is often just as important as that of the functions themselves. In some areas of dynamical system theory, continuous functions are assumed to have continuous derivatives of all orders, so that the subsequent analysis is simplified (see Chapter 11). On the other hand, some functions which in reality are continuous may be approximated by discontinuous functions for the purposes of the mathematical model. For example, when a rocket engine is switched on the thrust magnitude $T(t)$ may build up from zero to some maximum value T_{max}

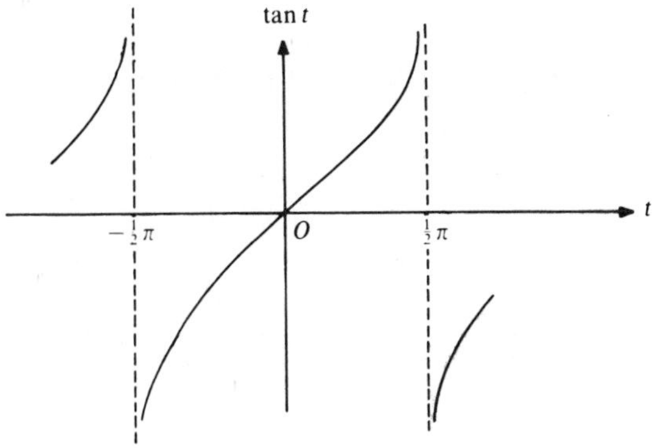

Fig. 10.4 The tangent function

in a very short but finite time. Nevertheless, for the mathematical model of the engine it is usually convenient to assume the maximum value T_{max} is attained instantaneously, immediately the engine is switched on. The real situation and the mathematical approximation are illustrated in Fig. 10.5(a) and (b) respectively. Similarly, when the engine is switched off, the thrust takes a finite time to reduce to zero but in the model it is often assumed that T changes from T_{max} to zero again instantaneously.

When the value of a control u is bounded, say $a \leqslant u(t) \leqslant b$ for all t, it is not unusual for the value of the control to be changed frequently from the lower bound a to the upper bound b and back. Many optimal-control profiles are of this form. Assuming the changes in value take place instantaneously, the graph of such a control u will take a form represented in Fig. 10.6 (see

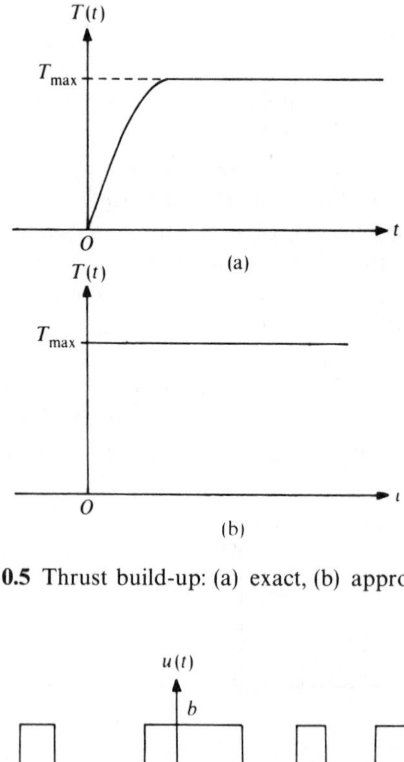

Fig. 10.5 Thrust build-up: (a) exact, (b) approximate

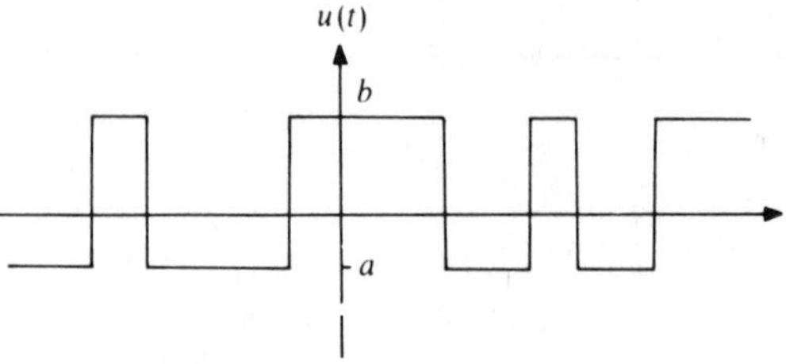

Fig. 10.6 Bang–bang control

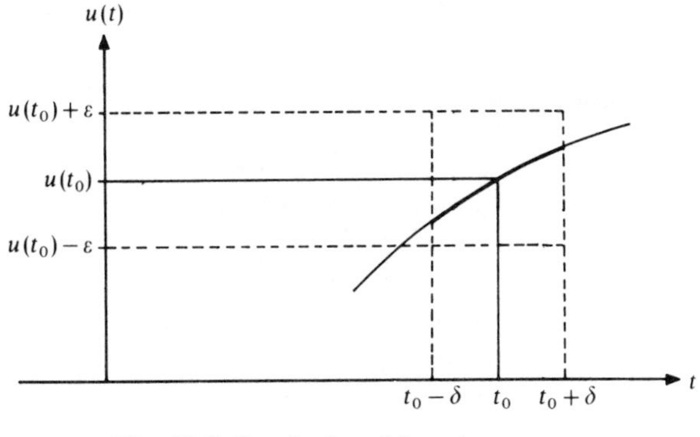

Fig. 10.7 Continuity of function u at t_0

Example 8.8). Such a control is called a *bang–bang* control in which the control is switched, or banged, from one bound to the other. Now this control strategy may not be very good for the equipment concerned, and hence not acceptable from a practical point of view. When a mathematical solution and practical considerations are in conflict in this way the engineer or scientist has to decide on some acceptable compromise.

Having looked at typical continuous and discontinuous functions which arise in dynamical system theory, the concept of continuity will now be presented in a more formal mathematical way. Suppose a function $u:I \to \mathbb{R}$, defined by $t \mapsto u(t)$, has a domain $I \subseteq \mathbb{R}$ which may be bounded or unbounded and is not necessarily open or closed. Then the function u is continuous at a point $t_0 \in I$ if, given any positive real number ε, there exists another positive real number δ (actually $\delta(\varepsilon)$ because δ will depend upon ε) such that changing the value of t_0 by less than δ causes $u(t_0)$ to change by less than ε (Fig. 10.7). Symbolically, given any $\varepsilon > 0$, there exists $\delta > 0$ such that

$$|t - t_0| < \delta \quad \Rightarrow \quad |u(t) - u(t_0)| < \varepsilon.$$

Alternatively, the condition of continuity requires that, for the point t_0, the limit equation

$$\lim_{t \to t_0} u(t) = u(t_0) \tag{10.1}$$

must be true. That is to say the value of the function u at the point t_0 has to be the same as the limit of the function values $u(t_n)$ for any arbitrary sequence of numbers t_n $(n \in \mathbb{N})$ converging to t_0. Notice that $t_0 \in I$, the domain of u, and so $u(t_0)$ does exist. Continuity of a function u can only be defined in the domain of u. [2]

Sometimes a function u which is discontinuous at a point t_0 can be changed to give a function continuous at t_0 by arranging for (10.1) to be satisfied. This process is illustrated in the next example.

Example 10.2

The signum function sgn: $\mathbb{R} \to \{-1, 0, 1\}$ is defined as

$$\operatorname{sgn} t \triangleq \begin{cases} 1 & \text{if } t>0, \\ 0 & \text{if } t=0, \\ -1 & \text{if } t<0. \end{cases}$$

It follows that the function $u:\mathbb{R} \to \{0, 1\}$, defined by

$$u(t) = \operatorname{sgn} |t| = \begin{cases} 1 & \text{if } t \neq 0, \\ 0 & \text{if } t=0, \end{cases}$$

is discontinuous at $t=0$ because $\lim_{t\to 0} u(t) = 1 \neq u(0)$. By redefining $u(0)=1$, then $u(t)=1$ for all $t \in \mathbb{R}$, and u is continuous at $t=0$.

It should be realized that not all discontinuities can be removed in the way of Example 10.2. The function $u:\mathbb{R} \to \mathbb{R}$, defined by $u(t)=1/t$ $(t \neq 0)$ and $u(0)=1$, has a discontinuity at $t=0$ [2] and no amount of redefining at $t=0$ will remove it.

A function $u:I \to \mathbb{R}$ is said to be continuous on the interval I if it is continuous at each point of I. This requires that, given $\varepsilon>0$, for each point $t_0 \in I$ there is some $\delta>0$, *depending on* ε *and* t_0, such that

$$|u(t)-u(t_0)|<\varepsilon \quad \text{whenever} \quad |t-t_0|<\delta \quad \text{and} \quad t \in I.$$

In this formal definition of a continuous function, if δ depends only on ε—so that, for any given ε, the same value of δ can be chosen for any $t_0 \in I$—then the continuity is said to be *uniform*. The function $u:I \to \mathbb{R}$ is then *uniformly continuous* on I if, given any $\varepsilon>0$, there exists δ such that

$$|u(t_1)-u(t_2)|<\varepsilon \quad \text{whenever} \quad |t_1-t_2|<\delta \text{ with } t_1,t_2 \in I.$$

The reader will notice that the idea of uniform continuity is much the same as that of uniform convergence of sequences of functions discussed in Section 8.4. In fact, an important result is that, if a sequence $(x^{(k)})$ of continuous functions converges uniformly on an interval I to a limit x, then this limit is also a continuous function on I. This result was required in the solution to Exercise 8.2. A proof of the result is given elsewhere. [3]

Obviously, a uniformly continuous function is necessarily continuous. Conversely, it can be shown that every function that is continuous on a closed interval $[a, b]$ is also uniformly continuous (Parzynski and Zipse 1987). The fact that the interval must be closed is crucial, as the next example shows.

Example 10.3

The function u: $(0\,,1) \to \mathbb{R}^+$ defined by $u(t)=1/t$ is continuous on the open interval $(0\,,1)$ but it is not uniformly continuous on $(0\,,1)$. To see this, suppose that $k>1$ is such that $1/k<\delta$, and let $t_1=1/k$ and $t_2=1/(k+1)$, so that $t_1,t_2 \in (0\,,1)$. Then

$$|t_1-t_2|=\left|\frac{1}{k}-\frac{1}{k+1}\right|=\frac{1}{k(k+1)}<\frac{1}{k}<\delta.$$

However,

$$|u(t_1)-u(t_2)|=|k-(k+1)|=1>\varepsilon.$$

So $|u(t_1)-u(t_2)|$ exceeds ε whenever $|t_1-t_2|<\delta$, thus violating the definition of uniform continuity. Hence u is not uniformly continuous on $(0\,,1)$.

The existence and uniqueness of a solution to the scalar differential equation

$$\dot{x}=f(x) \tag{10.2}$$

is ensured by the continuity of the function f and its derivative $\mathrm{d}f/\mathrm{d}x$ (Courant 1936). However, the continuity of the derivative can be replaced by the weaker Lipschitz condition. A function f: $[a\,,b] \to \mathbb{R}$ is said to satisfy a *Lipschitz condition* if there exists some positive real number K such that for all $t_1,t_2 \in [a\,,b]$,

$$|f(t_1)-f(t_2)| \leqslant K|t_1-t_2|.$$

A function that satisfies such a condition is sometimes called *K-Lipschitzian*. It follows immediately that a Lipschitzian function is also continuous. The Lipschitz condition can be generalized to cover the case when f in (10.2) is a vector-valued function (Lefschetz 1963). Continuity of function f in (10.2) is sufficient to ensure existence of a solution to the differential equation, but not uniqueness, as illustrated in the next example.

Example 10.4 (Coddington and Levinson 1955, Rosenbrock and Storey 1970)

$$\dot{x}=\tfrac{3}{2}x^{\frac{1}{3}} \quad (x \in \mathbb{R}), \qquad x(0)=0 \tag{10.3}$$

(note that $x \mapsto \frac{3}{2}x^{\frac{1}{3}}$ is continuous).

A function φ: $\mathbb{R} \to \mathbb{R}$ defined by

$$\varphi(t)=\begin{cases} 0 & \text{if } t \leqslant c, \\ (t-c)^{\frac{3}{2}} & \text{if } t>c, \end{cases} \tag{10.4}$$

for any real number $c \in (0, \infty)$, is a solution of (10.3). Certainly

$$\dot{\varphi}(t) = \begin{cases} 0 = \frac{3}{2}\varphi^{\frac{1}{3}}(t) & \text{if } t < c, \\ \frac{3}{2}(t-c)^{\frac{1}{2}} = \frac{3}{2}\varphi^{\frac{1}{3}}(t) & \text{if } t > c, \end{cases}$$

and furthermore,

$$\dot{\varphi}(c-) = \lim_{\varepsilon \to 0+} \frac{\varphi(c-\varepsilon) - \varphi(c)}{\varepsilon} = 0,$$

$$\dot{\varphi}(c+) = \lim_{\varepsilon \to 0+} \frac{\varphi(c+\varepsilon) - \varphi(c)}{\varepsilon} = \lim_{\varepsilon \to 0+} \frac{1}{\varepsilon}\varepsilon^{\frac{3}{2}} = \lim_{\varepsilon \to 0+} \varepsilon^{\frac{1}{2}} = 0.$$

Thus, $\dot{\varphi}(c)$ exists and is equal to zero. However, there are an infinite number of solutions to (10.3) because (10.4) is a solution for every positive real number c. So, even though $x \mapsto \frac{3}{2}x^{\frac{1}{3}}$ is continuous, there is no unique solution to (10.3).

The behaviour of control functions in dynamical systems can be quite complicated (e.g. the Fuller problem, Example 9.14). Such behaviour is exhibited in the next two examples.

Example 10.5

The function $u: \mathbb{R} \backslash \{0\} \to \mathbb{R}$ for which $u(t) = \sin 1/t$ takes values between -1 and 1 for t in the range $(2n - \frac{1}{2})\pi \leqslant 1/t \leqslant (2n + \frac{1}{2})\pi$ for any $n \in \mathbb{N}$. We have

$$u\left(\frac{2}{(4n-1)\pi}\right) = -1 \quad \text{and} \quad u\left(\frac{2}{(4n+1)\pi}\right) = 1,$$

so that, as n increases (and t decreases towards zero), the function values oscillate more and more rapidly between the values 1 and -1. Clearly an infinite number of oscillations occur in any small neighbourhood of the point $t = 0$. The function u cannot be extended to a continuous function, no matter what value is given to $u(0)$, [4] because u has no left or right limit at $t = 0$. Even if left and right limits did exist, u could not be extended to a continuous function unless both limits were equal.

Example 10.6

The function $u: \mathbb{R} \to \mathbb{R}$ satisfying $u(t) = t \sin 1/t$ for $t \neq 0$ is continuous at the point $t = 0$ provided that the value $u(0)$ is defined as zero, i.e. $u(0) = 0$. Again this function takes values that oscillate an infinite number of times in any small neighbourhood of $t = 0$ but, unlike Example 10.5, the amplitude of oscillations decrease to zero as the point $t = 0$ is approached.

Important continuity results are that, for any two functions $f, g: X \to \mathbb{R}$, with $X \subseteq \mathbb{R}$, both continuous at a point t_0, the sum $f+g$ and the product fg are continuous at t_0. The composite $g \circ f$ of two functions f and g is also continuous at t_0 provided that f is continuous at t_0 and g is continuous at $f(t_0)$. The quotient f/g of two such functions is also continuous at t_0 provided that $g(t_0) \neq 0$.

Example 10.7

The functions $f, g: \mathbb{R} \to \mathbb{R}$ defined by $f(s) = s+2$ and $g(s) = s^3 - 12s^2 + 48s - 64$ are both continuous everywhere on $\mathbb{R}$. The quotient function $q: \mathbb{R}\backslash\{4\} \to \mathbb{R}$ defined by $q(s) = f(s)/g(s) = (s+2)/(s-4)^3$ is continuous everywhere in the domain $\mathbb{R}\backslash\{4\}$. There is a discontinuity at $s=4$ because $q(s)$ becomes indefinitely large as $s \to 4$.

A function $f:[a, b] \to \mathbb{R}$ is said to be *absolutely continuous* on $[a, b]$ if, given $\varepsilon > 0$, there exists a $\delta > 0$ such that

$$\left| \sum_{i=1}^{r} (f(b_i) - f(a_i)) \right| < \varepsilon \quad \text{whenever} \quad \sum_{i=1}^{r} (b_i - a_i) < \delta,$$

where the open intervals (a_i, b_i) $(i = 1, \ldots, r)$ contained in $[a, b]$ are disjoint.

A function $f:[a, b] \to \mathbb{R}$ is said to be of *bounded variation* on $[a, b]$ if there is a bound $c \geqslant 0$ such that

$$\sum_{i=1}^{r} |f(t_i) - f(t_{i-1})| \leqslant c$$

for all partitions $a = t_0 < t_1 < t_2 < \cdots < t_r = b$ of the interval $[a, b]$. In this case, the least such bound c is called the *total variation* of f on $[a, b]$ and denoted by $\operatorname{var} f$, i.e.

$$\operatorname{var} f \triangleq \sup_{t \in T} \sum_{i=1}^{r} |f(t_i) - f(t_{i-1})|,$$

where $T = \{t = (t_0, \ldots, t_r): a = t_0 < t_1 < \cdots < t_r = b\}$.

It is obvious that the absolute continuity of f on $[a, b]$ implies the continuity of f on $[a, b]$. Although not so obviously, it also implies that f is of bounded variation on $[a, b]$. The set of functions of bounded variation on $[a, b]$ is a vector space under the usual addition and scalar multiplication. A normed vector space $\mathrm{BV}[a, b]$ is generated by the norm

$$\|f\| = |f(a)| + \operatorname{var} f.$$

In the proof of the existence of optimal controllers for nonlinear systems involving impulse controls, a control u is assumed to be of bounded variation on an open interval of $\mathbb{R}$ (Lee and Markus 1967). [5]

The concept of continuity has been described through a function on a subset of $\mathbb{R}$, but it can easily be generalized to a function on a subset of $\mathbb{R}^n$ by using a norm on $\mathbb{R}^n$, say the Euclidean norm

$$\|\boldsymbol{x}-\boldsymbol{\xi}\| = \sqrt{\sum_{i=1}^{n} (x_i-\xi_i)^2}.$$

Then, if A is a subset of $\mathbb{R}^n$, a function $f: A \to \mathbb{R}$ given by $(x_1, \ldots, x_n) \mapsto f(x_1, \ldots, x_n)$ is continuous at a point $\boldsymbol{\xi}=(\xi_1, \ldots, \xi_n) \in A$ if, given any positive real number ε, there exists another positive real number $\delta(\varepsilon)$ such that

$$\|\boldsymbol{x}-\boldsymbol{\xi}\| < \delta \quad \Rightarrow \quad |f(\boldsymbol{x})-f(\boldsymbol{\xi})| < \varepsilon.$$

Example 10.8

$f: \mathbb{R}^2 \to \mathbb{R}$ given by

$$f(x_1, x_2) = \begin{cases} (x_1+x_2)^3/(x_1^2+x_2^2) & \text{if } \boldsymbol{x} \neq \boldsymbol{0}, \\ 0 & \text{if } \boldsymbol{x} = \boldsymbol{0}. \end{cases}$$

The function f is continuous at $\boldsymbol{x}=\boldsymbol{0}$. Notice that the expression for $f(x_1, x_2)$ when $\boldsymbol{x} \neq \boldsymbol{0}$ is not defined at the origin of $\mathbb{R}^2$, where $x_1=0$ and $x_2=0$, and this is why it is necessary to define $f(x_1, x_2)$ separately at $\boldsymbol{x}=\boldsymbol{0}$.

Example 10.9

$f: \mathbb{R}^2 \to \mathbb{R}$ defined by

$$f(x_1, x_2) = \begin{cases} x_1 x_2/(x_1^2+x_2^2) & \text{if } \boldsymbol{x} \neq \boldsymbol{0}, \\ 0 & \text{if } \boldsymbol{x} = \boldsymbol{0}. \end{cases}$$

Here, $f(x_1, 0)=0$ for all x_1, including $x_1=0$ since $f(0, 0)=0$ by definition of f. The function restricted to the line $x_2=0$ is therefore continuous at the origin. Similarly, $f(x_1, x_2)$ is continuous in x_2 at the origin if x_1 is kept zero. However, f is discontinuous at the origin, since $f(x_1, x_2)=\frac{1}{2}$ at any point on the line $x_1=x_2$ except at the origin where $f(0, 0)=0$.

Continuity of a mapping from one metric space to another may be expressed in the following way. Given any mapping $f: X \to Y$ from one metric space (X, ρ_X) to another (Y, ρ_Y), then f is continuous at a point $x_0 \in X$ if, given any $\varepsilon > 0$, there exists $\delta > 0$ such that

$$\rho_X(x, x_0) < \delta \quad \Rightarrow \quad \rho_Y\big(f(x), f(x_0)\big) < \varepsilon. \tag{10.5}$$

A mapping f is continuous if it is continuous at every point of its domain X.

The definition of a continuous mapping (10.5) between two metric spaces X and Y is given above in terms of the chosen metrics ρ_X and ρ_Y. This definition therefore depends upon the choice of metrics and usually there are many different metrics to choose from. Again, some mappings will be between two spaces X and Y that are not metric spaces, and then the above definition is not applicable. Fortunately, continuity of general mappings can be studied using the open sets of a topology. As described in Section 9.5 these open sets enable topics to be discussed independently of the choice of metric or norm, even when these do exist. In order to move smoothly from metric spaces to topological spaces, continuity of a mapping will first be discussed in a metric space using the concept of open balls.

Given metric spaces (X, ρ_X) and (Y, ρ_Y), let $f: X \to Y$ be a map between them. Definition (10.5) is equivalent to (Fig. 10.8)

$$f(\mathrm{B}_\delta(x_0)) \subset \mathrm{B}_\varepsilon(f(x_0)). \tag{10.6}$$

In other words, choosing any open ball $\mathrm{B}_\varepsilon(f(x_0))$, no matter how small, an open ball $\mathrm{B}_\delta(x_0)$ can always be found such that all points of $\mathrm{B}_\delta(x_0)$ are mapped into $\mathrm{B}_\varepsilon(f(x_0))$. It follows from (10.6) that

$$\mathrm{B}_\delta(x_0) \subset f^{-1}(\mathrm{B}_\varepsilon(f(x_0)),$$

i.e. the open ball $\mathrm{B}_\delta(x_0)$ around x_0 lies in the subset $f^{-1}(\mathrm{B}_\varepsilon(f(x_0)))$ of X (Fig. 10.8). An extension of these results leads to an even simpler statement of continuity on the domain X: a map $f: X \to Y$ is continuous iff

$$V \subseteq Y \text{ is open} \quad \Rightarrow \quad f^{-1}(V) \subseteq X \text{ is open.}^{6} \tag{10.7}$$

All of this points to the importance of the open sets of X and Y rather than the chosen metrics ρ_X and ρ_Y. Indeed, by choosing two different pairs of metrics (ρ_{X_1}, ρ_{Y_1}) and (ρ_{X_2}, ρ_{Y_2}) that both yield the same family of open sets,

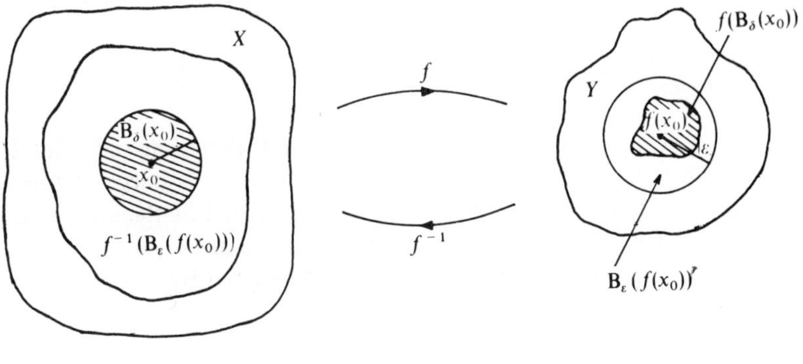

Fig. 10.8 Continuity of mapping f at x_0

any map that is continuous using (ρ_{X_1}, ρ_{Y_1}) will automatically be continuous using (ρ_{X_2}, ρ_{Y_2}).

It is now possible to define continuity for a mapping between topological spaces by using the concept expressed in (10.7). A map $f: X \to Y$ between two topological spaces X and Y is said to be continuous at a point $x_0 \in X$ if, given any open neighbourhood V of the point $f(x_0) \in Y$, there exists an open neighbourhood U of x_0 such that $f(U) \subseteq V$. The map f is said to be continuous if it is continuous at every point of X. The attraction of this definition of continuity is that it avoids any reference to metrics on X and Y. Many important topological spaces are such that no metrics can be chosen for them but, even when there are metrics available for a topological space (e.g. a function space), they are often rather artificial compared with the natural topological structure.

Mathematical analysis involving concepts such as continuity is often more straightforward using the topological structure of spaces. This is illustrated in the next example.

Example 10.10

Suppose $f: X \to Y$ and $g: Y \to Z$ are two bijective continuous maps between topological spaces such that the composite map $g \circ f: X \to Z$ is defined. Since g is continuous, then any open set $V \subseteq Z$ gives rise to an open set $g^{-1}(V) \subseteq Y$. Furthermore, because f is continuous and $g^{-1}(V)$ is an open subset of Y, it follows that the set $f^{-1}(g^{-1}(V))$ is open in X. Thus, for any open set V in Z, the set $(g \circ f)^{-1}(V)$ [7] is open in X, and so $g \circ f$ is a continuous map.

Two topological spaces X and Y are called *homeomorphic* or *topologically equivalent* if there exists a bijective mapping $f: X \to Y$ such that f and f^{-1} are continuous. The mapping f is then called a *homeomorphism*. Homeomorphic spaces all have the same topological properties and, from this point of view, are indistinguishable. This is exactly like, say, isomorphic groups discussed in Chapter 4.

Example 10.11

Let $X = \{(x_1, x_2): x_1^2 + x_2^2 = 1, x_2 \neq 1\}$, the unit circle excluding the point (0, 1). The function $f: X \to \mathbb{R}$ defined by $f(x_1, x_2) = x_1/(1 - x_2)$ is bijective and continuous (illustrated in Fig. 10.9, where f is represented as the stereographic projection of X from the point (0, 1) onto the x_1 axis which represents $\mathbb{R}$). Furthermore, the inverse function $f^{-1}: \mathbb{R} \to X$ defined by

$$f^{-1}(t) = \left(\frac{2t}{t^2+1}, \frac{t^2-1}{t^2+1}\right)$$

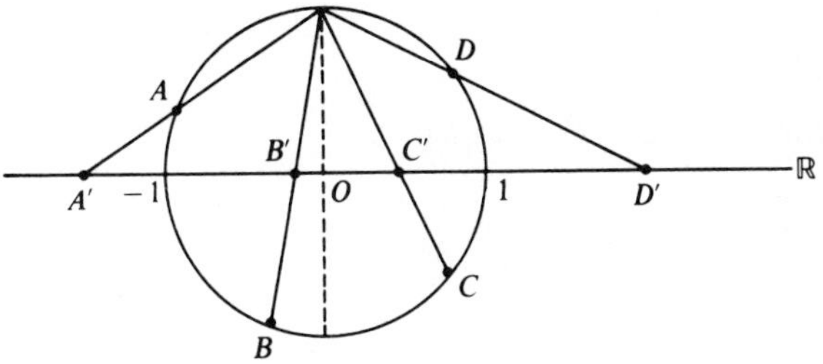

Fig. 10.9 Stereographic projection from (0 , 1) onto $\mathbb{R}$

is also continuous. Hence, f is a homeomorphism and so X and $\mathbb{R}$ are homeomorphic.

10.2 Differentiation

The formal definition of the derivative $\dot{x}$ of a function $x\colon \mathbb{R}\to\mathbb{R}$ at a point $t_0\in\mathbb{R}$ is usually given in the form

$$\dot{x}(t_0) \triangleq \lim_{h\to 0} \frac{x(t_0+h)-x(t_0)}{h}, \tag{10.8}$$

provided that the limit exists ($\pm\infty$ not admissible). Remember that the limit has to exist no matter how h approaches zero. Since $h\in\mathbb{R}$ there are two obvious approaches, either through negative values (the left-hand limit as $h\to 0^-$) or through positive values (the right-hand limit as $h\to 0^+$). The limit in both cases must have the same value. If the limit in (10.8) exists and is the same no matter how h approaches zero then the function x is said to be *differentiable* at t_0.

Similar definitions apply to functions $f\colon \mathbb{R}^n\to\mathbb{R}$, for which the partial derivative $\partial f/\partial x_i$ at a point $\boldsymbol{x}=(x_1 ,..., x_n)\in\mathbb{R}^n$ is defined as

$$\frac{\partial f}{\partial x_i}(\boldsymbol{x}) \triangleq \lim_{h\to 0} \frac{f(x_1 ,..., x_{i-1}, x_i+h, x_{i+1} ,..., x_n)-f(x_1 ,..., x_n)}{h} \qquad (i=1 ,..., n). \tag{10.9}$$

Example 10.12

$x\colon \mathbb{R}\to\mathbb{R}$ defined by $x(t)=t^k$, with $k\in\mathbb{N}$.

$$\lim_{h\to 0}\frac{x(t_0+h)-x(t_0)}{h}=\lim_{h\to 0}\frac{1}{h}[(t_0+h)^k-t_0^k]$$

$$=\lim_{h\to 0}\frac{1}{h}[t_0^k+kt_0^{k-1}h+O(h^2)-t_0^k]^{8}$$

$$=\lim_{h\to 0}[kt_0^{k-1}+O(h)]=kt_0^{k-1}=\dot{x}(t_0).$$

A function is said to be differentiable on an open interval $I\subset\mathbb{R}$ if it is differentiable at every point of I. If a function is differentiable on I, then it is also continuous on I. [9] However, the converse is not necessarily true, i.e. not every continuous function has a derivative at every point as the next examples demonstrate.

Example 10.13

The modulus function $u\colon\mathbb{R}\to\mathbb{R}$ defined by $t\mapsto|t|$ is continuous at $t=0$ (see Example 10.1). However,

$$\lim_{h\to 0^+}\frac{u(h)-u(0)}{h}=\lim_{h\to 0^+}\frac{h-0}{h}=1 \qquad \text{for } h>0$$

whereas

$$\lim_{h\to 0^-}\frac{u(h)-u(0)}{h}=\lim_{h\to 0^-}\frac{-h-0}{h}=-1 \qquad \text{for } h<0.$$

The limit is therefore different depending on whether h approaches zero through positive or negative values. Thus, the modulus function is not differentiable at $t=0$ even though it is continuous there (Example 10.1).

Example 10.14

$x\colon\mathbb{R}\to\mathbb{R}$ defined by $x(t)=t^{\frac{1}{3}}$ (Fig. 10.10). This function x is continuous at $t=0$, but

$$\lim_{h\to 0}\frac{x(h)-x(0)}{h}=\lim_{h\to 0}\frac{h^{\frac{1}{3}}-0}{h}=\lim_{h\to 0}1/h^{\frac{2}{3}}.$$

This limit does not exist since $h^{-\frac{2}{3}}\to\infty$ as $h\to 0$, and so x is not differentiable at $t=0$. Neither left-hand nor right-hand limit exists in this case. Geometri-

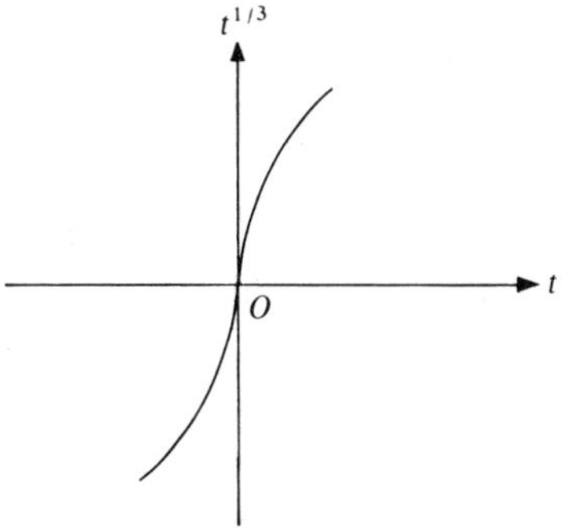

Fig. 10.10 Graph of function x given by $x(t)=t^{\frac{1}{3}}$

cally, the tangent to the graph representing x is the vertical at $t=0$ (Fig. 10.10), i.e. the gradient or slope is infinite.

If a function is continuous and has a derivative at every point, it is not necessarily true that the derivative of that function is continuous as the next example shows.

Example 10.15

$x:\mathbb{R}\to\mathbb{R}$ defined by

$$x(t)=\begin{cases} t^2 \sin 1/t & \text{if } t\neq 0, \\ 0 & \text{if } t=0. \end{cases}$$

This function x is continuous on $\mathbb{R}$ (Exercise 3). Furthermore, its derivative at any $t\neq 0$ is given by

$$\dot{x}(t)=-\cos 1/t+2t \sin 1/t$$

and, for the derivative of x at $t=0$, we have

$$\lim_{h\to 0}\frac{x(h)-x(0)}{h}=\lim_{h\to 0} h\sin 1/h=0.$$

Thus, x is differentiable everywhere. However, $\dot{x}$ is not continuous at $t=0$ since it does not have a limit as $t\to 0$. In fact, $\dot{x}(t)=-1$ when $t=1/2n\pi$, and $\dot{x}(t)=1$ when $t=1/(2n+1)\pi$ $(n\in\mathbb{Z})$.

The reader will be familiar with the differentiation of a composite function. In elementary mathematics, it is sometimes called the 'function of a function' rule. In general, it is known as the *chain rule* and is stated as follows. A composite function $F = g \circ f : \mathbb{R} \to \mathbb{R}$, defined by $x \mapsto g(f(x))$, of two differentiable functions f and g is differentiable, and its derivative is given by the formula

$$F'(x) = g'(f(x))f'(x).$$

Of course, care must be taken to restrict the domain of F to that for which it is defined (see comments following Example 3.15).

Example 10.16

$f:\mathbb{R}\backslash\{1\}\to\mathbb{R}$ defined by $f(x)=(1+x)/(1-x)$ and $g:(-\frac{1}{2}\pi, \frac{1}{2}\pi)\to\mathbb{R}$ defined by $g(x)=\tan x$. Then [10] the composite function

$$F:\left(\frac{\pi-2}{\pi+2}, \frac{\pi+2}{\pi-2}\right)\to\mathbb{R}$$

given by $F(x)=g(f(x))=\tan\dfrac{1+x}{1-x}$ has a derivative F' given by

$$F'(x) = -\frac{2x}{(1-x)^2}\sec^2\left(\frac{1+x}{1-x}\right).$$

On trying to generalize the concept of derivative to a general mapping, the definition (10.8) is not helpful. The definition is now rewritten in a form that can be generalized. Define a function $\lambda:\mathbb{R}\to\mathbb{R}$ as $\lambda(h)=\dot{x}(t_0)h$. (10.8) can then be reformulated as

$$\lim_{h\to 0}\frac{x(t_0+h)-x(t_0)-\dot{x}(t_0)h}{h}=0,$$

i.e.

$$\lim_{h\to 0}\frac{x(t_0+h)-x(t_0)-\lambda(h)}{h}=0. \tag{10.10}$$

Thus, a function $x:\mathbb{R}\to\mathbb{R}$ is differentiable at $t_0\in\mathbb{R}$ if there exists a homogeneous linear function $\lambda:\mathbb{R}\to\mathbb{R}$ defined by $\lambda(h)=Kh$ such that (10.10) is true. [11] Equation (10.10) can also be written as

$$x(t_0+h)-x(t_0)=\lambda(h)+\varepsilon|h| \tag{10.11}$$

where $\varepsilon\to 0$ as $h\to 0$. [12] The linear function λ is called the *derivative* of x at t_0, and the important point is that, in this new definition, the derivative is a linear function. In the present discussion, this linear function has a value which is a real number $\dot{x}(t_0)$. However, in the generalizations to follow, this will not be the case.

Example 10.17

Define the function $x:\mathbb{R}\to\mathbb{R}$ by $x(t)=t^3$. Let $t_0=2$. Then

$$x(2+h)-x(2)=(2+h)^3-2^3=12h+6h^2+h^3,$$

so that

$$\dot{x}(2)h+\varepsilon|h|=12h+6h^2+h^3.$$

Choose h positive. Then

$$\dot{x}(2)+\varepsilon=12+6h+h^2,$$

so that, in this case,

$$\dot{x}(2)=12 \quad \text{and} \quad \varepsilon=h(6+h)\to 0 \text{ as } h\to 0.$$

If h is chosen negative, then again $\dot{x}(2)=12$ and $\varepsilon=-h(6+h)\to 0$ as $h\to 0$.

The geometric significance of the derivative $\dot{x}(t_0)$ is shown in Fig. 10.11. The tangent to the graph of function x at the point t_0 touches the graph at the point O', a point with coordinates $(t_0, x(t_0))$ relative to axes TOX. Now change the origin from O to O' with new axes $T'O'X'$ as shown in Fig. 10.11. Then coordinates in the two sets of axes are related by the equations

$$T=t_0+T', \qquad X=x(t_0)+X'.$$

The equation $X=x(T)$ then becomes

$$x(t_0)+X'=x(t_0+T').$$

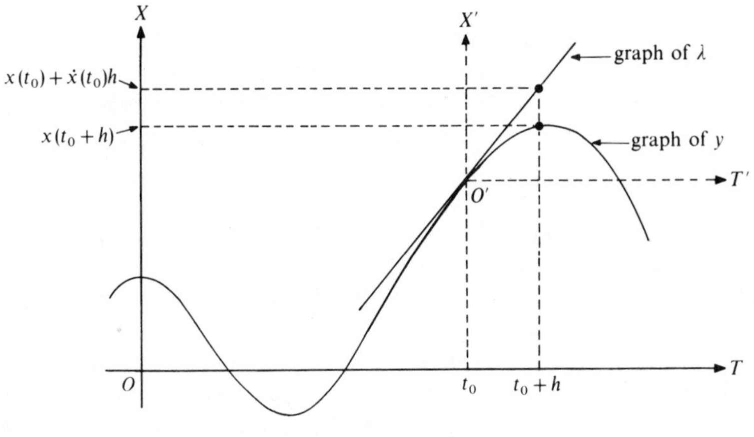

Fig. 10.11 Geometric significance of $\dot{x}(t_0)$

Now write $T'=h$; then the last equation is equivalent to

$$X'=x(t_0+h)-x(t_0).$$

Thus, the graph corresponding to $X=x(T)$ in the old axes now corresponds to $X'=x(t_0+h)-x(t_0)$ in the new axes $O'T'$ and $O'X'$.

With the functions y and λ defined by

$$y(h)=x(t_0+h)-x(t_0), \qquad \lambda(h)=\dot{x}(t_0)h,$$

the graphs of these two functions are tangential at $h=0$ (i.e. at $T'=0$, Fig. 10.11). Thus, λ may be described as a linear approximation to x at t_0. Put another way, the function defined by $h\mapsto x(t_0)+h\dot{x}(t_0)$ (the graph of λ being a straight line) is a first-order approximation to the function defined by $h\mapsto x(t_0+h)$ (see points marked • in Fig. 10.11).

From the above discussion, it will be clear that addition and subtraction of real numbers, linear functions, and limits all play an important role in the development of the derivative of a function. The generalization to the differentiation of a general mapping f from a domain A to a codomain B will involve vector spaces (for the addition and subtraction of elements in A and B and for linear mappings) and metric or normed spaces (to deal with limits). Before this generalization is discussed, it is convenient at this stage to consider some further topics concerning a function $x:\mathbb{R}\to\mathbb{R}$. These topics will also be generalized later.

The value $x(t_0+h)$ can be expressed exactly as a linear expression in h by the *mean-value theorem*. This states that, if a function $x:[t_0, t_0+h]\to\mathbb{R}$ is continuous in the closed interval $[t_0, t_0+h]$ and differentiable in the open interval (t_0, t_0+h), then there exists $\xi\in(t_0, t_0+h)$ such that

$$x(t_0+h)=x(t_0)+h\dot{x}(\xi). \tag{10.12}$$

Geometrically, this result means that the gradient of the graph of x at $t=\xi$ is equal to the slope of the chord joining the two points on the graph at $t=t_0$ and $t=t_0+h$ (Fig. 10.12). This is because (10.12) can be written as

$$\dot{x}(\xi)=\frac{x(t_0+h)-x(t_0)}{h},$$

and the left-hand side of this equation is the gradient of the graph at $t=\xi$, while the right-hand side is the gradient of the chord.

The class of all functions $x:U\to\mathbb{R}$ such that $U\subseteq\mathbb{R}$ and x is continuous on U is denoted by C^0. Similarly, x is said to be of class C^1 on U if its derivative λ, described earlier, exists and is continuous on U. Since the derivative of x is itself a function $\dot{x}$, the derivative of $\dot{x}$ (if it exists at some point t_0) is called the second derivative of x and denoted by $\ddot{x}$, and similarly for higher order derivatives $\dddot{x}$, $x^{(iv)}$,... . If $\ddot{x}$ is a continuous function, then $\dot{x}\in C^1$ and x is said to

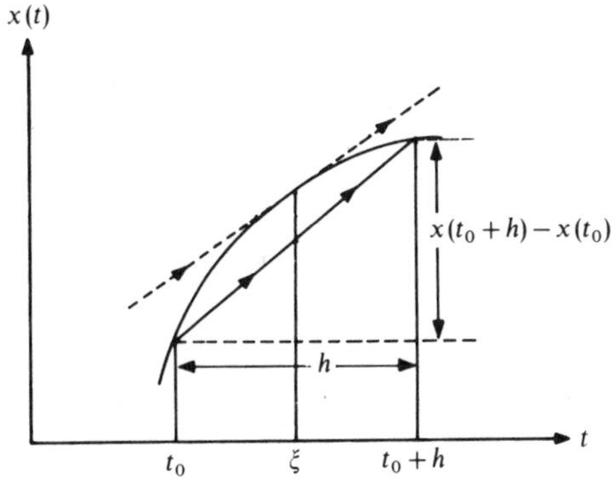

Fig. 10.12 The mean-value theorem

belong to class C^2. This idea generalizes to class C^r; that is, $x \in C^r$ (with $r \in \mathbb{N}$) means that $x, \dot{x}, \ddot{x}, \ldots, x^{(r)}$ are all continuous functions. If $x \in C^r$ for all $r \in \mathbb{N}$, so that x and all its derivatives are continuous, then one writes $x \in C^\infty$ and x is said to be *smooth*. Clearly, a C^∞ function (i.e. a function $x \in C^\infty$) is continuous. Authors often assume that r is sufficiently large for the purposes of their analysis without specifying the value of r explicitly. Alternatively, they assume all functions in the analysis are C^∞ (i.e. they all belong to class C^∞). The difficulty lies in the fact that C^r functions have derivatives which are only $C^{r-1}, C^{r-2}, \ldots$ functions. However, if functions are all assumed to be C^∞, then their derivatives are also C^∞ functions and the difficulty disappears.

The above discussion leads immediately to *Taylor's Theorem* which states that

$$x(t_0+h) = x(t_0) + h\dot{x}(t_0) + \frac{h^2}{2!}\ddot{x}(t_0) + \cdots + \frac{h^n}{n!}x^{(n)}(t_0) + \frac{h^{n+1}}{(n+1)!}x^{(n+1)}(\xi), \qquad (10.13)$$

where the last term on the right-hand side of (10.13) is the remainder term in which $t_0 < \xi < t_0 + h$. This remainder term does not necessarily tend to zero as $n \to \infty$ but, if it does for any particular x and t_0, then the convergent series

$$\sum_{n=0}^{\infty} \frac{h^n}{n!} x^{(n)}(t_0)$$

is called the *Taylor series for* x at t_0, and represents the value $x(t_0+h)$ exactly. If $x \in C^\infty$ and the Taylor series for x at t_0 converges to $x(t_0+h)$ for all h in some neighbourhood of t_0, then x is said to be (*real-*)*analytic* at t_0, written

$x \in C^\omega$. Real-analytic functions are usually satisfactory in practice but are somewhat restrictive in theory.

Example 10.18

The function $x: \mathbb{R} \to \mathbb{R}$ defined by $x(t) = \sin t$ is analytic for all $t \in \mathbb{R}$, and the Taylor series is given by

$$\sin t = t - \frac{1}{3!}t^3 + \frac{1}{5!}t^5 - \cdots .$$

Example 10.19

The function $x: \mathbb{R} \to \mathbb{R}$, defined by

$$x(t) = \begin{cases} \exp(-1/t^2) & \text{if } t \neq 0, \\ 0 & \text{if } t = 0, \end{cases}$$

is not analytic at $t=0$. Certainly $x \in C^\infty$ and all derivatives of x are zero at $t=0$. Every term in the Taylor series for x.expanded about 0 is therefore zero, and so the series converges to zero. But $x(h)$ is nonzero for any nonzero h and so x cannot be analytic at $t=0$.

10.3 Differentiable mappings

Generalization to the derivative of a general mapping, hinted at earlier, can now be described. Suppose $(X, \|\bullet\|_X)$ and $(Y, \|\bullet\|_Y)$ are two normed vector spaces and that A is an open subset of X. Let $f: A \to Y$ be a mapping that is not necessarily linear but is continuous on A. Then f is said to be differentiable at a point $x \in A$ iff there exists a linear map $\lambda: X \to Y$ which approximates f at x in the sense that

$$f(x+h) - f(x) = \lambda(h) + \|h\|_X \varepsilon(h)$$

for all $h \in X$, where $\varepsilon(h) \in Y$ with $\|\varepsilon(h)\|_Y \to 0$ as $\|h\|_X \to 0$. This is a direct generalization of (10.11) (function x is replaced by mapping f and t_0 by $x \in A$) with $\|h\|_X$ playing the role of $|h|$ and $\varepsilon(h)$ (thought of as small in some sense) replacing ε. Of course, $x, x+h \in A$ while $f(x), f(x+h), \lambda(h), \varepsilon(h) \in Y$.

If such a map λ exists, then it is unique and is called the derivative of f at x, denoted by $\mathsf{D}f(x)$ (or sometimes by $f^*(x)$). It is a linear map from X to Y and the value $\mathsf{D}f(x)(h)$ at any element $h \in X$ is not a real number in this more general setting. Note in particular that $\mathsf{D}f(x)$ is a linear map associated with the particular point $x \in A$. Thus, $\mathsf{D}f(x)$ maps all elements $h \in X$ to elements

$\mathrm{D}f(x)(h) \in Y$, for a fixed x. The element $\mathrm{D}f(x)(h)$ is usually abbreviated slightly, using operator-style notation, to $\mathrm{D}f(x)h$. Then

$$f(x+h)-f(x)=\mathrm{D}f(x)h+\|h\|\varepsilon(h),$$

where $\|h\|=\|h\|_X$ and $\varepsilon(h)\to 0$ in Y as $h\to 0$ in X. [13]

Example 10.20

$X=\mathbb{R}^n$, $Y=\mathbb{R}$, $f\colon \mathbb{R}^n\to\mathbb{R}$, $\boldsymbol{h}=h_1\boldsymbol{e}_1+ \cdots + h_n\boldsymbol{e}_n$, where $\{\boldsymbol{e}_1,\ldots, \boldsymbol{e}_n\}$ is the canonical basis in $\mathbb{R}^n$.

$$\begin{aligned}\mathrm{D}f(\boldsymbol{x})h&=\mathrm{D}f(\boldsymbol{x})(h_1\boldsymbol{e}_1+ \cdots + h_n\boldsymbol{e}_n)\\ &=h_1\mathrm{D}f(\boldsymbol{x})\boldsymbol{e}_1 + \cdots + h_n\mathrm{D}f(\boldsymbol{x})\boldsymbol{e}_n\end{aligned}$$

(because $\mathrm{D}f(\boldsymbol{x})$ is a linear map $\mathbb{R}^n\to\mathbb{R}$).

$$=h_1\alpha_1+h_2\alpha_2+ \cdots + h_n\alpha_n \quad \text{(say)},$$

where $\alpha_i\in\mathbb{R}$ $(i=1,\ldots, n)$. Now choose $\boldsymbol{h}=h_1\boldsymbol{e}_1$ so that $h_i=0$ $(i=2,\ldots, n)$. Substituting this $\boldsymbol{h}$ into the equation

$$f(\boldsymbol{x}+\boldsymbol{h})-f(\boldsymbol{x})=\mathrm{D}f(\boldsymbol{x})+\varepsilon\|\boldsymbol{h}\|$$

yields

$$f(x_1+h_1, x_2,\ldots, x_n)-f(x_1, x_2,\ldots, x_n)=h_1\alpha_1+\varepsilon h_1,$$

where $\varepsilon\to 0$ as $h_1\to 0$. Thus,

$$\lim_{h_1\to 0}\frac{f(x_1+h_1, x_2,\ldots, x_n)-f(x_1, x_2,\ldots, x_n)}{h_1}=\alpha_1,$$

i.e. $\alpha_1=\partial f/\partial x_1$ (cf. (10.9)). Similarly, $\alpha_i=\partial f/\partial x_i$ $(i=2,\ldots, n)$. Thus, $\mathrm{D}f(\boldsymbol{x})$ may be identified with the gradient row vector $[\partial f/\partial x_1,\ldots, \partial f/\partial x_n]$ evaluated at the point $\boldsymbol{x}$, and $\lambda(\boldsymbol{h})$ is the product $\mathrm{D}f(\boldsymbol{x})\boldsymbol{h}\in\mathbb{R}$.

Example 10.21

Let $X=\mathbb{R}$ and $Y=\mathbb{R}^n$. A mapping $\boldsymbol{\phi}\colon\mathbb{R}\to\mathbb{R}^n$ is called a *path* in $\mathbb{R}^n$. At each point $t\in\mathbb{R}$, the derivative of $\boldsymbol{\phi}$ at t (if it exists) is a linear map $\mathrm{D}\boldsymbol{\phi}(t)\colon\mathbb{R}\to\mathbb{R}^n$. Writing $\boldsymbol{\lambda}_t$ for $\mathrm{D}\boldsymbol{\phi}(t)$, we have

$$\boldsymbol{\lambda}_t(s)=\boldsymbol{\lambda}_t(s\cdot 1)=s\boldsymbol{\lambda}_t(1)$$

for all $s\in\mathbb{R}$, by the linearity of $\boldsymbol{\lambda}_t$. Thus, if $\boldsymbol{\lambda}_t(1)\in\mathbb{R}^n$ is known, then $\boldsymbol{\lambda}_t(s)$ is found simply by multiplying $\boldsymbol{\lambda}_t(1)$ by s. Obviously $\boldsymbol{\lambda}_t$ is the zero map if $\boldsymbol{\lambda}_t(1)=O$ but, if $\boldsymbol{\lambda}_t(1)$ is not the zero vector in $\mathbb{R}^n$, then the image of $\boldsymbol{\lambda}_t$ is the straight line through the origin and the point $\boldsymbol{\lambda}_t(1)$ in $\mathbb{R}^n$. When the origin is

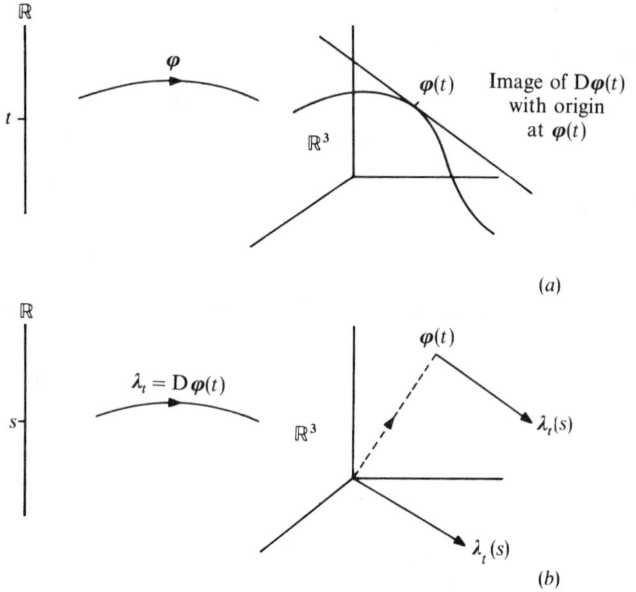

Fig. 10.13 Path of Example 10.21 ($n=3$)

translated to the point $\phi(t)$, this straight line is tangent to the image of the path ϕ at $\phi(t)$ (Fig. 10.13(a, b)). The element $\lambda_t(1) \in \mathbb{R}^n$ is what is usually denoted by $\dot{\phi}(t)$ (Exercise 12).

The reader will see from Example 10.21 that each point $\phi(t)$ on the image of the path ϕ gives rise to a linear map λ_t which in turn is associated with the tangent at $\phi(t)$ in the direction $\lambda_t(1)$. Thus, there is a one–one correspondence between the set $\mathrm{Lin}(\mathbb{R}, \mathbb{R}^n)$ of linear maps $\mathbb{R} \to \mathbb{R}^n$ and the elements $\lambda_t(s)$ in $\mathbb{R}^n$. Using pointwise addition and scalar multiplication, the set $\mathrm{Lin}(\mathbb{R}, \mathbb{R}^n)$ can be shown to be a vector space and there is a linear isomorphism between $\mathrm{Lin}(\mathbb{R}, \mathbb{R}^n)$ and $\mathbb{R}^n$ (Fig. 10.14).

Consider the differential equation

$$\dot{\boldsymbol{x}} = \boldsymbol{f}(\boldsymbol{x}), \qquad \boldsymbol{x}(0) = \boldsymbol{x}_0,$$

where the mapping $\boldsymbol{f}: U \to \mathbb{R}^n$ acts on an open set U of $\mathbb{R}^n$. A solution is a path $\boldsymbol{\phi}: I \to U$ defined by $t \mapsto \boldsymbol{\phi}(t)$ on an interval $I \subseteq \mathbb{R}$ with $\boldsymbol{\phi}(0) = \boldsymbol{x}_0$ and

$$\dot{\boldsymbol{\phi}}(t) = \boldsymbol{f}(\boldsymbol{\phi}(t)).$$

Now the differentiation of the composite function $\boldsymbol{f} \circ \boldsymbol{\phi}: I \to \mathbb{R}^n$ is given by a generalization of the chain rule (cf. p. 212).

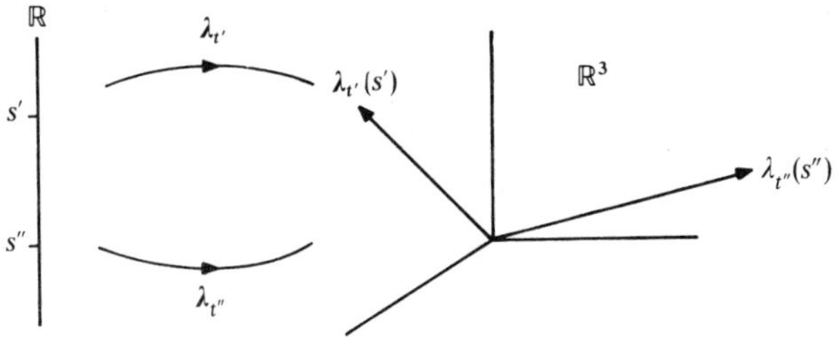

Fig. 10.14 Linear isomorphism Lin(ℝ, $\mathbb{R}^n$) ≃ $\mathbb{R}^n$

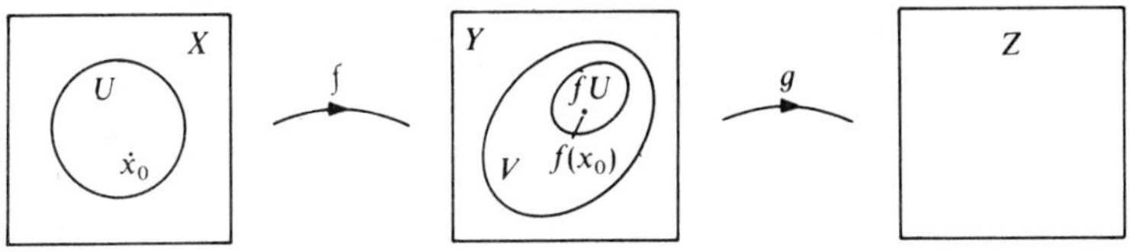

Fig. 10.15 A composite map

Looking at the general case first, let X, Y, Z be normed vector spaces and suppose that U and V are open sets in X and Y respectively (Fig. 10.15). If a mapping $f: U \to Y$ is differentiable at $x_0 \in U$ and a mapping $g: V \to Z$ is differentiable at $f(x_0) \in V$, then the composite mapping[14] $g \circ f: U \to Z$ is differentiable at x_0 and

$$\mathsf{D}(g \circ f)(x_0) = \mathsf{D}g(f(x_0)) \circ \mathsf{D}f(x_0). \tag{10.14}$$

In other words, the derivative of the composite map $g \circ f$ at x_0 is a continuous linear map from X to Z, being the composition of the two continuous linear maps $\mathsf{D}f(x): X \to Y$ and $\mathsf{D}g(f(x)): Y \to Z$.

In the special case of $\boldsymbol{f}(\boldsymbol{\phi}(t))$ mentioned above, (10.14) yields

$$\mathsf{D}(\boldsymbol{f} \circ \boldsymbol{\phi})(t) = \mathsf{D}\boldsymbol{f}(\boldsymbol{\phi}(t)) \circ \mathsf{D}\boldsymbol{\phi}(t),$$

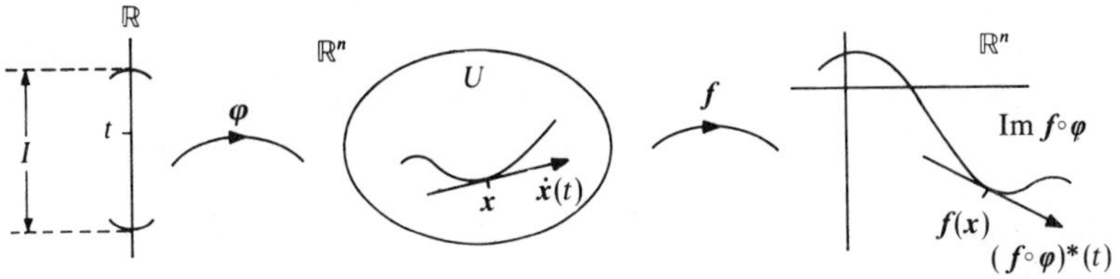

Fig. 10.16 Differentiation of a composite map

a map from $\mathbb{R}$ to $\mathbb{R}^n$. In an alternative notation, [15]

$$(\boldsymbol{f}\circ\boldsymbol{\phi})^*(t)=\mathsf{D}\boldsymbol{f}(\boldsymbol{x})\circ\dot{\boldsymbol{\phi}}(t). \tag{10.15}$$

This demonstrates that the derivative $\mathsf{D}\boldsymbol{f}(\boldsymbol{x})$ takes the tangent to the image of the path $\boldsymbol{\phi}$ at $\boldsymbol{x}$ to the tangent to the image of the path $\boldsymbol{f}\circ\boldsymbol{\phi}$ at $\boldsymbol{f}(\boldsymbol{x})$ (Fig. 10.16).

Example 10.22

$X=\mathbb{R}^n$, $Y=\mathbb{R}^m$, $Z=\mathbb{R}$; the map $\boldsymbol{f}\colon\mathbb{R}^n\to\mathbb{R}^m$ is defined by $\boldsymbol{f}(\boldsymbol{x})=\boldsymbol{y}$ and $g\colon\mathbb{R}^m\to\mathbb{R}$ is defined by $g(\boldsymbol{y})=z$. By Exercise 8 and Example 10.20, the linear maps $\mathsf{D}\boldsymbol{f}(\boldsymbol{x})\colon\mathbb{R}^n\to\mathbb{R}^m$ and $\mathsf{D}g(\boldsymbol{y})\colon\mathbb{R}^m\to\mathbb{R}$ are respectively represented by

$$\begin{bmatrix} \partial f_1/\partial x_1 & \cdots & \partial f_1/\partial x_n \\ \vdots & & \vdots \\ \partial f_m/\partial x_1 & \cdots & \partial f_m/\partial x_n \end{bmatrix} \quad \text{and} \quad [\partial g/\partial y_1, \ldots, \partial g/\partial y_m].$$

Again, Example 10.20 gives the representation of the linear map $\mathsf{D}(g\circ\boldsymbol{f})(x)\colon\mathbb{R}^n\to\mathbb{R}$ as $[\partial g/\partial x_1, \ldots, \partial g/\partial x_n]$ and (10.14) yields the result

$$[\partial g/\partial x_1, \ldots, \partial g/\partial x_n]=[\partial g/\partial y_1, \ldots, \partial g/\partial y_m]\begin{bmatrix} \partial f_1/\partial x_1 & \cdots & \partial f_1/\partial x_n \\ \vdots & & \vdots \\ \partial f_m/\partial x_1 & \cdots & \partial f_m/\partial x_n \end{bmatrix},$$

i.e.

$$\frac{\partial g}{\partial x_i}=\sum_{j=1}^{m}\frac{\partial g}{\partial y_j}\frac{\partial f_j}{\partial x_i} \quad (i=1, \ldots, n).$$

This is the usual 'function of a function' or chain rule given in elementary calculus texts (Apostol 1974).

As in the single variable case discussed in the previous section, classes C^r can be defined for maps whose domains are subsets of $\mathbb{R}^n$, with $n\geqslant 2$. A function f defined on an open set $U\subset\mathbb{R}^n$ is differentiable at a point $u\in U$ if the partial derivatives $\partial f/\partial x_i$ $(i=1, \ldots, n)$ are defined in a neighbourhood of u and are continuous at u. When the partial derivatives of f exist and are continuous on U, then f is differentiable at every point of U. If f has continuous partial derivatives of orders $1, \ldots, r$ on U then f is said to be of class C^r on U. f is said to be *smooth*, or f is of class C^∞, when f is of class C^r for all r. These classes of functions on U are denoted by $\mathrm{C}^r(U)$ or, in the latter case, by $\mathrm{C}^\infty(U)$.

The function x of Example 10.14 is the inverse function f^{-1} of the homeomorphism $f\colon\mathbb{R}\to\mathbb{R}$ defined by $t\mapsto t^3$. This function f is clearly differentiable on $\mathbb{R}$, but Example 10.14 shows that f^{-1} (i.e. x) is not differentiable at

$t=0$ ($f^{-1} \notin C^1$ on $\mathbb{R}$), and hence is not differentiable in a neighbourhood of $t=0$. In general, if a differentiable map f has an inverse which is also differentiable, then f is called a C^1-*diffeomorphism*. Thus, the function f above is not such a diffeomorphism in a neighbourhood of $t=0$. A mapping $f: A \to \mathbb{R}^n$, $A \subseteq \mathbb{R}^n$, is a C^r-*diffeomorphism* if (i) f is a homeomorphism and (ii) $f \in C^r$ and $f^{-1} \in C^r$ (with $r \geqslant 1$). When $f, f^{-1} \in C^\infty$, then f is referred to simply as a *diffeomorphism*.

Example 10.23

$\boldsymbol{f}: \mathbb{R}^n \to \mathbb{R}^n$ is defined by $\boldsymbol{x} \mapsto \boldsymbol{A}\boldsymbol{x}$. $\boldsymbol{f}$ is analytic and C^∞, so that, if $\det \boldsymbol{A} \neq 0$, then $\boldsymbol{f}$ is a diffeomorphism. However, if $\det \boldsymbol{A} = 0$, then $\boldsymbol{f}$ is not a 1–1 mapping; $\boldsymbol{f}$ maps at least a line to the origin (cf. Example 3.17). In this case, $\boldsymbol{f}$ is not a diffeomorphism, because the inverse mapping $\boldsymbol{f}^{-1}$ does not exist.

Let $\mathrm{Lin}(X, Y)$ be the set of all linear maps $f: X \to Y$ between two normed vector spaces X and Y. This set $\mathrm{Lin}(X, Y)$ is a vector space over $\mathbb{R}$ under pointwise addition and scalar multiplication defined as follows. For $f_1: X \to Y$ and $f_2: X \to Y$ belonging to $\mathrm{Lin}(X, Y)$, we define

$$(f_1 + f_2)(x) \triangleq f_1(x) + f_2(x), \qquad (\alpha f_1)(x) \triangleq \alpha f_1(x),$$

for all $\alpha \in \mathbb{R}$ and all $x \in X$. Alternatively, these two definitions can be incorporated into a single one:

$$(\alpha f + \beta g)(x) = \alpha f(x) + \beta g(x)$$

for all $\alpha, \beta \in \mathbb{R}$ and all $f, g \in \mathrm{Lin}(X, Y)$.

Not all linear maps are continuous (Exercise 15), but $\mathrm{Lin}(X, Y)$ contains a subspace $\mathfrak{L}(X, Y)$ consisting of all linear maps that are continuous. So, if $f \in \mathfrak{L}(X, Y)$, then f is a continuous linear map from X to Y. In this case, there exists a constant K, depending on f, such that

$$\| f(x) \|_Y < K$$

for all $x \in X$ satisfying $\| x \|_X = 1$ (Simmons 1963).

This enables a norm to be defined on $\mathfrak{L}(X, Y)$ as

$$\| f \| = \sup\{ \| f(x) \|_Y : \| x \|_X = 1 \}.$$

In the special case where $X = \mathbb{R}^n$ and $Y = \mathbb{R}^m$, bases can be chosen for $\mathbb{R}^n$ and $\mathbb{R}^m$. Then $\boldsymbol{f}$ can be represented by an $m \times n$ matrix $\boldsymbol{F} = [f_{ij}]$. Regarding $\boldsymbol{F}$ as an element of $\mathbb{R}^{m \times n}$, many different matrix norms can be chosen for $\mathfrak{L}(\mathbb{R}^n, \mathbb{R}^m)$ (Section 7.2), e.g.

$$\| \boldsymbol{f} \| = \| \boldsymbol{F} \| = \left(\sum_{i=1}^{m} \sum_{j=1}^{n} f_{ij}^2 \right)^{\frac{1}{2}} \quad \text{or} \quad \| \boldsymbol{f} \| = \max_{i,j} | f_{ij} |.$$

However, it makes no difference which norm is chosen on $\mathfrak{L}(\mathbb{R}^n, \mathbb{R}^m)$, because any two norms on a finite-dimensional vector space will induce the same topology (Simmons 1963). Every linear map $\mathbb{R}^n \to \mathbb{R}^m$ is continuous (Exercise 14), so that $\mathfrak{L}(\mathbb{R}^n, \mathbb{R}^m) = \mathrm{Lin}(\mathbb{R}^n, \mathbb{R}^m)$.

The reader already knows that, if $f \in \mathfrak{L}(A, Y)$, with $A \subseteq X$, [16] then, for every $x \in A$ at which f is differentiable, there is a linear map $\mathsf{D}f(x): X \to Y$ which is itself continuous and so belongs to $\mathfrak{L}(X, Y)$. Thus, a further map $\mathsf{D}f: A \to \mathfrak{L}(X, Y)$ can be defined by $x \mapsto \mathsf{D}f(x)$. If $\mathsf{D}f$ is continuous, then $f: A \to Y$ is said to be of class C^1. If $\mathsf{D}f$ is also differentiable at $x \in A$ then $\mathsf{D}(\mathsf{D}f)(x)$ will be a continuous linear map, called the *second derivative of f at x* and is denoted by $\mathsf{D}^2 f(x)$; it is a linear map $X \to \mathfrak{L}(X, Y)$. Just as continuity and differentiability of $f: A \to Y$ at $x \in A$ implies that $\mathsf{D}f(x) \in \mathfrak{L}(X, Y)$ is a continuous linear map, so the continuity and differentiability of $\mathsf{D}f: A \to \mathfrak{L}(X, Y)$ implies that $\mathsf{D}^2 f(x)$ is a continuous linear map. That is to say $\mathsf{D}^2 f(x) \in \mathfrak{L}(X, \mathfrak{L}(X, Y))$ and thus leads to a further map $\mathsf{D}^2 f: A \to \mathfrak{L}(X, \mathfrak{L}(X, Y))$. If $\mathsf{D}^2 f$ is continuous, then f is said to be of class C^2. The set $\mathfrak{L}(X, \mathfrak{L}(X, Y))$ is a normed vector space and so the procedure can be repeated to define classes C^r for $r = 3, 4, \ldots$. As in the earlier discussion, f is said to be of class C^∞ if it is of class C^r for all $r \in \mathbb{N}$.

The result of applying the map $\mathsf{D}^2 f(x): X \to \mathfrak{L}(X, Y)$ is to take any element $h_1 \in X$ into the linear map $\mathsf{D}^2 f(x)(h_1): X \to Y$, and this latter map takes any other element $h_2 \in X$ into the element $[\mathsf{D}^2 f(x)(h_1)](h_2)$ in Y. The notation rapidly gets out of hand as the process is repeated, but fortunately there is an isomorphism available which eases the situation. Suppose λ is a map $X \to \mathfrak{L}(X, Y)$ defined by $h_1 \mapsto \lambda(h_1)$, so that $\lambda(h_1)$ is a map $X \to Y$ belonging to $\mathfrak{L}(X, Y)$. This map $\lambda(h_1): X \to Y$ is defined by $h_2 \mapsto \lambda(h_1)h_2$. Writing $\tilde{\lambda}(h_1, h_2) = \lambda(h_1)h_2$ defines a *bilinear map* (linear in each factor separately [17])

$$\tilde{\lambda}: X \times X \to Y.$$

Since λ and $\lambda(h_1)$ are continuous linear maps, $\tilde{\lambda}$ is a continuous bilinear map (with the product topology on $X \times X$) and is a special case of a *multilinear map.* Such maps play an important role in multilinear algebra (Northcott 1984). Generally, let $X_1, \ldots, X_p$ (with $p \geqslant 1$) and Y be $\mathbb{R}$-vector spaces. Then the map $\phi: X_1 \times \cdots \times X_p \to Y$ is called a *p-multilinear map* if

$$\phi(x_1, \ldots, x_{i-1}, x+y, x_{i+1}, \ldots, x_p)$$
$$= \phi(x_1, \ldots, x_{i-1}, x, x_{i+1}, \ldots, x_p) + \phi(x_1, \ldots, x_{i-1}, y, x_{i+1}, \ldots, x_p),$$
$$\phi(x_1, \ldots, x_{i-1}, \alpha x, x_{i+1}, \ldots, x_p) = \alpha\phi(x_1, \ldots, x_{i-1}, x, x_{i+1}, \ldots, x_p),$$

for all $x, y \in X_i$, all $\alpha \in \mathbb{R}$, and all $x_j \in X_j$ $(j \neq i)$, for $i = 1, \ldots, p$. Note that a 1-multilinear map is simply an $\mathbb{R}$-vector space homomorphism.

It turns out that the set $\mathrm{M}_2(X \times X, Y)$ of continuous bilinear (2-multilinear) maps $X \times X \to Y$ is linearly isomorphic to the space $\mathfrak{L}(X, \mathfrak{L}(X, Y))$.

Thus, the element $[\mathsf{D}^2 f(x)(h_1)](h_2)$ in Y can be replaced by $\mathsf{D}^2 f(x)(h_1, h_2)$ with $\mathsf{D}^2 f$ playing the role of a map $A \to \mathrm{M}_2(X \times X, Y)$. All of this generalizes of course to $\mathsf{D}^p f(x)(h_1, \ldots, h_p)$, with $\mathsf{D}^p f$ a map $A \to \mathrm{M}_p(X^p, Y)$, and leads to a generalization of Taylor's Theorem given in Section 10.2 as follows.

Let $f: A \to \mathbb{R}$ be a C^∞ function, where A is an open subset of a normed vector space X. Suppose x and $x + h$ both belong to A with $x + h$ sufficiently close to x for the line segment l joining x and $x + h$ to lie wholly in A. Then

$$f(x+h) = f(x) + \mathsf{D}f(x)h + \frac{1}{2!}\mathsf{D}^2 f(x)(h, h) + \cdots + \frac{1}{n!}\mathsf{D}^n f(x)(h, \ldots, h) + R_n(h),$$

where the error term $R_n(h)$ has the form

$$\frac{1}{(n+1)!}\mathsf{D}^{n+1} f(\xi)(h, \ldots, h)$$

for some point ξ lying between x and $x + h$ on l and $\mathsf{D}^i f(x)$ is an i-multilinear map $X \times \cdots \times X$ (i terms) $\to \mathbb{R}$. If $\| R_n(h) \| \to 0$ as $n \to \infty$, then the Taylor series for f at x is given by

$$\sum_{n=0}^{\infty} \frac{1}{n!}\mathsf{D}^n f(x)(h, \ldots, h), \tag{10.16}$$

where $\mathsf{D}^0 f \triangleq f$. If $f \in \mathrm{C}^\infty$, then the Taylor series (10.16) always exists. If the series converges to $f(x + h)$ for all h in some neighbourhood of x then f is said to be *analytic* at x and $f \in \mathrm{C}^\omega$ (cf. Section 10.2).

An important result which gives local information concerning the differentiation of inverse functions is the *inverse-mapping theorem.* This result is restricted to Banach spaces rather than general topological spaces, and is as follows. Suppose X and Y are Banach spaces and U is an open set in X. Let $f: U \to Y$ be a C^r map (with $r \geqslant 1$) and $x \in U$ such that $\mathsf{D}f(x): X \to Y$ is an isomorphism. Then there exists an open neighbourhood V of x, with $V \subset U$, such that $f \upharpoonright V$ is a diffeomorphism from V to $f(V)$.

When X and Y are both $\mathbb{R}^n$ and $U \subseteq \mathbb{R}^n$, the above theorem can be stated as follows. Let $\boldsymbol{f}: U \to \mathbb{R}^n$ be a continuous differentiable mapping such that $\mathsf{D}\boldsymbol{f}(\boldsymbol{x})$ is nonsingular [18] for any point $\boldsymbol{x} \in U$. Then there exists an open neighbourhood V of $\boldsymbol{x}$ and an open neighbourhood W of $\boldsymbol{f}(\boldsymbol{x})$ such that $\boldsymbol{f}: V \to W$ has a continuous inverse $\boldsymbol{f}^{-1}: W \to V$ which is differentiable (i.e. $\boldsymbol{f}$ is a diffeomorphism). Furthermore, if $\boldsymbol{x} \in V$ and $\boldsymbol{y} = \boldsymbol{f}(\boldsymbol{x})$ then

$$\mathsf{D}\boldsymbol{f}^{-1}(\boldsymbol{y}) = [\mathsf{D}\boldsymbol{f}(\boldsymbol{x})]^{-1}.$$

Example 10.24

The mapping f: $\mathbb{R}^2 \to [0, \infty) \times (-\pi, \pi]$ given by $p = (x, y) \mapsto q = (r, \theta)$ takes the cartesian coordinates (x, y) of a point P to the polar coordinates (r, θ) of P, where $r = (x^2 + y^2)^{\frac{1}{2}}$ and θ is given by $x = r\cos\theta$ and $y = r\sin\theta$ $(-\pi < \theta \leqslant \pi)$.[19] So by Exercise 8, with $f_1(x, y) = (x^2 + y^2)^{\frac{1}{2}} = r$ and $f_2(x, y) = \arctan y/x = \theta$, we have

$$\mathsf{D}f(p) = \begin{bmatrix} \partial r/\partial x & \partial r/\partial y \\ \partial\theta/\partial x & \partial\theta/\partial y \end{bmatrix} = \begin{bmatrix} x/r & y/r \\ -y/r^2 & x/r^2 \end{bmatrix},$$

$\det \mathsf{D}f(p) = 1/r$ (i.e. $\mathsf{D}f(p)$ is nonsingular for all finite r). Therefore,

$$[\mathsf{D}f(p)]^{-1} = r\begin{bmatrix} x/r^2 & -y/r \\ y/r^2 & x/r \end{bmatrix} = \begin{bmatrix} x/r & -y \\ y/r & x \end{bmatrix}.$$

On the other hand, the inverse set of equations giving x and y in terms of r and θ, namely

$$x = r\cos\theta, \qquad y = r\sin\theta,$$

defines the inverse mapping f^{-1}: $\mathbb{R}^2 \to \mathbb{R}^2$, $(r, \theta) \mapsto (x, y)$. Here

$$\mathsf{D}f^{-1}(q) = \begin{bmatrix} \partial x/\partial r & \partial x/\partial\theta \\ \partial y/\partial r & \partial y/\partial\theta \end{bmatrix} = \begin{bmatrix} \cos\theta & -r\sin\theta \\ \sin\theta & r\cos\theta \end{bmatrix}$$
$$= \begin{bmatrix} x/r & -y \\ y/r & x \end{bmatrix}.$$

Thus, as the inverse-mapping theorem predicts,

$$\mathsf{D}f^{-1}(q) = [\mathsf{D}f(p)]^{-1}.$$

It is worth noting that an inverse function f^{-1} of a function f: $\mathbb{R} \to \mathbb{R}$ may exist even if $\mathsf{D}f(t) = 0$ (i.e. $\dot{f}(t) = 0$). For example, when f is defined by $f(t) = t^3$, the value of the derivative $\dot{f}(t)$ is $3t^2$ and $\dot{f}(0) = 0$. Nevertheless, the inverse function f^{-1} is defined as $f^{-1}(t) = t^{\frac{1}{3}}$. However, the value of the derivative of f^{-1} at t is $\frac{1}{3}t^{-\frac{2}{3}} = 1/3t^{\frac{2}{3}}$, and this does not exist at $t = 0$, so the inverse function f^{-1} is not differentiable at $t = 0$, i.e. f^{-1} is not C^1 on $\mathbb{R}$.

Generally, if f: $\mathbb{R}^n \to \mathbb{R}^n$, with $\det \mathsf{D}f(x) = 0$, then f^{-1} is not differentiable at $f(x)$. This is because $(f^{-1} \circ f)(x) = x$ and, if f^{-1} were differentiable at $f(x)$, then the chain rule (10.14) would give

$$\mathsf{D}f^{-1}(f(x)) \circ \mathsf{D}f(x) = \mathsf{D}(f^{-1} \circ f)(x) = I,$$

whence $\det \mathsf{D}f^{-1}(f(x)) \det \mathsf{D}f(x) = 1$, contradicting the fact that $\det \mathsf{D}f(x) = 0$.

A further generalization from Section 10.2 is the mean-value theorem for a differentiable function $f: U \to \mathbb{R}$, where U is an open subset of any normed linear space. Consider a line segment l joining elements x and $x+h$ belonging to U. Then there exists $\xi \in l$ such that

$$f(x+h) = f(x) + \mathsf{D}f(\xi)h.$$

This theorem is useful in estimating the difference between the two values $f(x)$ and $f(x+h)$ in $\mathbb{R}$ as, for example, when finding the maximum or minimum values of $f(x)$. If there is some $k \in \mathbb{R}$ such that $\| \mathsf{D}f(x) \| < k$ for all $x \in U$, then the theorem leads to

$$|f(x+h) - f(x)| < K \|h\|,$$

giving $K\|h\|$ as an upper bound on the difference between the two values of f.

10.4 Implicit functions

Suppose a dynamical system is governed by a differential equation and initial condition

$$\dot{x} = f(x, u), \qquad x(0) = x_0,$$

with $f: X \times U \to \mathbb{R}$ in which $X \subseteq \mathbb{R}$ and $U \subseteq \mathbb{R}$. If a control u is required to keep the system at the initial point x_0 for all $t \in [0, \infty)$ then it must be chosen so that $\dot{x} = 0$ everywhere in that time interval. This can be arranged, provided that u satisfies the implicit equation

$$f(x, u) = 0. \tag{10.17}$$

The function f is then called an *implicit function.* To actually find the required control it is necessary to solve the *implicit* equation (10.17) for u in terms of x so as to obtain an *explicit* equation $u = g(x)$. If such a function g exists then [20]

$$f(x, g(x)) \equiv 0.$$

The good news is that sometimes it is possible to find such a function g, and the fact that it may not be unique is not necessarily a difficulty—one is just spoilt for choice! The bad news is that it is not always possible; it all depends on the function f, as the following two examples illustrate.

Example 10.25

$f(x, u) = x^2 + u^2 - 1$. Equation (10.17) yields two explicit equations for $|x| \leqslant 1$: [21]

$$u = g_1(x) = \sqrt{(1-x^2)} \quad \text{or} \quad u = g_2(x) = -\sqrt{(1-x^2)}.$$

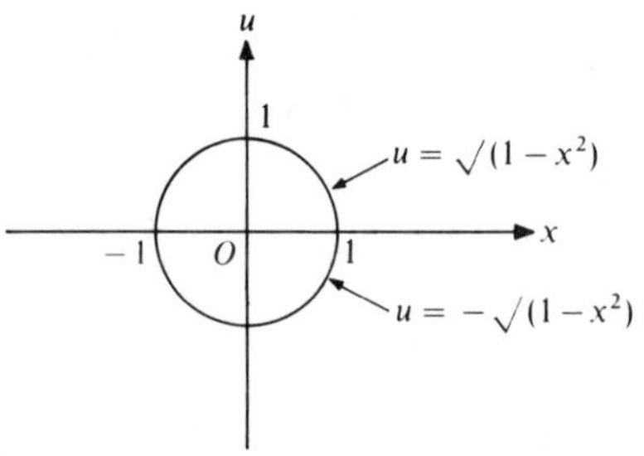

Fig. 10.17 Unit circle of Example 10.25

Geometrically, the two equations $u = g_1(x)$ and $u = g_2(x)$ represent the two parts of the unit circle in $\mathbb{R}^2$ above and below the x axis (Fig. 10.17).

Example 10.26

$f(x, u) = x^2 + u^2 + 1 = 0$. Although it is possible to write $u = \sqrt{(-x^2 - 1)}$ or $u = -\sqrt{(-x^2 - 1)}$, these equations yield values of $u \in \mathbb{C} \backslash \mathbb{R}$, since $-x^2 - 1 < 0$ for all $x \in \mathbb{R}$. But u is restricted to $U \subseteq \mathbb{R}$, so no function g exists for the given function f.

Implicit functions are intimately associated with the existence or otherwise of inverse functions. The question here is, given a function g defined by $u = g(x)$, does there exist an inverse function g^{-1} with $x = g^{-1}(u)$? Now $u = g(x)$ can be written as $f(x, u) = u - g(x) = 0$, and then the above question is whether this latter equation can be solved for x in terms of u.

Consider again the function $f: X \times U \to \mathbb{R}$ of Example 10.25 defined by $(x, u) \mapsto x^2 + u^2 - 1$. Suppose that the point (a, b) satisfies the equation $x^2 + u^2 - 1 = 0$, so that $f(a, b) = 0$, and assume that $a \neq \pm 1$. Then (a, b) is a point on the unit circle, say on the upper half-circle $u = \sqrt{(1 - x^2)}$. Fig. 10.18

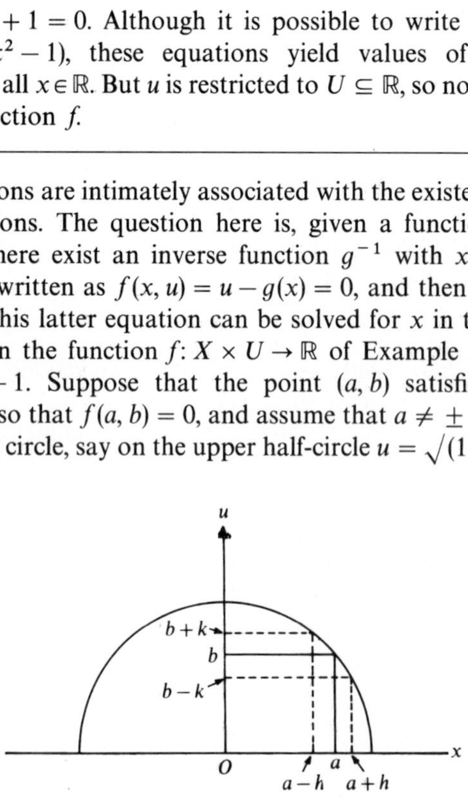

Fig. 10.18 Upper half-circle $u = \sqrt{(1 - x^2)}$

illustrates the existence of open intervals $(a-h, a+h)$ and $(b-k, b+k)$ such that, if $x \in (a-h, a+h)$, then there is a unique $u \in (b-k, b+k)$ with $f(x, u)=0$. This implies the existence of a function g: $(a-h, a+h) \to (b-k, b+k)$ defined by $x \mapsto g(x)$ such that $f(x, g(x)) = 0$. If $a = \pm 1$, no such function g can be found that is defined in an open interval containing a. Clearly, a criterion is needed for deciding when such a function can be found in the general case.

Geometrically, the problem of finding points $(x, u) \in X \times U \subset \mathbb{R}^2$ that satisfy the equation $f(x, u) = 0$ is one of determining the intersection of the surface defined by the equation $z = f(x, u)$ and the (x, u)-plane (which has the equation $z = 0$). If the surface and plane do not intersect at any point, then there is no solution to the equation $f(x, u) = 0$. Suppose that (a, b) is a point of intersection, and consider the tangent plane touching the surface at the point (a, b) (Fig. 10.19). Provided that this tangent plane is not parallel to the (x, u)-plane, it will intersect the plane $z = 0$ in a straight line. Intuitively, in a neighbourhood of the tangent plane, the surface will intersect the plane $z = 0$ in a single well-defined curve (Fig. 10.19).

The tangent plane will be parallel to the (x, u)-plane only if

$$\frac{\partial f}{\partial x}(a, b) = \frac{\partial f}{\partial u}(a, b) = 0.$$

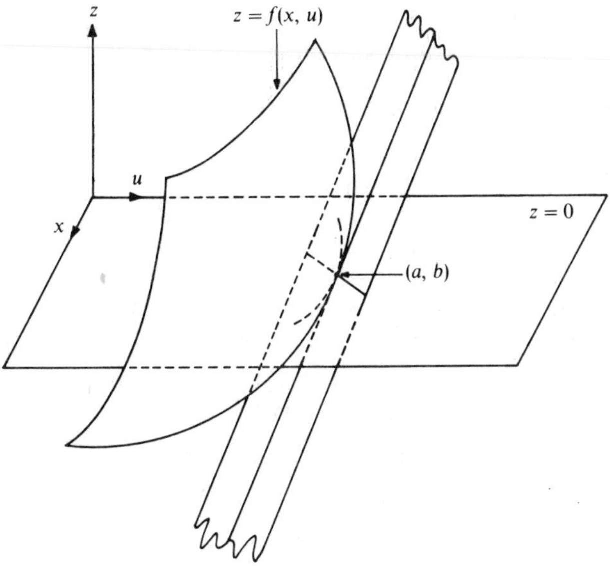

Fig. 10.19 Solution of $f(x, u) = 0$

Thus, if either one of these partial derivatives is nonzero, then the existence of an explicit solution of the form $u = g(x)$ or $x = g^{-1}(u)$ can be expected. These considerations lead to the following *implicit-function theorem*. Suppose that $f: X \times U \to \mathbb{R}$ is a function defined by $(x, u) \mapsto f(x, u)$, continuous in some neighbourhood of a point $(a, b) \in X \times U$ and having continuous derivatives represented by $\partial f/\partial x$ and $\partial f/\partial u$. Furthermore, assume that $f(a, b) = 0$ and $(\partial f/\partial u)(a, b) \neq 0$. Then there exists a function $g: X \to \mathbb{R}$ such that

(i) $g(a) = b$.
(ii) g is defined, continuous, and differentiable in some neighbourhood of a,
(iii) $f(x, g(x)) \equiv 0$ whenever g is defined.

Moreover, g is uniquely determined by these conditions and its derivative is given by the equation

$$\frac{du}{dx} = g'(x) = -\frac{\partial f}{\partial x}\bigg/\frac{\partial f}{\partial u}.$$

Example 10.27

$f(x, u) = x^2 + xu + u^2 - 7$. The equation $f(x, u) = 0$ has a unique solution for u near the point $(2, 1)$, because $f(2, 1) = 0$ and $\partial f/\partial u(2, 1) = 4 \neq 0$.

All the conditions of the implicit-function theorem are satisfied, and so there exists a function g such that $u = g(x)$ in a neighbourhood of the point $(2, 1)$. By completing the square with respect to u in $f(x, u)$, it is not difficult to find an explicit expression for $g(x)$:

$$f(x, u) = (u + \tfrac{1}{2}x)^2 + \tfrac{3}{4}x^2 - 7.$$

Thus, $f(x, u) = 0$ gives $(u + \frac{1}{2}x)^2 = 7 - \frac{3}{4}x^2$, i.e.

$$u + \tfrac{1}{2}x = \pm\sqrt{(7 - \tfrac{3}{4}x^2)},$$

so that

$$u = g(x) = -\tfrac{1}{2}x + \sqrt{(7 - \tfrac{3}{4}x^2)},$$

the positive sign being chosen because $u = 1$ when $x = 2$.

Since $\partial f/\partial x(2, 1) = 5 \neq 0$, the inverse function g^{-1} is also defined in a neighbourhood of the point $(2, 1)$ such that $x = g^{-1}(u)$ within that neighbourbood. By symmetry,[22] g^{-1} is given by

$$g^{-1}(u) = -\tfrac{1}{2}u + \sqrt{(7 - \tfrac{3}{4}u^2)}.$$

For the dynamical system governed by

$$\dot{x} = x^2 + xu + u^2 - 7, \qquad x(0) = 2,$$

the defining equation for the equilibrium point $x(t) = 2$ for all $t \in [0, \infty)$, is

$f(x, u) = 0$, i.e.

$$\begin{aligned} f(2, u) &= 4 + 2u + u^2 - 7 \\ &= (u+3)(u-1) \\ &= 0, \end{aligned}$$

whence $u = -3$ or $u = 1$. With $u = 1$, the expression for $g(x)$ is as given above. The control $u = -3$ corresponds to the choice of the negative sign for $\sqrt{(7 - \frac{3}{4}x^2)}$ so that

$$g(x) = -\tfrac{1}{2}x - \sqrt{(7 - \tfrac{3}{4}x^2)}.$$

There now follows a generalization of the above implicit-function theorem.

Suppose that $X \subseteq \mathbb{R}^n$, $U \subseteq \mathbb{R}$, and $f: X \times U \to \mathbb{R}$ is a function given by $(x_1, \ldots, x_n, u) \mapsto f(x_1, \ldots, x_n, u)$ which is defined and continuous in some neighbourhood of a point $(a_1, \ldots, a_n, b) \in X \times U$ and which has continuous partial derivatives $f_{x_1}, \ldots, f_{x_n}, f_u$. Furthermore, assume that $f(a_1, \ldots, a_n, b) = 0$ and $f_u(a_1, \ldots, a_n, b) \neq 0$. Then there exists a function $g: X \to \mathbb{R}$ such that

(i) $g(a_1, \ldots, a_n) = b$,
(ii) g is defined, continuous, and differentiable in some neighbourhood of $(a_1, \ldots, a_n)$,
(iii) $f(x_1, \ldots, x_n, g(x_1, \ldots, x_n)) \equiv 0$ whenever g is defined.

Moreover, g is uniquely determined by these conditions, and its continuous partial derivatives are given by the equations

$$\begin{aligned} f_{x_1} + f_u g_{x_1} &= 0, \\ f_{x_2} + f_u g_{x_2} &= 0 \\ \vdots \quad \vdots \quad &\quad \vdots \\ f_{x_n} + f_u g_{x_n} &= 0. \end{aligned} \tag{10.18}$$

These equations are obtained from the fact that the differential df is given by

$$df(x_1, \ldots, x_n, u) = f_{x_1} dx_1 + \cdots + f_{x_n} dx_n + f_u du = 0, \tag{10.19}$$

and the differentials $dx_1, \ldots, dx_n$, du are dependent, since the variables $x_1, \ldots, x_n$, u are dependent through the equation $f(x_1, \ldots, x_n, u) = 0$. Writing $u = g(x_1, \ldots, x_n)$, expressing du in the form

$$du = g_{x_1} dx_1 + g_{x_2} dx_2 + \cdots + g_{x_n} dx_n,$$

and substituting into (10.19) yields

$$(f_{x_1} + f_u g_{x_1}) dx_1 + \cdots + (f_{x_n} + f_u g_{x_n}) dx_n = 0. \tag{10.20}$$

The system (10.18) now follows from the fact that the differentials $dx_1, ..., dx_n$ are independent and hence each of their coefficients in (10.20) must vanish.[23]

Example 10.28

$n=2$, $X \subseteq \mathbb{R}^2$, $U \subseteq \mathbb{R}$, $f(x_1, x_2, u) = x_1 + x_2 + u - \sin x_1 x_2 u$, $a = (0, 0)$, $b = 0$, $f(0, 0, 0) = 0$, $f_u(x_1, x_2, u) = 1 - x_1 x_2 \cos x_1 x_2 u$, $f_u(0, 0, 0) = 1 \neq 0$, $f_{x_1} = 1 - x_2 u \cos x_1 x_2 u$, $f_{x_2} = 1 - x_1 u \cos x_1 x_2 u$.
Then (10.18) yields

$$\frac{\partial g}{\partial x_1} = -\frac{f_{x_1}}{f_u} = \frac{x_2 u \cos x_1 x_2 u - 1}{1 - x_1 x_2 \cos x_1 x_2 u},$$

$$\frac{\partial g}{\partial x_2} = -\frac{f_{x_2}}{f_u} = \frac{x_1 u \cos x_1 x_2 u - 1}{1 - x_1 x_2 \cos x_1 x_2 u}.$$

The most general form of the implicit-function theorem concerns the mapping $\boldsymbol{f}: X \times U \to \mathbb{R}^m$ with $X \subseteq \mathbb{R}^n$, $U \subseteq \mathbb{R}^m$, and $\boldsymbol{f}$ defined by

$$(x_1, ..., x_n; u_1, ..., u_m) \mapsto \boldsymbol{f}(x_1, ..., x_n; u_1, ..., u_m).$$

Such a case arises when m algebraic equations are to be solved for m unknowns $u_1, ..., u_m$ in terms of n parameters $x_1, ..., x_n$:

$$\begin{aligned} f^1(x_1, ..., x_n; u_1, ..., u_m) &= 0, \\ \vdots \qquad\qquad & \vdots \\ f^m(x_1, ..., x_n; u_1, ..., u_m) &= 0. \end{aligned} \tag{10.21}$$

Suppose that a solution to (10.21) exists at the point $(\boldsymbol{a}, \boldsymbol{b}) \in X \times U$. For each $\boldsymbol{x}$ in a neighbourhood of the point $\boldsymbol{a}$ the conditions which guarantee a solution for $\boldsymbol{u}$ in a neighbourhood of $\boldsymbol{b}$ are given by the following form of the implicit-function theorem which includes all the previous special cases.

Suppose that $\boldsymbol{f}: X \times U \to \mathbb{R}^m$, with $X \subseteq \mathbb{R}^n$ and $U \subseteq \mathbb{R}^m$, is a continuous differentiable mapping in an open neighbourhood of a point $(\boldsymbol{a}, \boldsymbol{b}) \in X \times U$ satisfying $\boldsymbol{f}(\boldsymbol{a}, \boldsymbol{b}) = \boldsymbol{0}$. If

$$\det \left[\frac{\partial \boldsymbol{f}}{\partial \boldsymbol{u}}\right]_{\substack{x=a \\ u=b}} \neq 0, \tag{10.22}$$

then there exist open neighbourhoods A and B of $\boldsymbol{a}$ and $\boldsymbol{b}$ respectively such that, for each $\boldsymbol{x} \in A$, there is a unique mapping $\boldsymbol{g}: X \to \mathbb{R}^m$ with the properties

(i) $\boldsymbol{g}(\boldsymbol{a}) = \boldsymbol{b}$,
(ii) $\boldsymbol{f}$ is defined, continuous, and differentiable in A,
(iii) $\boldsymbol{f}(\boldsymbol{x}, \boldsymbol{g}(\boldsymbol{x})) \equiv \boldsymbol{0}$ in A.

As before, equations can be found from which the partial derivatives of $\boldsymbol{g}=(g^1,\ldots,g^m)$ with respect to $x_1,\ldots,x_n$ can be found. For example, differentiation with respect to x_1 gives

$$\begin{aligned} f^1_{x_1} + f^1_{u_1} g^1_{x_1} + \cdots + f^1_{u_m} g^m_{x_1} &= 0, \\ f^2_{x_1} + f^2_{u_1} g^1_{x_1} + \cdots + f^2_{u_m} g^m_{x_1} &= 0, \\ &\vdots \\ f^m_{x_1} + f^m_{u_1} g^1_{x_1} + \cdots + f^m_{u_m} g^m_{x_1} &= 0. \end{aligned} \tag{10.23}$$

In matrix notation (10.23) can be written as

$$\frac{\partial \boldsymbol{f}}{\partial \boldsymbol{u}} \begin{bmatrix} g^1_{x_1} \\ \vdots \\ g^m_{x_1} \end{bmatrix} = - \begin{bmatrix} f^1_{x_1} \\ \vdots \\ f^m_{x_1} \end{bmatrix},$$

so that the partial derivatives of $g^1,\ldots,g^m$ with respect to x_1 are given by

$$\begin{bmatrix} g^1_{x_1} \\ \vdots \\ g^m_{x_1} \end{bmatrix} = \left(\frac{\partial \boldsymbol{f}}{\partial \boldsymbol{u}}\right)^{-1} \begin{bmatrix} f^1_{x_1} \\ \vdots \\ f^m_{x_1} \end{bmatrix},$$

and the inverse matrix $(\partial \boldsymbol{f}/\partial \boldsymbol{u})^{-1}$ exists in A because of (10.22). Similar equations can be obtained for the derivatives $g^1_{x_i},\ldots,g^m_{x_i}$ $(i = 2,\ldots,n)$ by differentiating with respect to x_i.

The partial derivatives $g^j_{x_i}$ $(j = 1,\ldots,m;\, i = 1,\ldots,n)$, obtained by the above procedure, depend on the function $\boldsymbol{g}$, and there may be several such functions, as was seen in Example 10.25. However, for any particular $\boldsymbol{g}$, the above equations yield its partial derivatives. To illustrate this, the function f of Example 10.25 will again be examined.

Example 10.29

$f\colon X \times U \to \mathbb{R}$, with $X \subseteq \mathbb{R}$ and $U \subseteq \mathbb{R}$, defined by $f(x, u) = x^2 + u^2 - 1$. From Example 10.25, two possible solutions for u are

$$u = g^1(x) = \sqrt{(1-x^2)} \quad \text{or} \quad u = g^2(x) = -\sqrt{(1-x^2)}.$$

For either g, we have $f_x + f_u g_x = 0$, i.e. $2x + 2ug_x = 0$, whence

$$g_x = -x/u = -x/g(x),$$

and so g_x clearly depends on $g(x)$. Now

$$g^1_x = -x/\sqrt{(1-x^2)} = -x/g^1(x), \qquad g^2_x = x/\sqrt{(1-x^2)} = -x/g^2(x).$$

Thus, for the two different g's, the partial derivative g_x is different, but the formula $g_x = -x/g(x)$ is valid in both cases.

Exercises

1. Prove from first principles that the function $x\colon \mathbb{R} \to \mathbb{R}^+$ defined by $x(t) = t^2$ is continuous but not uniformly continuous on $\mathbb{R}$.

2. Use the basic definition of continuity to show that the function $u\colon [-1, 1] \to \mathbb{R}$ given by $u(t) = 1/(1+t^2)$ is uniformly continuous at any point $t_0 \in [-1, 1]$.

3. Show that the function $u\colon \mathbb{R} \to \mathbb{R}$, given by

$$u(t) = \begin{cases} t^2 \sin 1/t & \text{if } t \neq 0, \\ 0 & \text{if } t = 0, \end{cases}$$

is continuous at $t = 0$.

4. Prove that the function $f\colon \mathbb{R}^2 \to \mathbb{R}$, given by

$$f(x_1, x_2) = \begin{cases} x_1 x_2^2/(x_1^2 + x_2^4) & \text{if } x \neq 0, \\ 0 & \text{if } x = 0, \end{cases}$$

is not continuous at $x = 0$.

5. Given the set $X = \{(x_1, x_2)\colon x_1^2 + x_2^2 = 1,\ x_2 \neq -1\}$, show that the function $f\colon X \to \mathbb{R}$ defined by $f(x_1, x_2) = x_1/(1+x_2)$ is a homeomorphism.

6. Show that the function $f\colon (-1, 1) \to \mathbb{R}$ defined by $f(t) = \tan \frac{1}{2}\pi t$ is a homeomorphism.

7. Prove that the function $u\colon \mathbb{R} \to \mathbb{R}$, given by

$$u(t) = \begin{cases} t \sin 1/t & \text{if } t \neq 0, \\ 0 & \text{if } t = 0, \end{cases}$$

is not differentiable at $t = 0$.

8. Find a representation for the derivative $\mathsf{D}\boldsymbol{f}(\boldsymbol{x})\colon \mathbb{R}^n \to \mathbb{R}^m$ of a function $\boldsymbol{f}\colon X \to \mathbb{R}^m$, where $X \subseteq \mathbb{R}^n$.

9. $\boldsymbol{f}\colon X \to \mathbb{R}^2$, with $X \subseteq \mathbb{R}^2$, is defined by $(x_1, x_2) \mapsto (x_1^2 + x_2^2, 2x_1x_2)$. Calculate the derivative $\mathsf{D}\boldsymbol{f}(\boldsymbol{x})$ for $\boldsymbol{x} \in X$, and show that its rank is 2 in the whole of $\mathbb{R}^2$ except on lines $x_2 = \pm x_1$.

10. A mapping $\boldsymbol{f}\colon \mathbb{R}^2 \to \mathbb{R}^2$ is defined by the equations

$$u = x^2 - y^2, \qquad v = 2xy,$$

while a mapping $g\colon \mathbb{R}^2 \to \mathbb{R}$ is given by $g(u, v) = u^2 - v^2$. Use the chain rule (see Example 10.22) to calculate the partial derivatives $\partial g/\partial x$ and $\partial g/\partial y$.

11. If $f\colon [a, b] \to \mathbb{R}$, defined on the interval $[a, b] \subset \mathbb{R}$, is differentiable on (a, b) and is such that $\dot{f}(t) > 0$ for all $t \in (a, b)$, then show that $f^{-1}\colon [f(a), f(b)] \to \mathbb{R}$ is also differentiable on $\big(f(a), f(b)\big)$.

12. Describe the image of the path $\varphi\colon \mathbb{R} \to \mathbb{R}^2$ arising from the dynamical system

$$\dot{x}_1 = x_2, \quad x_1(0) = 0; \qquad \dot{x}_2 = -x_1, \quad x_2(0) = 1.$$

Find the 'value' of the derivative $\dot{\varphi}(t)$ when $t = \frac{1}{4}\pi$.

13. The mapping $\boldsymbol{f}\colon \mathbb{R} \times \mathbb{R}^2 \to \mathbb{R}^2$ defined by $\boldsymbol{f}(t, (x, y)) = (x + t, y)$ transfers each point $(x, y) \in \mathbb{R}^2$ to the point $(x + t, y)$. For a fixed t, denote the appropriate mapping by $\boldsymbol{f}_t$. Show that $\boldsymbol{f}_t\colon \mathbb{R}^2 \to \mathbb{R}^2$ is a diffeomorphism with $(\boldsymbol{f}_t)^{-1} = \boldsymbol{f}_{-t}$.

14. Prove that an arbitrary linear map $\boldsymbol{f}\colon \mathbb{R}^n \to \mathbb{R}^m$, defined by $\boldsymbol{x} \mapsto \boldsymbol{A}\boldsymbol{x}$ with $\boldsymbol{A} \in \mathbb{R}^{m \times n}$, is continuous on $\mathbb{R}^n$.

15. Consider the normed vector space X of all infinite sequences $\boldsymbol{a} = (a_i)$, where $a_i \in \mathbb{R}$ $(i = 1, 2, \dots)$ such that $\sum_{i=1}^{\infty} |a_i|$ is finite, with norm defined as

$$\|\boldsymbol{a}\| = \sum_{i=1}^{\infty} \frac{1}{i!} |a_i|.$$

Prove that the map $f\colon X \to X$ defined by $f(a_1, a_2, a_3, \dots) = (a_2, a_3, a_4, \dots)$ is linear but that it is not continuous at $\mathbf{0} = (0, 0, \dots)$.

16. Prove that the equation $x \cos xu = 0$ has a unique solution for u in a neighbourhood of the point $(1, \frac{1}{2}\pi)$, and show that the derivative $\mathrm{d}u/\mathrm{d}x$ vanishes when $\tan xu = 1/xu$ and $xu \neq k\pi$ $(k \in \mathbb{Z})$.

Notes for Chapter 10

1. The value of the function u at the point of discontinuity can be chosen as 1 in Fig. 10.1(b). This is purely a matter of convenience (see Section 9.4 in particular).

2. $u\colon \mathbb{R}\setminus\{0\} \to \mathbb{R}$ defined by $u(t) = 1/t$ is continuous at all points of its domain. However, if $u\colon \mathbb{R} \to \mathbb{R}$ is defined by $u(t) = 1/t$ for $t \neq 0$, and $u(0) = 1$, then u is discontinuous at $t = 0$.

3. To prove that the limit of a sequence of continuous functions is continuous if the convergence is uniform, one may proceed as follows. Let the sequence $(x^{(k)}(t))$ converge to $x(t)$ uniformly on an interval I. Let $t_0 \in I$ be an arbitrary point in I. Because the convergence is uniform, given $\varepsilon > 0$, there exists a sufficiently large k such that

$$|x(t) - x^{(k)}(t)| < \tfrac{1}{3}\varepsilon \quad \forall\, t \in I.$$

But $x^{(k)}(t)$ is continuous at t_0 so there is some $\delta > 0$ such that

$$|x^{(k)}(t) - x^{(k)}(t_0)| < \tfrac{1}{3}\varepsilon \quad \text{whenever} \quad |t - t_0| < \delta.$$

Thus, if $|t - t_0| < \delta$, we have

$$\begin{aligned} |x(t) - x(t_0)| &= |x(t) - x^{(k)}(t) + x^{(k)}(t) - x^{(k)}(t_0) + x^{(k)}(t_0) - x(t_0)| \\ &\leqslant |x(t) - x^{(k)}(t)| + |x^{(k)}(t) - x^{(k)}(t_0)| + |x^{(k)}(t_0) - x(t_0)| \\ &< \tfrac{1}{3}\varepsilon + \tfrac{1}{3}\varepsilon + \tfrac{1}{3}\varepsilon = \varepsilon. \end{aligned}$$

So x is continuous at t_0 and, since t_0 is an arbitrary point in I, it follows that x is continuous on I.

4. This new function $u: \mathbb{R} \to \mathbb{R}$ has a different domain to the original function $u: \mathbb{R} \setminus \{0\} \to \mathbb{R}$ and so they are actually different functions. The fact that they are denoted by the same letter u should not cause confusion in this instance. See also Note 2.
5. Integration theory, involving impulses, point loads, and other generalized functions, also uses the concept of bounded variation. Both Riemann–Stieltjes and Lebesgue integration involve this concept (Rosenbrock and Storey 1970, Weir 1973).
6. It is not generally true that, if U is open in X, then $f(U)$ is open in Y. For example, $f: \mathbb{R} \to \mathbb{R}$ defined by $f(t) = t^2$; the set $U = \mathbb{R}$ is open but $f(U) = \mathbb{R}^+$ is not open.
7. By (3.1), $f^{-1}(g^{-1}(V)) = (gf)^{-1}(V)$.
8. The remaining terms $\varphi(h)$ after $kt_0^{k-1}h$ in the binomial expansion of $(t_0 + h)^k$ are denoted by $O(h^2)$ (O for order) which means that $\varphi(h)$ vanishes to at least the same order h but not necessarily to a higher order. Thus,

$$O(h^2) = \varphi(h) = \binom{n}{2} t_0^{k-2} h^2 + \binom{n}{3} t_0^{k-3} h^3 + \cdots + h^k,$$

where $\binom{n}{r}$ is the binomial coefficient $n(n-1) \cdots (n-r+1)/r!$. The notation $O(h^2)$ means that $|O(h^2)/h^2|$ is bounded as $h \to 0$. Clearly, $|O(h^2)/h^2| \to \binom{n}{2} t_0^{k-2}$ as $h \to 0$.

Alternatively, $\varphi(h)$ can be denoted by $o(h)$ which means that $o(h)/h \to 0$ as $h \to 0$, i.e.

$$\varphi(h)/h = \binom{n}{2} t_0^{k-2} h + \binom{n}{3} t_0^{k-3} h^2 + \cdots + h^{k-1} \to 0 \quad \text{as } h \to 0.$$

In general, the *Landau order symbols O* and *o* are defined as follows. Given two functions $\varphi(h)$ and $\psi(h)$, one writes:

$$\varphi = O(\psi) \text{ as } h \to 0 \quad \text{if } |\varphi(h)/\psi(h)| \text{ is bounded as } h \to 0$$

and

$$\varphi = o(\psi) \text{ as } h \to 0 \quad \text{if } \varphi(h)/\psi(h) \to 0 \text{ as } h \to 0.$$

9. If x is differentiable at t_0 then $\dot{x}(t_0)$ exists, which means that $h^{-1}[x(t_0 + h) - x(t_0)]$ tends to a definite limit (say λ) as $h \to 0$. Thus,

$$x(t_0 + h) - x(t_0) \to \lambda \lim_{h \to 0} h = 0.$$

But $x(t_0 + h) - x(t_0) \to 0$ as $h \to 0$ implies that x is continuous at t_0.
10. The bounds on x to ensure that the image of f is equal to the domain of g are given by $-\frac{1}{2}\pi < (1 + x)/(1 - x) < \frac{1}{2}\pi$. These inequalities are equivalent to $(\pi - 2)/(\pi + 2) < x < (\pi + 2)/(\pi - 2)$.
11. The derivative is being defined in terms of a linear approximation. If the scale of the graph of a function x that is differentiable at a point t_0 is blown up without limit, then the graph at t_0 becomes flat. In symbols,

$$x(t_0 + h) = x(t_0) + Kh + \varepsilon|h|,$$

where K is a constant and $\varepsilon \to 0$ as $|h| \to 0$. Then $K = \dot{x}(t_0)$ defines the derivative.

A *homogeneous* linear function λ: $\mathbb{R} \to \mathbb{R}$ is defined by $\lambda(h) = Kh$, where $K(=\dot{x}(t_0))$ is a constant. To call λ just a linear function could mislead, since this could be interpreted as meaning $\lambda(h) = a + bh$ for constants a and b. Nevertheless, despite this danger, λ is usually called a linear function.

12. (10.11) is equivalent to (10.10), because the former can be written as

$$\frac{x(t_0 + h) - x(t_0)}{h} = \dot{x}(t_0) + \frac{\varepsilon|h|}{h} \to \dot{x}(t_0) \quad \text{as } h \to 0,$$

since $\varepsilon \to 0$ as $h \to 0$.

13. Remember that the element 0 in '$\varepsilon(h) \to 0$' is the zero element in Y, whereas the element 0 in '$h \to 0$' is the zero element in X.

14. The composite mapping $g \circ f$: $U \to Z$ will exist, provided that the image of f lies in V, i.e. provided that $f(U) \subseteq V$.

15. The derivative of a function ψ at x is sometimes denoted by $\psi^*(x)$ or $\psi^*_{(x)}$ (Lobry 1970).

16. Clearly f need only be defined on some open set A of X for differentiability to be defined at $x \in A$.

17. Remember λ and $\lambda(h_1)$ are linear maps.

18. The notation $\mathsf{D}\boldsymbol{f}(\boldsymbol{x})$ is also being used here to denote the matrix representation of the linear map $\mathsf{D}\boldsymbol{f}(\boldsymbol{x})$: $\mathbb{R}^n \to \mathbb{R}^n$. This is commonly done and should not lead to any confusion.

19. Notice that the function defined by $\theta = \arctan y/x$ has range $(-\frac{1}{2}\pi, \frac{1}{2}\pi)$. This indicates that $\theta = \arctan y/x$ is an unsatisfactory definition for θ, since

$$\arctan\frac{y}{x} = \arctan\frac{-y}{-x},$$

whereas (x, y) and $(-x, -y)$ each has a different polar coordinate θ. This can be remedied by taking for θ the interval $-\pi < \theta \leqslant \pi$, with

$$\cos\theta = x/r, \qquad \sin\theta = y/r,$$

and then, of course, θ is discontinuous on the negative x axis, i.e. $\{(x, y): y = 0, x < 0\}$.

20. The sign $\equiv$ in $f(x, g(x)) \equiv 0$ means that the expression $f(x, g(x))$ is identically zero, i.e. all terms occurring in $f(x, g(x))$ actually cancel out so that the value of f is zero for all $x \in X$. This is to be distinguished from the situation in which, because of the particular time behaviour of a function x in an interval $t_1 \leqslant t \leqslant t_2$, we have $f(x(t)) = 0$. These considerations are particularly important in the discussion of Poisson and Lie brackets in Chapter 11.

21. $x^2 + u^2 - 1 = 0$ gives $u^2 = 1 - x^2$ and, since $u \in \mathbb{R}$, we have $u^2 \geqslant 0$. It follows that $1 - x^2 \geqslant 0$ and hence $|x| \leqslant 1$.

22. $f(x, u)$ is unchanged with x written as u and u as x.

23. For example, since $\mathrm{d}x_1, \ldots, \mathrm{d}x_n$ are independent, $\mathrm{d}x_1$ may be chosen nonzero and $\mathrm{d}x_2, \ldots, \mathrm{d}x_n$ all zero. Then (10.20) becomes $(f_{x_1} + f_u g_{x_1})\mathrm{d}x_1 = 0$ and, since $\mathrm{d}x_1 \neq 0$, we get $f_{x_1} + f_u g_{x_1} = 0$. Now choose $\mathrm{d}x_2$ nonzero and $\mathrm{d}x_1, \mathrm{d}x_3, \ldots, \mathrm{d}x_n$ zero. As before,

$$(f_{x_2} + f_u g_{x_2})\mathrm{d}x_2 = 0 \quad \Rightarrow \quad f_{x_2} + f_u g_{x_2} = 0,$$

and so on.

11

MANIFOLDS AND LIE ALGEBRAS

For some dynamical systems, the state space turns out not to be a vector space but is instead a type of topological space known as a manifold. Since a system in this case is still governed in general by a set of nonlinear differential equations, a calculus on manifolds becomes essential. Such a calculus can be described using the ideas of differentiability on $\mathbb{R}^n$ from Chapter 10. Again, special types of dynamical systems known as linear–analytic may be analysed by means of a binary operation called a Lie bracket. This operation in turn leads to the idea of a Lie algebra, which has played an important role in the theory of nonlinear systems (Brockett 1973). Many of the concepts from linear system theory such as controllability, observability, and realization have been generalized in the study of nonlinear systems.

This final chapter very briefly describes these rather sophisticated mathematical ideas and leads the reader to the point where he or she can (hopefully) move on to read some of the early literature on nonlinear system theory. Many of the basic papers are referenced in this chapter.

11.1 Vector fields

Many dynamical systems are governed by the vector differential equation [1]

$$\dot{\boldsymbol{x}} = \boldsymbol{X}(\boldsymbol{x}), \qquad \boldsymbol{x}(t_0) = \boldsymbol{x}_0, \qquad \boldsymbol{x}(t) \in \mathbb{R}^n, \tag{11.1}$$

where $\boldsymbol{X}$ is a map from $\mathbb{R}^n$ to $\mathbb{R}^n$. The solution of (11.1) is a path (see Example 10.21) $\varphi: I \to \mathbb{R}^n$, with $I \subseteq \mathbb{R}$, defined by $t \mapsto \varphi(t)$, satisfying $\varphi(t_0) = \boldsymbol{x}_0$ and

$$\dot{\varphi}(t) = \boldsymbol{X}(\varphi(t)) \quad \forall\, t \in I.$$

The image of each point $\boldsymbol{x} \in \mathbb{R}^n$ under the map $\boldsymbol{X}$ is a vector $\boldsymbol{X}(\boldsymbol{x}) \in \mathbb{R}^n$ which is equal to the velocity vector $\dot{\boldsymbol{x}}$ at $\boldsymbol{x}$ according to the differential equation (11.1).

The reader will be familiar with the usual idea of a velocity vector being tangent to the curve traced out in state space by a dynamical system. Therefore, it is more natural to look upon $\boldsymbol{X}(\boldsymbol{x})$ in (11.1) as a vector radiating from the point $\boldsymbol{x}$ and tangent to the curve at $\boldsymbol{x}$, rather than as a vector in $\mathbb{R}^n$ (which would radiate from the origin of $\mathbb{R}^n$ as shown in Fig. 11.1 for $n = 3$). The set of all vectors $\boldsymbol{X}(\boldsymbol{x})$ localized at $\boldsymbol{x}$ (i.e. origin at the point $\boldsymbol{x} \in \mathbb{R}^n$) is called the *tangent space* to $\mathbb{R}^n$ at $\boldsymbol{x}$ and is denoted by $\mathrm{T}_{\boldsymbol{x}}(\mathbb{R}^n)$. Geometrically, this is

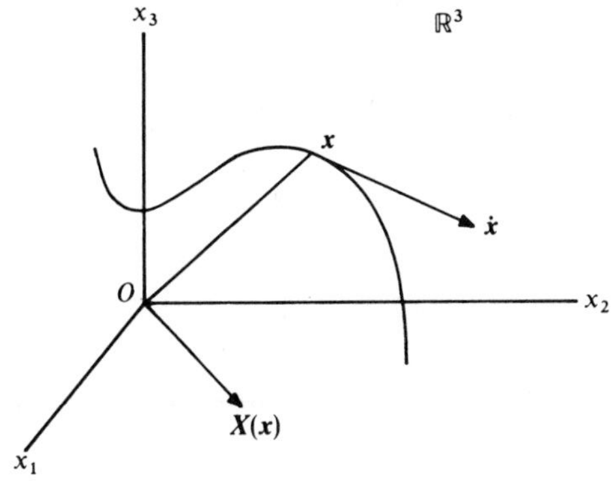

Fig. 11.1 Velocity vector $\dot{x}$

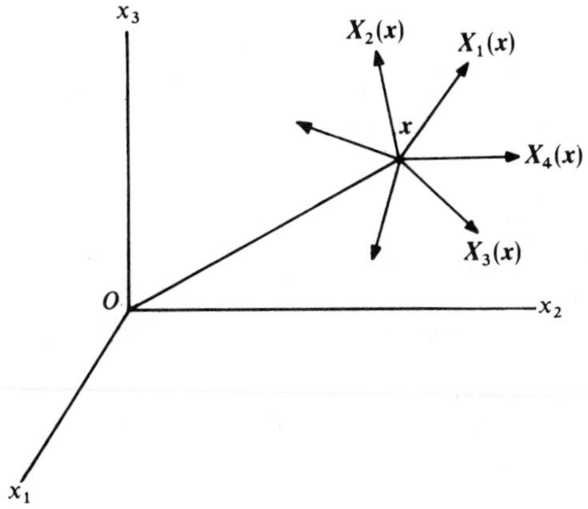

Fig. 11.2 Tangent space to $\mathbb{R}^3$ at x

illustrated in Fig. 11.2 for $n = 3$. Obviously the tangent space $T_x(\mathbb{R}^n)$ is a vector space, since the sums and scalar multiples of vectors localized at x are also vectors localized at x. [2] The union

$$\bigcup_{x \in \mathbb{R}^n} T_x(\mathbb{R}^n)$$

is the union of the tangent spaces to all points $x \in \mathbb{R}^n$, called the *tangent bundle* to $\mathbb{R}^n$ and denoted by $T(\mathbb{R}^n)$. [3]

It now follows that (11.1) may be considered as defining a map $X: \mathbb{R}^n \to T(\mathbb{R}^n)$ which assigns to each x belonging to $\mathbb{R}^n$ a vector $X(x)$ in the tangent space $T_x(\mathbb{R}^n) \subseteq T(\mathbb{R}^n)$. Such a map is called a *vector field* on $\mathbb{R}^n$.

Example 11.1

Given a function $f: \mathbb{R}^3 \to \mathbb{R}$ defined by $(x, y, z) \mapsto f(x, y, z)$, the gradient vector [4]

$$\nabla f = \frac{\partial f}{\partial x} \boldsymbol{i}_1 + \frac{\partial f}{\partial y} \boldsymbol{i}_2 + \frac{\partial f}{\partial z} \boldsymbol{i}_3$$

defines a vector field on $\mathbb{R}^3$. Any point $\boldsymbol{p} \in \mathbb{R}^3$ is mapped to the vector $\nabla f \in \mathrm{T}_{\boldsymbol{p}}(\mathbb{R}^3)$ and this vector is in the direction of the normal to the surface $f(x, y, z) = C$ that passes through the point $\boldsymbol{p}$ (Figs 11.3 and 11.4). [5]

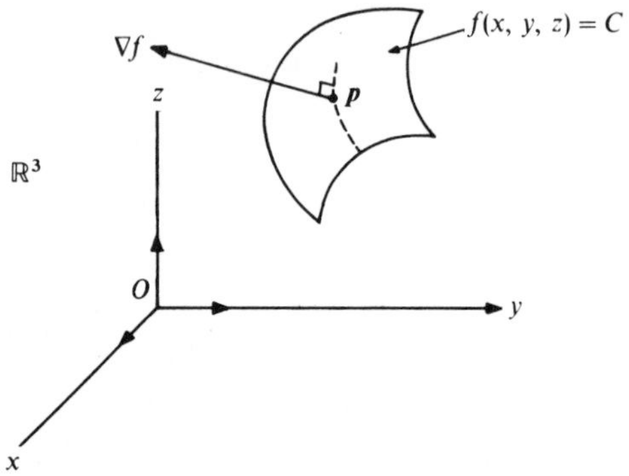

Fig. 11.3 Vector field on $\mathbb{R}^3$

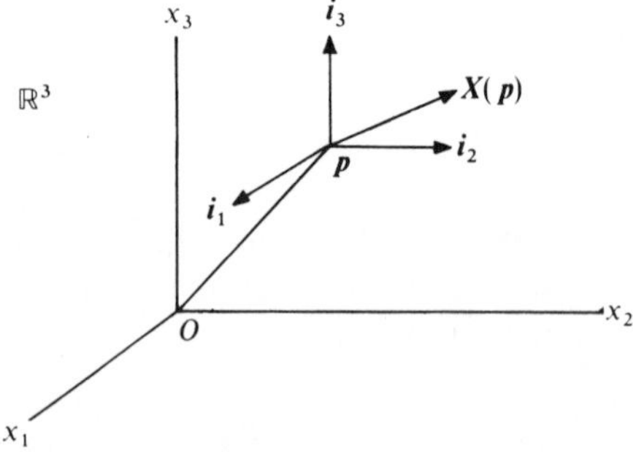

Fig. 11.4 Canonical basis for $\mathrm{T}_p(\mathbb{R}^3)$

The natural projection map $\pi: \mathrm{T}(\mathbb{R}^n) \to \mathbb{R}^n$ associates any vector in the tangent bundle $\mathrm{T}(\mathbb{R}^n)$ with a particular point in $\mathbb{R}^n$ to which it is attached. Then the vector field $\boldsymbol{X}: \mathbb{R}^n \to \mathrm{T}(\mathbb{R}^n)$ is such that the composite map $\pi \circ \boldsymbol{X}$ is simply the identity map $\mathrm{id}_{\mathbb{R}^n}$ on $\mathbb{R}^n$.

As the state $\boldsymbol{x}$ of the dynamical system (11.1) moves about in $\mathbb{R}^n$, its value $\boldsymbol{x}(t)$ at each time t will be associated with a particular tangent space $\mathrm{T}_{\boldsymbol{x}(t)}(\mathbb{R}^n)$. Therefore, the solution φ of (11.1) may be associated with the ordered pair $(\varphi(t), \dot{\varphi}(t))$, belonging to $\mathbb{R}^n \times \mathbb{R}^n$. With this interpretation, the tangent bundle can be defined as

$$\bigcup_{\boldsymbol{x} \in \mathbb{R}^n} \mathrm{T}_{\boldsymbol{x}}(\mathbb{R}^n) \triangleq \mathrm{T}_{\mathbb{R}^n}(\mathbb{R}^n) = \mathbb{R}^n \times \mathbb{R}^n.$$ [6]

For any $\boldsymbol{p} \in \mathbb{R}^n$, consider a vector $\boldsymbol{X}(\boldsymbol{p}) \in \mathrm{T}_{\boldsymbol{p}}(\mathbb{R}^n)$ and let unit vectors $\boldsymbol{i}_1, \ldots, \boldsymbol{i}_n$ constitute the canonical basis in the tangent space (Fig. 11.4, $n = 3$). Suppose that

$$\boldsymbol{X}(\boldsymbol{p}) = \alpha_1 \boldsymbol{i}_1 + \cdots + \alpha_n \boldsymbol{i}_n. \tag{11.2}$$

Let $\mathrm{F}_{\boldsymbol{p}}(\mathbb{R}^n)$ be the set of all real-valued C^∞ functions whose domains are open subsets of $\mathbb{R}^n$ containing the point $\boldsymbol{p}$. No distinction is made between functions that are equal on some neighbourhood of $\boldsymbol{p}$, since only derivatives of these functions at $\boldsymbol{p}$ are needed. For any $f \in \mathrm{F}_{\boldsymbol{p}}(\mathbb{R}^n)$, the directional derivative $X_{\boldsymbol{p}} f$ of f at the point $\boldsymbol{p}$ in the direction of $\boldsymbol{X}(\boldsymbol{p})$ is defined as

$$\begin{aligned} X_{\boldsymbol{p}} f &\triangleq \alpha_1 \left(\frac{\partial f}{\partial x_1}\right)_{\boldsymbol{p}} + \cdots + \alpha_n \left(\frac{\partial f}{\partial x_n}\right)_{\boldsymbol{p}} \\ &= \left(\sum_{i=1}^{n} \alpha_i \frac{\partial}{\partial x_i}\right) f \in \mathbb{R} \end{aligned} \tag{11.3}$$

in which the partial derivatives are all evaluated at the point $\boldsymbol{p}$ (Boothby 1975). Thus, the directional derivative $X_{\boldsymbol{p}} f$ is the scalar product $\boldsymbol{X}(\boldsymbol{p}) \cdot (\nabla f)_{\boldsymbol{p}}$.[7]

Example 11.2

$$\begin{bmatrix} \dot{x}_1 \\ \dot{x}_2 \\ \dot{x}_3 \end{bmatrix} = \boldsymbol{X}(\boldsymbol{x}) = \begin{bmatrix} 2x_3 - x_1^2 \\ x_1^3 - x_3^2 \\ x_2^4 \end{bmatrix}.$$

If $\boldsymbol{p} = (2, 0, 3)$, then $\boldsymbol{X}(\boldsymbol{p}) = (2, -1, 0)$, i.e. $\boldsymbol{X}(\boldsymbol{p})$ is the vector $2\boldsymbol{i}_1 - \boldsymbol{i}_2$ in $T_{\boldsymbol{p}}(\mathbb{R}^3)$. Consider the C^∞ function f defined by $f(x_1, x_2, x_3) = x_1^2 + x_2^2 + x_3^2$. Then $\nabla f = (2x_1, 2x_2, 2x_3)$ which, at the given point $\boldsymbol{p}$, is $(\nabla f)_{\boldsymbol{p}} = (4, 0, 6)$. The directional derivative of f at the point $(2, 0, 3)$ in the direction $2\boldsymbol{i}_1 - \boldsymbol{i}_2$ is

$$X_{\boldsymbol{p}} f = \boldsymbol{X}(\boldsymbol{p}) \cdot (\nabla f)_{\boldsymbol{p}} = 8.$$

From (11.3) it will be apparent that X_p can be thought of as a function from $F_p(\mathbb{R}^n)$ to $\mathbb{R}$ and expressed as

$$\sum_{i=1}^{n} \alpha_i \frac{\partial}{\partial x_i}, \tag{11.4}$$

but remember that the derivatives must be evaluated at $\boldsymbol{p}$. The functions $X_p: F_p(\mathbb{R}^n) \to \mathbb{R}$ are called *derivations* on $F_p(\mathbb{R}^n)$ into $\mathbb{R}$. The set L of functions X_p for a given $\boldsymbol{p} \in \mathbb{R}^n$ is actually a vector space over $\mathbb{R}$ because any two functions $X_p, Y_p \in L$ can be added together and multiplied by real numbers under the definitions

$$(X_p + Y_p)f \triangleq X_p f + Y_p f, \qquad (\alpha X_p)f \triangleq \alpha(X_p f). \tag{11.5}$$

Furthermore, [8] elements of L are linear functions because

$$X_p(\alpha f + \beta g) = \alpha(X_p f) + \beta(X_p g) \tag{11.6}$$

for all $X_p \in L$, all $f, g \in F_p(\mathbb{R}^n)$, and all $\alpha, \beta \in \mathbb{R}$. Any $X_p \in L$ also satisfies the *Leibniz rule* for the product of two C^∞ functions f and g:

$$X_p(fg) = (X_p f)g(\boldsymbol{p}) + f(\boldsymbol{p})(X_p g). \tag{11.7}$$

It should be clear to the reader that to every vector $X(\boldsymbol{p}) \in T_p(\mathbb{R}^n)$ is associated a function $X_p \in L$, and it turns out that the vector space L is isomorphic to the set of tangent vectors at the point $\boldsymbol{p}$. Thus, the tangent space $T_p(\mathbb{R}^n)$ can be considered either as a set of tangent vectors $\boldsymbol{X}(\boldsymbol{p})$ or as a set of derivations X_p of $F_p(\mathbb{R}^n)$ into $\mathbb{R}$. In particular, the canonical basis vectors $\boldsymbol{i}_1, \ldots, \boldsymbol{i}_n$ in (11.2) are associated with $\partial/\partial x_1, \ldots, \partial/\partial x_n$ in (11.4).

The set $F_p(\mathbb{R}^n)$ is a ring with identity (see Example 5.2) with $f: \mathbb{R}^n \to \mathbb{R}$ and with the identity element the function 1 for which $1(\boldsymbol{x}) = 1$ for all $\boldsymbol{x} \in \mathbb{R}^n$. The set $V(\mathbb{R}^n)$ of vector fields on $\mathbb{R}^n$ is an additive abelian group (Exercise 11.2). Furthermore, an element $\boldsymbol{X} \in V(\mathbb{R}^n)$ can be multiplied by a function $f \in F_p(\mathbb{R}^n)$ under the definition

$$(f\boldsymbol{X})_p \triangleq f(\boldsymbol{p})\boldsymbol{X}(\boldsymbol{p}) \quad (\boldsymbol{p} \in \mathbb{R}^n).$$

Thus, $V(\mathbb{R}^n)$ forms a module over the ring $F_p(\mathbb{R}^n)$ with module mapping $\varphi_p: F_p(\mathbb{R}^n) \times V(\mathbb{R}^n) \to V(\mathbb{R}^n)$ defined by $\varphi_p(f, \boldsymbol{X}) = f(\boldsymbol{p})\boldsymbol{X}(\boldsymbol{p})$ (Fig. 11.5). This is quite analogous to the way $\mathbb{R}^m[z]$ forms a module over the ring of polynomials $\mathbb{R}[z]$, illustrated in Fig. 6.6.

There is yet another way of describing a tangent space. Suppose a mapping $f: A \to \mathbb{R}^m$ over a domain $A \subseteq \mathbb{R}^n$ is differentiable at a point $\boldsymbol{p} \in A$. Let $\varphi: I \to A$ be a smooth (i.e. C^∞) path defined on an open interval $I \subseteq \mathbb{R}$ which contains zero and such that $\varphi(0) = \boldsymbol{p}$. Consider the set of all paths with images that pass through $\boldsymbol{p}$. As described in Example 2.15, an equivalence relation can be set up to give equivalence classes $[\varphi(\cdot)]$. To each class $[\varphi(\cdot)]$ there corresponds a value $\dot{\varphi}(0)$. Such equivalence classes are called *tangency*

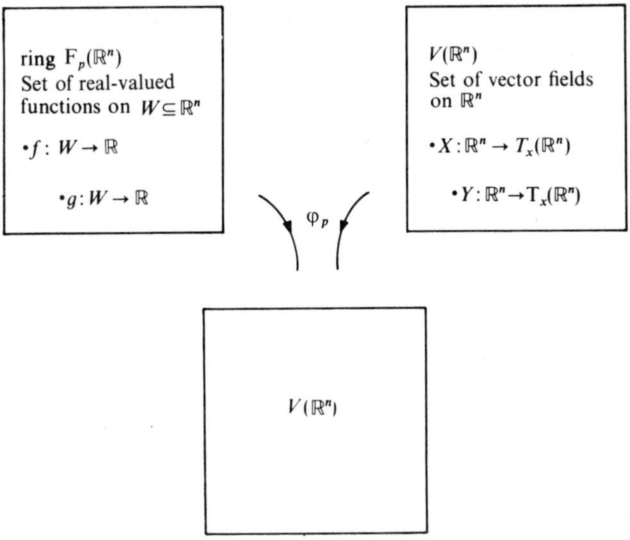

Fig. 11.5 $F_p(\mathbb{R}^n)$-module, $V(\mathbb{R}^n)$

classes. Then the tangent space $T_p(A)$ to A at the point $\boldsymbol{p}$ can be represented as the set of all tangency classes of smooth paths with images through $\boldsymbol{p}$. Now consider the map $\boldsymbol{\alpha}(\boldsymbol{p})\colon T_p(A) \to \mathbb{R}^n$ defined by $[\varphi(\cdot)] \mapsto \dot{\varphi}(0)$. This is an isomorphism (Exercise 3.7), so that $T_p(A)$ is isomorphic to $\mathbb{R}^n$. Similarly, $T_{f(p)}(B)$ with $B \subseteq \mathbb{R}^m$ is isomorphic to $\mathbb{R}^m$.

The composite map $\boldsymbol{f} \circ \varphi\colon I \to B$ is a smooth path with image through $\boldsymbol{f}(\boldsymbol{p})$ (Fig. 11.6). By the chain rule,

$$\mathsf{D}(\boldsymbol{f} \circ \boldsymbol{\varphi})(0) = \mathsf{D}\boldsymbol{f}\big(\boldsymbol{\varphi}(0)\big) \circ \mathsf{D}\boldsymbol{\varphi}(0)$$

i.e.

$$\mathsf{D}(\boldsymbol{f} \circ \boldsymbol{\varphi})(0) = \mathsf{D}\boldsymbol{f}(\boldsymbol{p}) \circ \dot{\boldsymbol{\varphi}}(0).$$

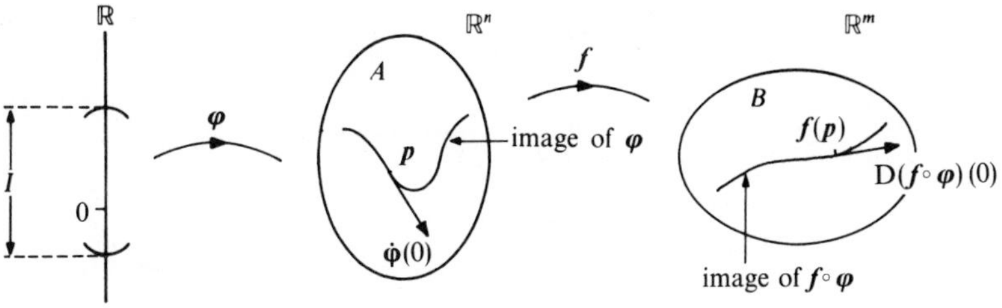

Fig. 11.6 Differentiation of a composite map

In other words, the derivative $\mathsf{D}f(p)$ takes the tangent to the image of the path φ at p to the tangent to the image of the path $f \circ \varphi$ at $f(p)$ (Fig. 11.6 and Fig. 10.16). Thus, the linear map $\mathsf{D}f(p)$: $A \to B$ can now be represented as the linear map

$$\mathrm{T}_p f \colon \mathrm{T}_p(A) \to \mathrm{T}_{f(p)}(B)$$

defined by $[\varphi(\cdot)] \mapsto [f \circ \varphi]$.

11.2 Manifolds

Introductory texts on dynamical systems (e.g. Hirsch and Smale 1974) usually assume that the state space for the system is a vector space. But there are many problems for which this assumption cannot be made (Example 11.4 below is a case in point). However, when the state space is not a vector space it may still be a space known as a *manifold*. In this case the system equations can be treated analytically but on a manifold instead of on a vector space. Examples of manifolds include hypersurfaces in $\mathbb{R}^n$, the simplest cases of curves in $\mathbb{R}^2$ and $\mathbb{R}^3$ and surfaces in $\mathbb{R}^3$ having been studied originally in classical differential geometry (Vaisman 1984). For example, the surface of the unit sphere, with centre the origin and equation

$$x_1^2 + x_2^2 + x_3^2 = 1,$$

is a two-dimensional manifold in $\mathbb{R}^3$ (Fig. 11.7), usually denoted by S^2. [9] Another example is the torus T^2 (Fig. 11.8). $\mathbb{R}^3$ itself is actually a manifold (and also a vector space of course) so that strictly speaking S^2 is a two-dimensional *submanifold* of $\mathbb{R}^3$ (or the Euclidean space E^3 in the sense of Euclidean geometry).

Manifolds arise quite naturally in dynamical systems. Sometimes solutions of (11.1) lie completely on one of the simple manifolds mentioned above. In other cases, as mentioned earlier, the state of a system is such that the state

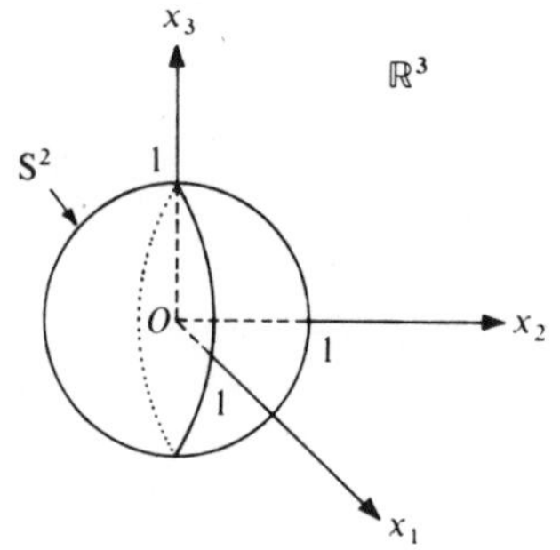

Fig. 11.7 Manifold S^2

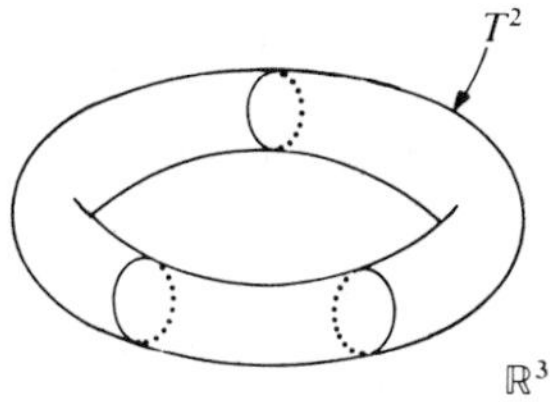

Fig. 11.8 Manifold T^2

space is not a vector space but is a manifold. Such possibilities are illustrated in the next two examples.

Example 11.3

Consider the system

$$\dot{x}_1 = 0, \qquad x_1(0) = 1/\sqrt{3}, \tag{11.8}$$

$$\dot{x}_2 = x_3, \qquad x_2(0) = 1/\sqrt{3}, \tag{11.9}$$

$$\dot{x}_3 = -x_2, \qquad x_3(0) = 1/\sqrt{3}, \tag{11.10}$$

From (11.8), $x_1(t) = 1/\sqrt{3}$ for all $t \in [0, \infty)$. Equations (11.9) and (11.10) imply that

$$\frac{dx_2}{dx_3} = -\frac{x_3}{x_2},$$

and integration of this last equation yields the solution

$$x_2^2 + x_3^2 = \text{constant}.$$

Since $x_2(0) = 1/\sqrt{3} = x_3(0)$, the constant has the value $\frac{2}{3}$. Thus, at any time t, the state vector $(x_1(t), x_2(t), x_3(t))$ of the system satisfies the equations

$$x_1 = 1/\sqrt{3}, \qquad x_2^2 + x_3^2 = \tfrac{2}{3},$$

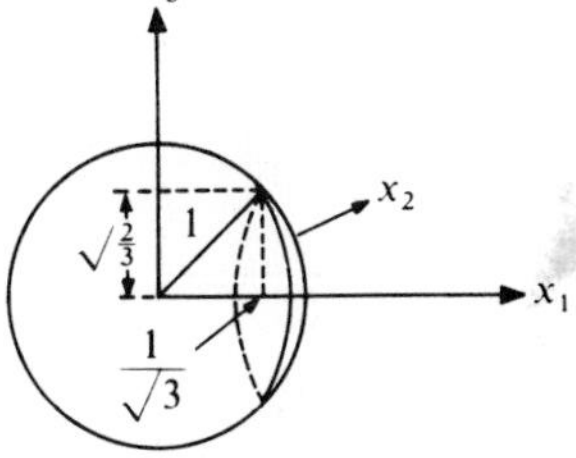

Fig. 11.9 Trajectory of Example 11.3

so that $x_1^2 + x_2^2 + x_3^2 = 1$ for all $t \in [0, \infty)$. Thus, the trajectory representing the time behaviour of the system lies wholly on the manifold S^2, in fact, on the circle $x_2^2 + x_3^2 = \frac{2}{3}$ lying in the plane $x_1 = 1/\sqrt{3}$, with centre at $(1/\sqrt{3}, 0, 0)$ and radius $\sqrt{(\frac{2}{3})}$ (Fig. 11.9).

Example 11.4 (Brockett 1972)

The output-feedback problem, discussed in Section 2.2 for a linear, time-invariant system

$$\dot{\boldsymbol{x}} = \boldsymbol{Ax} + \boldsymbol{Bu}, \qquad \boldsymbol{x}(0) = \boldsymbol{x}_0,$$

$$\boldsymbol{y} = \boldsymbol{Cx},$$

is to find a matrix $\boldsymbol{F}$ that minimizes an objective function $\boldsymbol{x}_0^{\mathrm{T}}\boldsymbol{K}(\boldsymbol{F})\boldsymbol{x}_0$ subject to the system equation $\dot{\boldsymbol{x}} = (\boldsymbol{A} + \boldsymbol{BFC})\boldsymbol{x}$. Since the objective function depends on the initial state $\boldsymbol{x}_0$ as well as $\boldsymbol{F}$, choosing an optimal $\boldsymbol{F}$ for a given $\boldsymbol{x}_0$ will not be satisfactory for any other initial condition. When the initial state is not known, then it is better to pick a collection of initial state vectors and to pick $\boldsymbol{F}$ to minimize a weighted average of the individual performances. If exactly n initial states $\boldsymbol{x}_1(0), \ldots, \boldsymbol{x}_n(0) \in \mathbb{R}^n$ are chosen, then $\boldsymbol{F}$ should be considered as controlling the solution of the matrix equation

$$\dot{\boldsymbol{\Phi}}(t) = (\boldsymbol{A} + \boldsymbol{BFC})\boldsymbol{\Phi}(t), \qquad \boldsymbol{\Phi}(0) = [\boldsymbol{x}_1(0), \ldots, \boldsymbol{x}_n(0)].$$

The state space is therefore $G\ell(n, \mathbb{R})$ which is a group (Exercise 4.3) and also a manifold, but not a vector space.

The manifolds referred to already are actually topological manifolds, being topological spaces which locally look like $\mathbb{R}^n$. That is to say, for any point p belonging to a manifold M, there is an open neighbourhood U of p in M and a homeomorphism $\varphi\colon U \to U'$ which maps U onto an open neighbourhood U' of the point $\varphi(p) \in \mathbb{R}^n$ (Fig. 11.10). M is said to be *locally* homeomorphic to $\mathbb{R}^n$ and to be of dimension n. [10] Formally, an n-dimensional manifold M is a

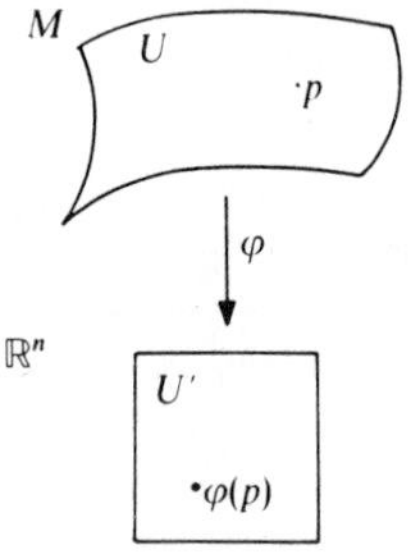

Fig. 11.10 Homeomorphism $\phi\colon U \to U'$

Hausdorff topological space which is locally homeomorphic to $\mathbb{R}^n$ and has a countable basis of open sets for the topology.

The open neighbourhood U with an associated homeomorphism φ together form a pair (U, φ) which is called a *chart*. If the dimension of M is finite, then any $p \in U$ gives rise to a corresponding image $\varphi(p) \in \mathbb{R}^n$, and n coordinates $x_1(p), \ldots, x_n(p)$ [11] may be assigned to $\varphi(p)$, which then leads to coordinates in U through the homeomorphism φ. This is why charts are sometimes called *local coordinate charts*. A collection of charts $\{(U_i, \varphi_i): i \in \mathscr{I}\}$ such that

$$\bigcup_{i \in \mathscr{I}} U_i = M$$

is called an *atlas* for M.

Example 11.5

$M = S^1$ where S^1 is the unit circle in $\mathbb{R}^2$ given by

$$\{(x_1, x_2) \in \mathbb{R}^2 : x_1^2 + x_2^2 = 1\}.$$

A possible choice of charts is (U_1, φ_1) and (U_{-1}, φ_{-1}) where the two open sets U_1 and U_{-1} are

$$U_1 = \{(x_1, x_2) \in \mathbb{R}^2 : x_1^2 + x_2^2 = 1, x_2 \neq 1\},$$
$$U_{-1} = \{(x_1, x_2) \in \mathbb{R}^2 : x_1^2 + x_2^2 = 1, x_2 \neq -1\},$$

and the homeomorphisms φ_1 and φ_{-1} (see Example 10.11 and Exercise 10.5) are

$$\varphi_1 : U_1 \to \mathbb{R} : (x_1, x_2) \mapsto x_1/(1 - x_2), \qquad \varphi_{-1} : U_{-1} \to \mathbb{R} : (x_1, x_2) \mapsto x_1/(1 + x_2).$$

$\{(U_1, \varphi_1), (U_{-1}, \varphi_{-1})\}$ is an atlas for S^1 with $U_1 \cup U_{-1} = S^1$.

If the state x of a dynamical system $\dot{x} = X(x)$ belongs to a manifold such as $G\ell(n, \mathbb{R})$, then X is a mapping from the manifold to some (tangent) space containing $X(x)$. Therefore, it is necessary to define the tangent space to some point on a manifold and also to discuss the question of differentiable mappings between manifolds. The latter topic will be dealt with first.

Continuity of a map from one manifold M to another one N is no problem, since manifolds are topological spaces, and continuous maps between these types of spaces have already been investigated in Section 10.1. Differentiable maps have also been discussed in Section 10.3, but only mappings between normed vector spaces, $\mathbb{R}^n \to \mathbb{R}^m$ in particular. To extend the ideas of differentiation to maps between manifolds, local coordinate charts are used to obtain representative maps between normed vector spaces (usually Euclidean spaces). Since manifolds are locally like $\mathbb{R}^n$, the calculus on manifolds (Spivak 1965) can actually be carried out in the normed vector spaces just as in Chapter 10. The approach is now described more formally.

Suppose $f: M \to N$ is a continuous map between two manifolds M and N. Let (U, φ) be a chart for M, as described earlier, with $p \in U$ and $\varphi: U \to U'$, where U' is an open neighbourhood of $\varphi(p) \in \mathbb{R}^n$. Similarly, let (V, ψ) be a chart for N with $f(p) \in V$ and $\psi: V \to V'$, where V' is an open neighbourhood of $\psi(f(p)) \in \mathbb{R}^m$ (Fig. 11.11). The composite map $\psi \circ f \circ \varphi^{-1}: U' \to \mathbb{R}^m$ is a continuous map, and locally (in U and V) this map can be used to represent the mapping f. Differentiation of f can then be formulated through a map $\mathbb{R}^n \to \mathbb{R}^m$ for which the machinery is already available in Chapter 10. The map f from M to N is called a *smooth* (or C^∞) mapping when the composite map $\psi \circ f \circ \varphi^{-1}$ is of class C^∞ as defined for mappings between normed vector spaces in Section 10.3.

The fact that there are many choices for the charts of a manifold M does raise a difficulty with regard to differentiation of a mapping from M to some other manifold N. Consider two charts (U_1, φ_1) and (U_2, φ_2) for M, and suppose that the point $p \in M$ belongs to both U_1 and U_2. Then, for a given ψ in a chart (V, ψ) for N, it might be that $\psi \circ f \circ \varphi_1^{-1}$ is differentiable at $\varphi_1(p)$ and yet $\psi \circ f \circ \varphi_2^{-1}$ not differentiable at $\varphi_2(p)$ (Fig. 11.12). The difficulty is resolved by obtaining an overlap mapping v through the commutative diagram shown in Fig. 11.13. The mapping $v = \varphi_2 \circ \varphi_1^{-1}$ is a homeomorphism between open subsets of $\mathbb{R}^n$, namely $\varphi_1(U_{12}) \subseteq U_1' \subseteq \mathbb{R}^n$ and $\varphi_2(U_{12}) \subseteq U_2' \subseteq \mathbb{R}^n$. The map

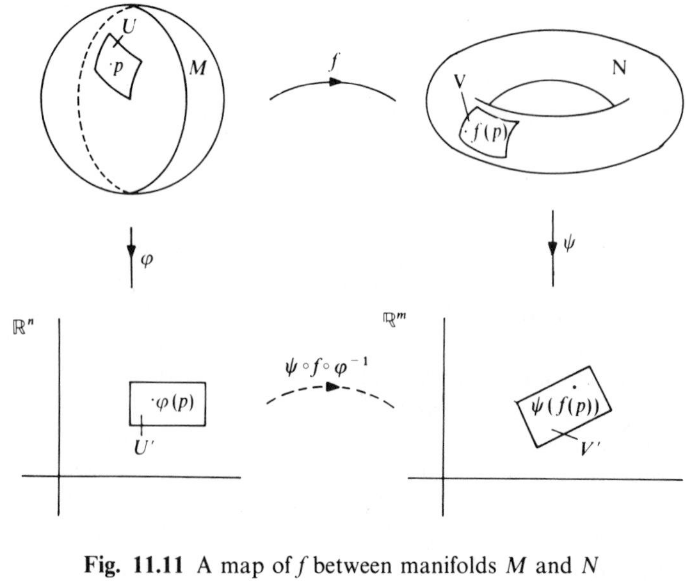

Fig. 11.11 A map of f between manifolds M and N

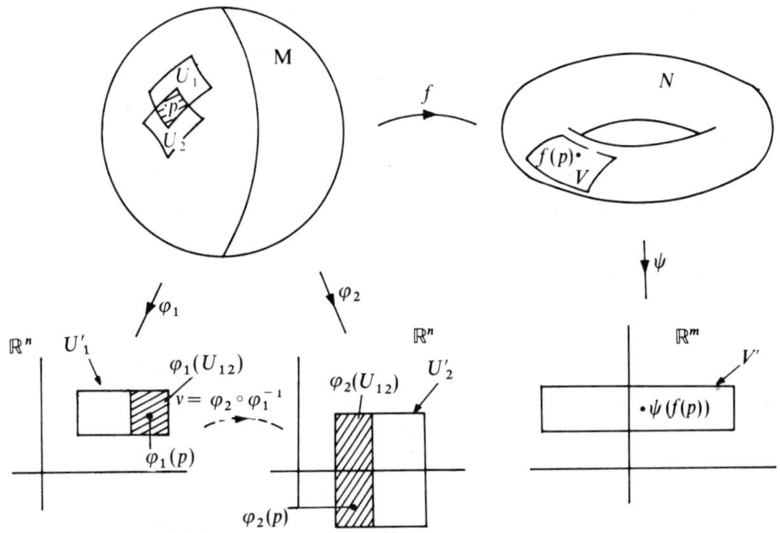

Fig. 11.12 Choice of charts in mapping $f: M \to N$

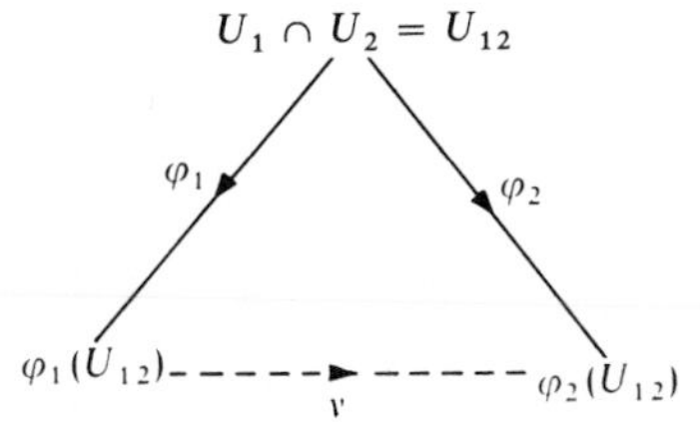

Fig. 11.13 Overlap mapping $v = \varphi_2 \circ \varphi_1^{-1}$

$\psi \circ f \circ \varphi_2^{-1}: \varphi_2(U_{12}) \to \mathbb{R}^m$ can now be represented by the composition of the two maps

$$v^{-1}: \varphi_2(U_{12}) \to \varphi_1(U_{12}) \text{ and } \psi \circ f \circ \varphi_1^{-1}: \varphi_1(U_{12}) \to \mathbb{R}^m,$$

i.e. $\psi \circ f \circ \varphi_2^{-1} = (\psi \circ f \circ \varphi_1^{-1}) \circ v^{-1}$ where $v^{-1} = (\varphi_2 \circ \varphi_1^{-1})^{-1} = \varphi_1 \circ \varphi_2^{-1}$.

The chain rule states that the derivative of a composition is the composition of the derivatives. Thus, if every overlap map v within the given atlas is of class C^∞, then $\psi \circ f \circ \varphi_1^{-1}$ is C^∞ whenever $\psi \circ f \circ \varphi_2^{-1}$ is C^∞, and the original difficulty is resolved.

There is a further difficulty when a given set of coordinate charts on a manifold is being used to investigate the time behaviour of a dynamical system. For one reason or another, a change of coordinates is often desired and, in order for such transformations to be C^∞, the maps v and v^{-1} defined above must be diffeomorphisms of the open subsets $\varphi_1(U_{12})$ and $\varphi_2(U_{12})$ of $\mathbb{R}^n$. In this case, the charts (U_1, φ_1) and (U_2, φ_2) are said to be C^∞-*compatible*. Consider a family of coordinate charts (U_i, φ_i) for which the U_i's cover a topological manifold M and such that (i) for any i and j, the charts (U_i, φ_i) and (U_j, φ_j) are C^∞-compatible and (ii) any coordinate chart (V, ψ) compatible with all members of the family (U_i, φ_i) is also a member. Then such a family is called a C^∞ *structure*. A topological manifold with a C^∞ structure is called a C^∞ (or smooth) manifold.

Return now to the set of differential equations

$$\dot{x} = X(x), \tag{11.11}$$

where $x(t)$ now evolves on a general manifold M rather than on a vector space. It is possible to generalize the concept of a tangent space to a manifold M at a point $p \in M$. However, the important manifolds that arise in dynamical system theory are usually those, such as $G\ell(n, \mathbb{R})$ and $So(n, \mathbb{R})$, that are isomorphic to hypersurfaces in $\mathbb{R}^n$ for some n (e.g. S^2 shown in Fig. 11.14), and so the previous description also covers these cases. Therefore, in the remaining part of this section, we only give a very brief indication of this generalization to more abstract manifolds with more general tangent spaces.

The tangent space to M at p, denoted by $T_p(M)$, is the set of all functions $X_p\colon F_p(M) \to \mathbb{R}$ of the form

$$X_p f = \sum_{i=1}^{n} \alpha_i \left(\frac{\partial f}{\partial x_i}\right)_p,$$

where now $F_p(M)$ is the set of all real-valued C^∞ functions defined on domains that are open subsets of M which contain the point p. These derivations X_p again satisfy the vector space operations (11.5), linearity (11.6), and the Leibniz rule (11.7). A tangent vector to M at p is any $X_p \in T_p(M)$. The tangent bundle $T(M)$ to M is the union of all the tangent spaces $T_p(M)$ for all $p \in M$. The map X in (11.11) can now be thought of again as a vector field defined as $X\colon M \to T(M)$. It satisfies $\pi \circ X = \mathrm{id}_M$, where π is the natural projection $T(M) \to M$, so that $(\pi \circ X)(x) = x$ (Fig. 11.15).

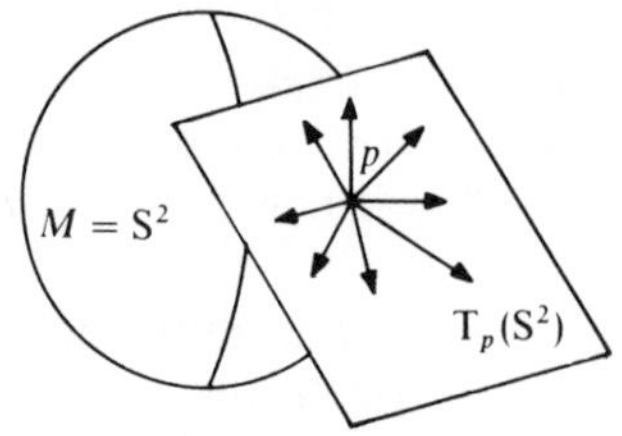

Fig. 11.14 Tangent space to S^2 at p

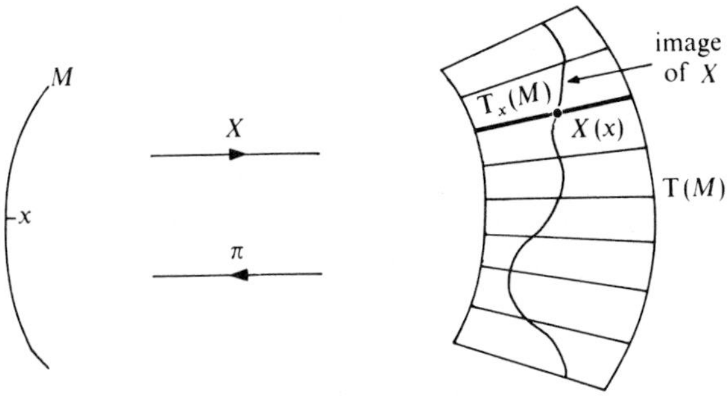

Fig. 11.15 Natural projection map $\pi: \mathrm{T}(M) \to M$

11.3 Lie groups

Let G be a group that is also a differentiable manifold. Then G is a *Lie group*, provided that the mapping $G \times G \to G$ defined by $(x, y) \mapsto xy$ and the mapping $G \to G$ given by $x \mapsto x^{-1}$ are both C^∞ mappings.

Example 11.6

$G\ell(n, \mathbb{R})$ is a group, and both the maps $(\boldsymbol{A}, \boldsymbol{B}) \mapsto \boldsymbol{AB}$ and $\boldsymbol{A} \mapsto \boldsymbol{A}^{-1}$ are smooth. The product $\boldsymbol{AB}$ has entries that are polynomials in the entries of $\boldsymbol{A}$ and $\boldsymbol{B}$, and these entries are exactly the expressions in local coordinates of the product map, which is thus C^∞. The inverse of $\boldsymbol{A} = [a_{ij}]$ may be written as

$$\boldsymbol{A}^{-1} = \frac{1}{\det A} [A_{ij}]^{\mathrm{T}},$$

where A_{ij} are the cofactors of $\boldsymbol{A}$ (and so are polynomials in the entries of $\boldsymbol{A}$) and where $\det \boldsymbol{A}$ is a polynomial in these entries that does not vanish on $G\ell(n, \mathbb{R})$. It follows that the entries of $\boldsymbol{A}^{-1}$ are rational functions on $G\ell(n, \mathbb{R})$ with nonvanishing denominators and hence they are C^∞. Therefore $G\ell(n, \mathbb{R})$ is a Lie group.

Consider again a dynamical system of the form (11.11), with initial condition $x(0) = x_0$, which evolves on a manifold M. As usual, the solution of equation (11.11) is represented by a path φ through x_0 (written φ_{x_0}) defined by $t \mapsto \varphi_{x_0}(t)$ and where $\varphi_{x_0}(0) = x_0$. If M is compact, then it turns out that φ_{x_0} is the global solution for the vector field X through x_0. The path φ_{x_0} is a map $\mathbb{R} \to M$ but, if $\varphi_{x_0}(t)$ is written in the form $\varphi(t, x_0)$, then this implies a map $\varphi: \mathbb{R} \times M \to M$, called a *flow* on M or an *action* of $\mathbb{R}$ on M. The images of the solution curves of the vector field X are called the *orbits* of this action.

Again, $\varphi(t, x_0)$ can be expressed as $\varphi_t(x_0)$, in which case another map $\varphi_t: M \to M$ is defined which satisfies

$$\varphi_t \circ \varphi_s = \varphi_{t+s} \quad \forall\ s,t \in \mathbb{R}, \qquad \varphi_0 = \mathrm{id}_M, \tag{11.12}$$

the last equation resulting from the fact that $\varphi_0(x_0) = \varphi_{x_0}(0) = x_0$. This map φ_t is the map discussed at some length in Chapter 4 following (4.16). Also,

$$\varphi_t \circ \varphi_{-t} = \varphi_{-t} \circ \varphi_t = \varphi_0 = \mathrm{id}_M,$$

so that φ_{-t} is the inverse of φ_t. Hence, φ_t is a diffeomorphism from M to M.

Suppose that $\mathrm{D}(M)$ is the set of all C^∞-diffeomorphisms $M \to M$. Composition of two diffeomorphisms is a diffeomorphism and, under this operation of composition, $\mathrm{D}(M)$ is a group. Then (11.12) shows that the map $\mathbb{R} \to \mathrm{D}(M)$ defined by $t \mapsto \varphi_t$ is a group homomorphism.

In the above discussion, the time parameter t belongs to $\mathbb{R}$ which is an additive Lie group acting on the manifold M. This idea of a group (not necessarily a Lie group) acting on a manifold can be generalized. Suppose S is a set and G a group. Then G is said to act on S [12] if there exists $\theta: G \times S \to S$ such that

(i) if e is the identity in G, then $\theta(e, s) = s$ for all $s \in S$,

(ii) if $g_1, g_2 \in G$, then $\theta(g_1 g_2, s) = \theta(g_1, \theta(g_2, s))$ for all $s \in S$.

When G is a Lie group and S is a C^∞ manifold M, and θ is C^∞, the above set-up defines a C^∞ action. As with the path φ discussed earlier, for a fixed $g \in G$, we can write $\theta(g, x)$ as $\theta_g(x) \in M$, so that $\theta_g: M \to M$ is a C^∞-diffeomorphism belonging to $\mathrm{D}(M)$. Then a C^∞ action of G on M can be defined alternatively as a map $\alpha: G \to \mathrm{D}(M)$ in which $\theta_{g_2} \circ \theta_{g_1} = \theta_{g_1 g_2}$ for all $g_1, g_2 \in G$. Here θ_g is an alternative way of writing $\alpha(g)$. For a given $x \in M$, the set of points $\theta_g(x)$ in M, as g runs through the whole of G, is called the *orbit of x under the action of* α. A point $x \in M$ is called a *fixed point* of G if $Gx = x$. [13] If $Gx = M$ for some $x \in M$, then G is said to be *transitive* on M in which case $Gx = M$ for all $x \in M$.

Example 11.7

The *natural action* of $\mathrm{G}\ell(n, \mathbb{R})$ on $\mathbb{R}^n$ is $\theta: G \times \mathbb{R}^n \to \mathbb{R}^n$ defined by $\theta(\boldsymbol{A}, \boldsymbol{x}) = \boldsymbol{A}\boldsymbol{x}$. Here $M = \mathbb{R}^n$ and G is the multiplicative group $\mathrm{G}\ell(n, \mathbb{R})$. The identity in G is the unit matrix $\mathbf{I}_n$ and $\theta(\mathbf{I}_n, \boldsymbol{x}) = \mathbf{I}_n \boldsymbol{x} = \boldsymbol{x}$ for all $\boldsymbol{x} \in \mathbb{R}^n$; this is (i) above. If $\boldsymbol{A}, \boldsymbol{B} \in G$, then $\theta(\boldsymbol{AB}, \boldsymbol{x}) = (\boldsymbol{AB})\boldsymbol{x}$ and $\theta(\boldsymbol{A}, \theta(\boldsymbol{B}, \boldsymbol{x})) = \theta(\boldsymbol{A}, \boldsymbol{Bx}) = \boldsymbol{A}(\boldsymbol{Bx})$ for all $\boldsymbol{x} \in \mathbb{R}^n$. So (ii) above is equivalent to $(\boldsymbol{AB})\boldsymbol{x} = \boldsymbol{A}(\boldsymbol{Bx})$, the associative law for matrix multiplication ($\boldsymbol{x}$ being an $n \times 1$ matrix).

If $\boldsymbol{x} = \mathbf{0} \in \mathbb{R}^n$ and $\boldsymbol{A} \in \mathrm{G}\ell(n, \mathbb{R})$, then $\boldsymbol{A}\mathbf{0} = \mathbf{0}$ so $\mathbf{0}$ is a fixed point of $\mathrm{G}\ell(n, \mathbb{R})$. If $\boldsymbol{e}_1$ is the first element of the canonical basis in $\mathbb{R}^n$, then the orbit of

$e_1 \in \mathbb{R}^n$ is the set $\{Ae_1 : A \in \mathrm{G}\ell(n, \mathbb{R})\}$, namely the first columns of all matrices in $\mathrm{G}\ell(n, \mathbb{R})$. Since a first column can be any non-null vector in $\mathbb{R}^n$, it follows that $\{Ae_1 : A \in \mathrm{G}\ell(n, \mathbb{R})\} = \mathbb{R}^n \setminus \{0\}$ and so $\mathrm{G}\ell(n, \mathbb{R})$ is transitive on $\mathbb{R}^n \setminus \{0\}$.

Points on a manifold M that belong to the same orbit define an equivalence relation on M. If $x_1, x_2 \in M$, then $x_1 \sim x_2$ when there exists a $g \in G$ such that $gx_1 = x_2$. The points on the orbit of x form the equivalence class $[x]$ and, from Section 2.2, these equivalence classes partition the manifold M into the orbits of the points of M.

11.4 The singular-control problem

The mapping $\boldsymbol{f}$ in (1.27) is a vector field and, when the state x belongs to a manifold, the topics discussed so far in this Chapter are of particular relevance. In fact, classical mechanics can be presented in terms of vector fields, manifolds, and tangent spaces (Arnold 1978; Abraham and Marsden 1978).

Consider the so-called linear–analytic system given by

$$\dot{\boldsymbol{x}} = \boldsymbol{f}(\boldsymbol{x}) + u\boldsymbol{g}(\boldsymbol{x}), \qquad u(t) \in U \subseteq \mathbb{R}, \quad \boldsymbol{x}(t) \in M \subseteq \mathbb{R}^n, \tag{11.13}$$

where $\boldsymbol{f}(\boldsymbol{x})$ is known as the drift term, u is a scalar control, and M is the state space. The vector fields $\boldsymbol{f}$ and $\boldsymbol{g}$ are both assumed C^∞; in fact f_i and g_i ($i = 1, ..., n$), the elements of $\boldsymbol{f}$ and $\mathbf{g}$ respectively, are assumed analytic functions. Suppose that an optimal-control problem of *Mayer* type [14] is formulated for system (11.13). The *Hamiltonian* (Athans and Falb 1966) for this problem is given by

$$H(\boldsymbol{x}, \lambda, u) = \boldsymbol{\lambda}^{\mathrm{T}}[\boldsymbol{f}(\boldsymbol{x}) + u\boldsymbol{g}(\boldsymbol{x})], \tag{11.14}$$

in which the adjoint vector $\boldsymbol{\lambda}(t) \in \mathbb{R}^n$ satisfies the differential equation

$$\dot{\boldsymbol{\lambda}}^{\mathrm{T}} = -H_{\boldsymbol{x}} = -\boldsymbol{\lambda}^{\mathrm{T}}(\boldsymbol{f}_{\boldsymbol{x}} + u\boldsymbol{g}_{\boldsymbol{x}}). \tag{11.15}$$

$\boldsymbol{f}_x$ and $\boldsymbol{g}_x$ are the Jacobian matrices [15] of $\boldsymbol{f}$ and $\boldsymbol{g}$ with respect to $\boldsymbol{x}$. Using (11.14), the system equation (11.13) can be written as [16]

$$\dot{\boldsymbol{x}} = H_{\lambda}^{\mathrm{T}}. \tag{11.16}$$

If the *switching function* $\boldsymbol{\lambda}^{\mathrm{T}}\boldsymbol{g}$ is such that $\boldsymbol{\lambda}^{\mathrm{T}}(t)\boldsymbol{g}(\boldsymbol{x}(t))$ is zero for a finite interval of time (t_1, t_2), the corresponding trajectory and control are called *singular* (Bell and Jacobson 1975). It follows that, along a singular trajectory, the total time derivative of the switching function also vanishes along the singular arc. Now

$$\begin{aligned}\frac{\mathrm{d}}{\mathrm{d}t}\boldsymbol{\lambda}^{\mathrm{T}}\boldsymbol{g} &= \dot{\boldsymbol{\lambda}}^{\mathrm{T}}\boldsymbol{g} + \boldsymbol{\lambda}^{\mathrm{T}}\boldsymbol{g}_x\dot{\boldsymbol{x}}\\ &= -\boldsymbol{\lambda}^{\mathrm{T}}(\boldsymbol{f}_x + u\boldsymbol{g}_x)\boldsymbol{g} + \boldsymbol{\lambda}^{\mathrm{T}}\boldsymbol{g}_x(\boldsymbol{f} + u\boldsymbol{g}) \qquad \text{(by (11.13) and (11.15))}\\ &= \boldsymbol{\lambda}^{\mathrm{T}}(\boldsymbol{g}_x\boldsymbol{f} - \boldsymbol{f}_x\boldsymbol{g}) + u\boldsymbol{\lambda}^{\mathrm{T}}(\boldsymbol{g}_x\boldsymbol{g} - \boldsymbol{g}_x\boldsymbol{g})\\ &= \boldsymbol{\lambda}^{\mathrm{T}}[\boldsymbol{f}, \boldsymbol{g}],\end{aligned}$$

where

$$[\boldsymbol{f}, \boldsymbol{g}] = \boldsymbol{g}_x\boldsymbol{f} - \boldsymbol{f}_x\boldsymbol{g} \tag{11.17}$$

is called the *Lie bracket* of $\boldsymbol{f}$ and $\boldsymbol{g}$ [17] and is itself a vector field.

Example 11.8

$\boldsymbol{f}(\boldsymbol{x}) = (-x_3, 0, x_1), \qquad \boldsymbol{g}(\boldsymbol{x}) = (x_2, -x_1, 0).$

$$\boldsymbol{f}_x = \begin{bmatrix} 0 & 0 & -1 \\ 0 & 0 & 0 \\ 1 & 0 & 0 \end{bmatrix}, \qquad \boldsymbol{g}_x = \begin{bmatrix} 0 & 1 & 0 \\ -1 & 0 & 0 \\ 0 & 0 & 0 \end{bmatrix},$$

$$[\boldsymbol{f}, \boldsymbol{g}](\boldsymbol{x}) = \begin{bmatrix} 0 & 1 & 0 \\ -1 & 0 & 0 \\ 0 & 0 & 0 \end{bmatrix}\begin{bmatrix} -x_3 \\ 0 \\ x_1 \end{bmatrix} - \begin{bmatrix} 0 & 0 & -1 \\ 0 & 0 & 0 \\ 1 & 0 & 0 \end{bmatrix}\begin{bmatrix} x_2 \\ -x_1 \\ 0 \end{bmatrix} = \begin{bmatrix} 0 \\ x_3 \\ -x_2 \end{bmatrix}.$$

Since the expression $\boldsymbol{\lambda}^{\mathrm{T}}[\boldsymbol{f}, \boldsymbol{g}](\boldsymbol{x}(t))$ is zero along a singular trajectory, its total time derivative also remains zero along such a trajectory. It can be shown (Bell 1984) that

$$\frac{\mathrm{d}}{\mathrm{d}t}\boldsymbol{\lambda}^{\mathrm{T}}[\boldsymbol{f}, \boldsymbol{g}] = \boldsymbol{\lambda}^{\mathrm{T}}[\boldsymbol{f}, [\boldsymbol{f}, \boldsymbol{g}]] + u\boldsymbol{\lambda}^{\mathrm{T}}[\boldsymbol{g}, [\boldsymbol{f}, \boldsymbol{g}]]. \tag{11.18}$$

Lie brackets have the following properties [18]: for all $f, g, h \in \mathrm{V}(M)$ (the set of all C^∞ vector fields on a manifold M),

$$[f, f] = 0, \tag{11.19}$$

$$[f, 0] = 0, \tag{11.20}$$

$$[g, f] = -[f, g] \quad \text{(skew symmetry)}, \tag{11.21}$$

$$[(f+g), h] = [f, h] + [g, h], \tag{11.22}$$

$$[f, [g, h]] + [g, [h, f]] + [h, [f, g]] = 0. \tag{11.23}$$

Equation (11.23) is called the *Jacobi identity*.

The Lie bracket is a binary operation on the space $V(M)$ but this operation is not associative. This is because, by (11.21), the Jacobi identity (11.23) can be written as

$$[f, [g, h]] = [[f, g], h] + [[h, f], g].$$

The term $[[h, f], g]$ will not generally be the null vector field 0, in which case

$$[f, [g, h]] \neq [[f, g], h].$$

Thus, the Lie bracket is a nonassociative operation.

Associated with an expression $\boldsymbol{\lambda}^{\mathrm{T}}[\boldsymbol{f}, \boldsymbol{g}]$ is an alternative formulation in terms of Poisson brackets. Consider two functions $p, q \in C^{\infty}(M \times \mathbb{R}^n)$, with $M \subseteq \mathbb{R}^n$. The *Poisson bracket* of p and q is defined by [19]

$$\{p, q\} \triangleq \frac{\partial q}{\partial x_i}\frac{\partial p}{\partial \lambda_i} - \frac{\partial p}{\partial x_i}\frac{\partial q}{\partial \lambda_i}.$$

Poisson brackets have the same properties as Lie brackets and, in particular,

$$\{q, p\} = -\{p, q\} \quad \text{and} \quad \{p, 0\} = 0. \tag{11.24}$$

A relationship between Lie and Poisson brackets (Bell 1984) is

$$\boldsymbol{\lambda}^{\mathrm{T}}[\boldsymbol{f}, \boldsymbol{g}] = \{\boldsymbol{\lambda}^{\mathrm{T}}\boldsymbol{f}, \boldsymbol{\lambda}^{\mathrm{T}}\boldsymbol{g}\}. \tag{11.25}$$

Example 11.9

$$p(\boldsymbol{x}, \boldsymbol{\lambda}) = x_2\lambda_3 - x_3\lambda_2, \qquad q(\boldsymbol{x}, \boldsymbol{\lambda}) = x_3\lambda_1 - x_1\lambda_3,$$

$$\{p, q\}(\boldsymbol{x}, \boldsymbol{\lambda}) = \sum_{i=1}^{3}\left(\frac{\partial q}{\partial x_i}\frac{\partial p}{\partial \lambda_i} - \frac{\partial p}{\partial x_i}\frac{\partial q}{\partial \lambda_i}\right) = x_2\lambda_1 - x_1\lambda_2.$$

Given any $p \in C^{\infty}(M \times \mathbb{R}^n)$, with $M \subseteq \mathbb{R}^n$, and using (11.15–16), we obtain

$$\begin{aligned}\frac{\mathrm{d}p}{\mathrm{d}t} &= \frac{\partial p}{\partial x_i}\dot{x}_i + \frac{\partial p}{\partial \lambda_i}\dot{\lambda}_i \\ &= \frac{\partial p}{\partial x_i}\frac{\partial H}{\partial \lambda_i} - \frac{\partial p}{\partial \lambda_i}\frac{\partial H}{\partial x_i} \\ &= \{H, p\}. \text{ [20]}\end{aligned}$$

Thus, the Poisson bracket $\{H, p\}$ is zero iff $p(\boldsymbol{x}, \boldsymbol{\lambda})$ is constant, i.e. p is a constant of motion.

Jacobi's identity (11.23) is also true for Poisson brackets. If $p, q, r \in C^{\infty}(M \times \mathbb{R}^n)$, with $M \subseteq \mathbb{R}^n$ then

$$\{p, \{q, r\}\} + \{q, \{r, p\}\} + \{r, \{p, q\}\} = 0. \tag{11.26}$$

Now take r as the Hamiltonian H. If p and q are constants of motion, then both $\{q, H\}$ and $\{H, p\}$ are zero in (11.26), which reduces to

$$\{H, \{p, q\}\} = 0.$$

This implies that $\{p, q\}$ is a constant of motion. In other words, the Poisson bracket of two constants of motion is itself a constant of motion (Goldstein 1950). This result is known as *Poisson's theorem.*

Example 11.10 (Gabasov and Kirillova 1972)

For a singular control of system (11.13),

$$\boldsymbol{\lambda}^{\mathsf{T}}\boldsymbol{g}(\boldsymbol{x}) = 0 \tag{11.27}$$

and so p, defined by $p(\boldsymbol{x}, \boldsymbol{\lambda}) = \boldsymbol{\lambda}^{\mathsf{T}}\boldsymbol{g}(\boldsymbol{x})$, is a constant of motion. If the Hamiltonian H does not depend explicitly on the time t, then H is also a constant of motion (Citron 1969). In this case, because of (11.14) and (11.27),

$$\boldsymbol{\lambda}^{\mathsf{T}}\boldsymbol{f}(\boldsymbol{x}) = \text{constant}.$$

Thus, the function q, defined by $q(\boldsymbol{x}, \boldsymbol{\lambda}) = \boldsymbol{\lambda}^{\mathsf{T}}\boldsymbol{f}(\boldsymbol{x})$, is also a constant of motion. Poisson's theorem then ensures that $\{\boldsymbol{\lambda}^{\mathsf{T}}\boldsymbol{f}, \boldsymbol{\lambda}^{\mathsf{T}}\boldsymbol{g}\}$ is also a constant of motion. Because of (11.25) this result is equivalent to the vanishing of the derivative in (11.18).

11.5 Linear algebras

In any vector space V over $\mathbb{R}$, the product of a vector $v \in V$ and a scalar $\alpha \in \mathbb{R}$ yields a vector αv belonging to V. In an inner product space over $\mathbb{R}$ the inner product of two vectors belongs to $\mathbb{R}$. What has now been found in Section 11.4 is a product of two vector fields (through the Lie bracket) which is itself another vector field. The set $\mathrm{V}(M)$ of all C^{∞} vector fields on a manifold M, together with the Lie bracket operation, form what is known as a linear algebra.

In general, a *linear algebra* L over the field $\mathbb{R}$ is a finite-dimensional vector space V over $\mathbb{R}$ in which there is defined a vector product $V \times V \rightarrow V$: $(u, v) \mapsto uv$ such that

$$(u + v)w = uw + vw, \qquad w(u + v) = wu + wv,$$

$$(\alpha u)v = \alpha(uv) = u(\alpha v),$$

for all $u,v,w \in V$ and all $\alpha \in \mathbb{R}$. Note that neither the commutative nor associative laws necessarily hold in a linear algebra. However, in some linear algebras, both laws are valid.

Example 11.11

The set $\mathbb{C}$ of all complex numbers over the field $\mathbb{R}$ with ordinary multiplication form a linear algebra which is both commutative and associative.

A *Lie algebra* is a nonassociative linear algebra in which the Jacobi identity (11.23) replaces the associative law. The nonassociative Lie bracket operation on the set $V(M)$ of all C^∞ vector fields on M yields a Lie algebra. A Lie algebra can be constructed from any associative linear algebra by defining

$$a \circ b = ba - ab,$$

in which $a \circ b$ represents the product in the Lie algebra, while ba and ab represent products in the associative algebra. Of course, the resulting Lie algebra is trivial if the original associative algebra is commutative, because then every Lie product $a \circ b$ is null.

Example 11.12

In the bilinear system

$$\dot{\boldsymbol{x}} = \boldsymbol{A}\boldsymbol{x} + u\boldsymbol{B}\boldsymbol{x}, \qquad u(t) \in \mathbb{R}, \quad \boldsymbol{x}(t) \in \mathbb{R}^n, \quad \boldsymbol{A}, \boldsymbol{B} \in \mathbb{R}^{n \times n},$$

the vector fields $\boldsymbol{f}$ and $\boldsymbol{g}$ of (11.13) are given by

$$\boldsymbol{f}(\boldsymbol{x}) = \boldsymbol{A}\boldsymbol{x}, \qquad \boldsymbol{g}(\boldsymbol{x}) = \boldsymbol{B}\boldsymbol{x}.$$

Then $[\boldsymbol{f}, \boldsymbol{g}](\boldsymbol{x}) = \boldsymbol{g}_x \boldsymbol{f}(\boldsymbol{x}) - \boldsymbol{f}_x \boldsymbol{g}(\boldsymbol{x}) = \boldsymbol{B}(\boldsymbol{A}\boldsymbol{x}) - \boldsymbol{A}(\boldsymbol{B}\boldsymbol{x}) = (\boldsymbol{B}\boldsymbol{A} - \boldsymbol{A}\boldsymbol{B})\boldsymbol{x}$. The expression $\boldsymbol{B}\boldsymbol{A} - \boldsymbol{A}\boldsymbol{B}$ is known as the *commutator product*, usually denoted by $[\boldsymbol{A}, \boldsymbol{B}]$. If $\boldsymbol{A}$ and $\boldsymbol{B}$ commute, then $[\boldsymbol{A}, \boldsymbol{B}]$ is the null matrix.

The set of all vector fields $\boldsymbol{f}$ defined by $\boldsymbol{f}(\boldsymbol{x}) = \boldsymbol{A}\boldsymbol{x}$ is an n^2-dimensional Lie algebra, because $\mathbb{R}^{n \times n}$ is an n^2-dimensional vector space (see remark following Example 6.10). Similarly, $\mathbb{R}^{n \times n}$ is a Lie algebra with the Lie bracket defined by $[\boldsymbol{A}, \boldsymbol{B}] = \boldsymbol{B}\boldsymbol{A} - \boldsymbol{A}\boldsymbol{B}$.

The linear-analytic system (11.13) has only one control variable u but it is easy to generalize to a vector control by

$$\dot{\boldsymbol{x}} = \boldsymbol{f}(\boldsymbol{x}) + \boldsymbol{G}(\boldsymbol{x})\boldsymbol{u} \tag{11.28}$$

where $\boldsymbol{u}(t) \in U \subseteq \mathbb{R}^m$ and $\boldsymbol{x}(t) \in M$. Then $\boldsymbol{G}(\boldsymbol{x}(t)) \in \mathbb{R}^{n \times m}$ and can be written in terms of its columns as $\boldsymbol{G}(\boldsymbol{x}) = [\boldsymbol{g}_1(\boldsymbol{x}), \ldots, \boldsymbol{g}_m(\boldsymbol{x})]$ so that (11.28) may also be expressed in the form

$$\dot{\boldsymbol{x}} = \boldsymbol{f}(\boldsymbol{x}) + \sum_{j=1}^{m} u_j \boldsymbol{g}_j(\boldsymbol{x}). \tag{11.29}$$

Example 11.13

Consider the system with no drift term:

$$\dot{\boldsymbol{x}}=\boldsymbol{G}(\boldsymbol{x})\boldsymbol{u}, \qquad \boldsymbol{u}(t)\in\mathbb{R}^3, \qquad \boldsymbol{G}(\boldsymbol{x})=\begin{bmatrix} 0 & -x_3 & x_2 \\ x_3 & 0 & -x_1 \\ -x_2 & x_1 & 0 \end{bmatrix},$$

Thus,

$$\boldsymbol{g}_1(\boldsymbol{x})=\begin{bmatrix} 0 \\ x_3 \\ -x_2 \end{bmatrix}, \qquad \boldsymbol{g}_2(\boldsymbol{x})=\begin{bmatrix} -x_3 \\ 0 \\ x_1 \end{bmatrix}, \qquad \boldsymbol{g}_3(\boldsymbol{x})=\begin{bmatrix} x_2 \\ -x_1 \\ 0 \end{bmatrix}.$$

It follows that

$$\begin{aligned}[\boldsymbol{g}_1, \boldsymbol{g}_2](\boldsymbol{x}) &= \boldsymbol{g}_{2x}\boldsymbol{g}_1(\boldsymbol{x})-\boldsymbol{g}_{1x}\boldsymbol{g}_2(\boldsymbol{x})=(x_2, 0, 0)-(0, x_1, 0)\\ &=\boldsymbol{g}_3(\boldsymbol{x}),\end{aligned}$$

$[\boldsymbol{g}_2, \boldsymbol{g}_3](\boldsymbol{x})=\boldsymbol{g}_1(\boldsymbol{x})$ (see Example 11.8), and $[\boldsymbol{g}_3, \boldsymbol{g}_1](\boldsymbol{x})=\boldsymbol{g}_2(\boldsymbol{x})$ (Exercise 10).

The set $D=\{\boldsymbol{g}_1, \boldsymbol{g}_2, \boldsymbol{g}_3\}$ of vector fields in Example 11.13 is said to be *involutive* because, for any $\boldsymbol{g}_i, \boldsymbol{g}_j \in D$ $(i,j=1, 2, 3)$, either $[\boldsymbol{g}_j, \boldsymbol{g}_i]$ or $[\boldsymbol{g}_i, \boldsymbol{g}_j]$ also belongs to D. [21] More generally, a set of vector fields $\boldsymbol{g}_1, \ldots, \boldsymbol{g}_m$ is called an involutive set if there exist functions γ_{ijk} such that

$$[\boldsymbol{g}_i, \boldsymbol{g}_j](x)=\sum_{k=1}^{m} \gamma_{ijk}(x)\boldsymbol{g}_k(x).$$

The set $\mathrm{V}(M)$ of all C^∞ vector fields on M can be regarded as a real (infinite dimensional) Lie algebra over $\mathbb{R}$ under the Lie bracket operation $(\boldsymbol{f}, \boldsymbol{g})\mapsto[\boldsymbol{f},\boldsymbol{g}]$ for all $\boldsymbol{f},\boldsymbol{g}\in\mathrm{V}(M)$. In Example 11.13,

$$\gamma_{121}=0=\gamma_{122}, \qquad \gamma_{123}=1,$$

$$\gamma_{232}=0=\gamma_{233}, \qquad \gamma_{231}=1,$$

$$\gamma_{311}=0=\gamma_{313}, \qquad \gamma_{312}=1.$$

Example 11.3 is the special case of Example 11.13 in which $u_1(\cdot)=1$, $u_2(\cdot)=0$, $u_3(\cdot)=0$, and integration is carried out along the vector field $\boldsymbol{g}_1$ which is tangent to the manifold S^2 in $\mathbb{R}^3$. In fact, in Example 11.13 the vector fields $\boldsymbol{g}_1, \boldsymbol{g}_2, \boldsymbol{g}_3$ are all tangent to the manifold S^2. So, if u_1, u_2, u_3 are controls that can be turned on and off, reversed, etc., then the state $\boldsymbol{x}$ can be moved anywhere on S^2. This sort of analysis is clearly important for controllability of systems (Sussmann and Jurdjevic 1972). The main concepts of this part of dynamical system theory will be given here only for the state space $\mathbb{R}^n$. However, these concepts readily generalize to a more general manifold M.

For controlling the system (11.29), the chosen class of admissible control functions $\boldsymbol{u}:[0, T] \to U$ given by $t \mapsto \boldsymbol{u}(t)$, with $U \subseteq \mathbb{R}^m$, usually includes all piecewise constant controls. It is also assumed that, for any given initial condition, the corresponding solution of (11.29) for such admissible controls will also be defined on $[0, T]$ and be absolutely continuous.

The reachable set at time T for system (11.29), with initial condition $\boldsymbol{x}(0)=\boldsymbol{x}_0$, is defined as the set of all states $\boldsymbol{x}_1 \in \mathbb{R}^n$ such that there exists an admissible control function $\boldsymbol{u}$ on $[0, T]$ for which the corresponding solution of (11.29) satisfies $\boldsymbol{x}(0)=\boldsymbol{x}_0$ and $\boldsymbol{x}(T)=\boldsymbol{x}_1$. This set is usually denoted by $R(T, \boldsymbol{x}_0)$. The set

$$R(\boldsymbol{x}_0) \triangleq \bigcup_{T \geqslant 0} R(T, \boldsymbol{x}_0)$$

is called the *reachable set from* $\boldsymbol{x}_0$.

System (11.29) is said to be *accessible from* $\boldsymbol{x}_0$ if $R(\boldsymbol{x}_0)$ contains a neighbourhood of some state $\boldsymbol{x} \in \mathbb{R}^n$; in other words if $R(\boldsymbol{x}_0)$ has a non-empty interior. The system is *accessible* if it is accessible from each $\boldsymbol{x}_0 \in \mathbb{R}^n$ and controllable if $R(\boldsymbol{x}_0) = \mathbb{R}^n$ for all $\boldsymbol{x}_0 \in \mathbb{R}^n$.

Necessary and sufficient conditions can be given (Krener 1974) under which (11.29) is accessible from $\boldsymbol{x}_0$. These conditions are in terms of the Lie algebra $\mathrm{L}(\boldsymbol{f}, \boldsymbol{g}_1, \ldots, \boldsymbol{g}_m)$ generated by the vector fields $\boldsymbol{f}, \boldsymbol{g}_1, \ldots, \boldsymbol{g}_m$. This Lie algebra is the smallest vector space of vector fields containing $\boldsymbol{f}, \boldsymbol{g}_1, \ldots, \boldsymbol{g}_m$ and which is involutive.[22] Denote by $\mathrm{L}(\boldsymbol{f}, \boldsymbol{g}_1, \ldots, \boldsymbol{g}_m)(\boldsymbol{x})$ the subspace of $\mathbb{R}^n$ defined by

$$\mathrm{L}(\boldsymbol{f}, \boldsymbol{g}_1, \ldots, \boldsymbol{g}_m)(\boldsymbol{x}) \triangleq \{\boldsymbol{h}(\boldsymbol{x}) : \boldsymbol{h} \in \mathrm{L}(\boldsymbol{f}, \boldsymbol{g}_1, \ldots, \boldsymbol{g}_m)\}.$$ [23]

Then system (11.29) is accessible from $\boldsymbol{x}_0$ iff $\mathrm{L}(\boldsymbol{f}, \boldsymbol{g}_1, \ldots, \boldsymbol{g}_m)(\boldsymbol{x}_0)=\mathbb{R}^n$. It is accessible if this condition is satisfied for all $\boldsymbol{x}_0 \in \mathbb{R}^n$.

Example 11.14

$$\dot{x}_1 = x_1 + u, \quad x_1(0) = x_1^0, \qquad \dot{x}_2 = x_1^2, \quad x_2(0) = x_2^0. \tag{11.30}$$

$$\boldsymbol{f}(\boldsymbol{x}) = \begin{bmatrix} x_1 \\ x_1^2 \end{bmatrix}, \quad \boldsymbol{g}(\boldsymbol{x}) = \begin{bmatrix} 1 \\ 0 \end{bmatrix}, \quad [\boldsymbol{f}, \boldsymbol{g}](\boldsymbol{x}) = \begin{bmatrix} -1 \\ -2x_1 \end{bmatrix},$$

$$[\boldsymbol{f}, [\boldsymbol{f}, \boldsymbol{g}]](\boldsymbol{x}) = \begin{bmatrix} 1 \\ 0 \end{bmatrix} = \boldsymbol{g}(\boldsymbol{x}), \qquad [\boldsymbol{g}, [\boldsymbol{f}, \boldsymbol{g}]](\boldsymbol{x}) = \begin{bmatrix} 0 \\ -2 \end{bmatrix}.$$

All other Lie brackets are null. Thus

$$\mathrm{L}(\boldsymbol{f}, \boldsymbol{g})(\boldsymbol{x}) \supset \left\{ \begin{bmatrix} 0 \\ -2 \end{bmatrix}, \begin{bmatrix} 1 \\ 0 \end{bmatrix} \right\}$$

for all $\boldsymbol{x} \in \mathbb{R}^2$, and so system (11.30) is accessible.

Although system (11.30) is accessible, it is not controllable. To see this, integrate the second differential equation in (11.30), with the given initial condition, to obtain

$$x_2(t)=x_2^0+\int_0^t x_1^2(\tau)\mathrm{d}\tau,$$

so that $x_2(t) \geqslant x_2^0$ for all controls u. Thus, no point in $\mathbb{R}^2$ below the line $x_2=x_2^0$ can be reached from $\boldsymbol{x}(0)$, no matter what control is used. Therefore, (11.30) is not controllable. This illustrates the fact that nonlinear systems can behave very differently from linear ones. To see this, consider the linear system

$$\dot{\boldsymbol{x}}(t)=\boldsymbol{A}\boldsymbol{x}(t)+\sum_{j=1}^{m} u_j(t)\boldsymbol{b}_j, \qquad \boldsymbol{A}\in\mathbb{R}^{n\times n}, \qquad \boldsymbol{b}_j\in\mathbb{R}^n \quad (j=1,\ldots,m),$$

i.e.

$$\dot{\boldsymbol{x}}(t)=\boldsymbol{A}\boldsymbol{x}(t)+\boldsymbol{B}\boldsymbol{u}(t), \tag{11.31}$$

where $\boldsymbol{B}=[\boldsymbol{b}_1,\ldots,\boldsymbol{b}_m]$. If the rank of the matrix $[\boldsymbol{B}, \boldsymbol{AB}, \boldsymbol{A}^2\boldsymbol{B},\ldots, \boldsymbol{A}^{n-1}\boldsymbol{B}]$ is less than n, then the reachable set never contains a neighbourhood of any state, so that the linear system is not accessible from any point $\boldsymbol{x}_0$. On the other hand, if the rank is n, then the system is controllable (Barnett and Cameron 1985).

The necessary and sufficient condition quoted above for the accessibility of system (11.29) is equivalent to the rank condition just quoted for system (11.31). This is easily seen since, in (11.31),

$$\boldsymbol{f}(\boldsymbol{x})=\boldsymbol{A}\boldsymbol{x},$$

$$\boldsymbol{g}_j(\boldsymbol{x})=\boldsymbol{b}_j, \qquad [\boldsymbol{f},\boldsymbol{g}_j](\boldsymbol{x})=-\boldsymbol{A}\boldsymbol{b}_j, \qquad [\boldsymbol{f},-\boldsymbol{A}\boldsymbol{b}_j]=\boldsymbol{A}^2\boldsymbol{b}_j, \ldots,$$
$$[\boldsymbol{f},(-1)^{n-1}\boldsymbol{A}^{n-1}\boldsymbol{b}_j](\boldsymbol{x})=(-1)^n\boldsymbol{A}^n\boldsymbol{b}_j \quad (n=2,3,\ldots),$$

for $j=1,\ldots,m$. It follows that

$$\mathrm{L}(A, b_1,\ldots, b_m)(\boldsymbol{x})=\{\boldsymbol{A}\boldsymbol{x}\pm\boldsymbol{b}_j, \pm\boldsymbol{A}\boldsymbol{b}_j,\ldots, \pm\boldsymbol{A}^{n-1}\boldsymbol{b}_j \quad (j=1,\ldots,m)\}.$$

The Cayley–Hamilton theorem ensures that these are all the terms which are needed. Then the system (11.31) is accessible iff the vector space generated by the vectors $\pm\boldsymbol{b}_j, \pm\boldsymbol{A}\boldsymbol{b}_j,\ldots, \pm\boldsymbol{A}^{n-1}\boldsymbol{b}_j$ is of dimension n. That is to say, iff the rank of matrix $[\boldsymbol{B}, \boldsymbol{AB}, \boldsymbol{A}^2\boldsymbol{B},\ldots, \boldsymbol{A}^{n-1}\boldsymbol{B}]$ is n.

Exercises

1. A particle of mass m_1 is fixed at the origin of $\mathbb{R}^3$. A second particle of mass m_2 is free to move in $\mathbb{R}^3$. Explain how the gravitational attraction defines a vector field on

$\mathbb{R}^3\backslash\{0\}$ given that the force between the two particles is inversely proportional to the square of the distance between them.

2. Show that the set of vector fields on $\mathbb{R}^n$ is an additive abelian group.

3. Find the directional derivatives of a function $f:\mathbb{R}^3 \to \mathbb{R}$ at a point $\boldsymbol{p}\in\mathbb{R}^3$ in the direction of the canonical coordinate axes in $\mathbb{R}^3$.

4. A dynamical system is given by

$$\dot{x}_1 = x_1^2 - x_3,$$
$$\dot{x}_2 = x_1^3 + x_2^2 - x_3^2,$$
$$\dot{x}_3 = x_1^4 + 9x_2 - x_3^3.$$

Find the directional derivative of the function $f:\mathbb{R}^3 \to \mathbb{R}$ defined by $f(x_1, x_2, x_3) = x_1^2 + x_2^2 + x_3^2$ at the point (2, 1, 3) in the direction of the vector $\boldsymbol{X}(\boldsymbol{x})$.

5. Demonstrate that the space E^2 of Euclidean geometry is a topological manifold.

6. Show that the group of rigid motions is a Lie group.

7. (a_1, a_2, a_3) is a fixed point distinct from the origin of the manifold $\mathbb{R}^3$. Show that $\varphi_t:\mathbb{R}^3 \to \mathbb{R}^3$ given by $\varphi_t(\boldsymbol{x}) = (x_1 + a_1 t, x_2 + a_2 t, x_3 + a_3 t)$ defines a C^∞ action of $\mathbb{R}$ on $\mathbb{R}^3$. Find the orbits of $\boldsymbol{x} = (x_1, x_2, x_3)$.

8. Demonstrate that the vector product $\boldsymbol{a}\times\boldsymbol{b}$ of two vectors $\boldsymbol{a}$ and $\boldsymbol{b}$ is a non-associative binary operation on $\mathbb{R}^3$ similar to the Lie bracket.

9. Prove that (11.19)$\Rightarrow$(11.21). (Hint: consider $[(f+g), (f+g)]$ and use $[(f+g), f+g] = [(f+g), f] + [(f+g), g]$ by a slight modification of (11.22)).

10. For the system of Example 11.13, carry out the calculations to show that $[g_3, g_1](\boldsymbol{x}) = g_2(\boldsymbol{x})$.

11. Given the system

$$\dot{x}_1 = u_1, \qquad \dot{x}_2 = u_2, \qquad \dot{x}_3 = x_1 x_2,$$

calculate the Lie brackets $[f, g_1]$, $[f, g_2]$, $[g_2, [f, g_1]]$, $[g_1, [f, g_2]]$.

12. Show that the set $\mathbb{R}^{2\times 2}$ with ordinary matrix multiplication forms a linear algebra. Demonstrate that this algebra forms a ring over $\mathbb{R}$.

Notes for Chapter 11

1. $\boldsymbol{X}$ is used here in place of the usual $\boldsymbol{f}$ in $\dot{\boldsymbol{x}} = \boldsymbol{f}(\boldsymbol{x})$ since f will be used later as a C^∞ function.

2. Note that $\mathrm{T}_{\boldsymbol{x}}(\mathbb{R}^n)$ is not a subspace of $\mathbb{R}^n$ but is the hyperplane $\boldsymbol{x} + \mathbb{R}^n$. This is a translation of a subspace, sometimes called a *linear variety* (Luenberger 1969). Notice that the plus sign in $\boldsymbol{x} + \mathbb{R}^n$ is 'structural'. (See also Note 3.11.)

3. At this stage $\mathrm{T}(\mathbb{R}^n)$ cannot be called a vector space because there is no rule to add a vector localized at $\boldsymbol{x}$ (i.e. in $\mathrm{T}_{\boldsymbol{x}}(\mathbb{R}^n)$) to a vector localized at $\boldsymbol{y}$ (i.e. in $\mathrm{T}_{\boldsymbol{y}}(\mathbb{R}^n)$).

4. $\boldsymbol{i}, \boldsymbol{j}, \boldsymbol{k}$ are unit vectors in the directions of the axes in the chosen coordinate system for $\mathbb{R}^3$. Any coordinate system can be chosen; $\nabla\varphi$ depends only on the domain of definition for ∇ and will be the same vector whatever coordinate system is chosen. $\nabla\varphi$ is said to be *invariant* under a change of axes.

5. Notice that no two surfaces $\varphi(x, y, z)=C_1$ and $\varphi(x, y, z)=C_2$, for two different constants C_1 and C_2, can intersect. If they did, it would mean that the value $\varphi(\alpha, \beta, \gamma)$ at a point of intersection (α, β, γ) could assume the two values C_1 and C_2. This would violate the fact that φ is a function and hence single-valued (cf. Chapter 3).
6. Note that this interpretation of the tangent bundle is now a vector space.
7. This is a slight deviation from the usual definition of directional derivative (Kreyszig 1979) in which $\boldsymbol{X}(\boldsymbol{x})$ is a unit vector.
8. Any two C^∞ functions $f, g \in F_p(\mathbb{R}^n)$ can be added together and multiplied by any $\alpha \in \mathbb{R}$ to give $f+g$ and αf in $F_p(\mathbb{R}^n)$ (cf. Exercise 6.1).
9. S^2 is two-dimensional because only two coordinates are required to specify a point on its surface, e.g. two spherical polar angles (radius $r=1$).
10. S^2 is compact, but $\mathbb{R}^2$ is not, so that S^2 and $\mathbb{R}^2$ cannot be homeomorphic; hence the reason for a *local* homeomorphism.
11. The ith coordinate function x_i is a real-valued function on U.
12. As quoted in the text, G acts on S *on the left* which is the usual case. For a right action, the value of θ is written as $\theta(s, g)$ instead of $\theta(g, s)$.
13. From Section 4.5

$$GM \triangleq \{gx : g \in G,\ x \in M\}$$

and so the orbit of any point $x \in M$ is the set Gx. This set is the same as $\{\theta_g(x)\}$.
14. An optimal-control problem with a performance index that involves only final values of state variables is called a *Mayer* problem. A transformation to such a problem is given in Exercise 1.5.
15. $\boldsymbol{f_x}$ involves the first partial derivatives $\partial f_i/\partial x_j$ (see solution to Exercise 10.8). Thus, if $\boldsymbol{f}=(f_1, f_2)$ and $\boldsymbol{x}=(x_1, x_2)$, then

$$\boldsymbol{f_x}=\begin{bmatrix} \partial f_1/\partial x_1 & \partial f_1/\partial x_2 \\ \partial f_2/\partial x_1 & \partial f_2/\partial x_2 \end{bmatrix},$$

and similarly for $\boldsymbol{g_x}$.
16. $\boldsymbol{x}$ and $\boldsymbol{\lambda}$ are column vectors so that $H_{\boldsymbol{x}}$ and $H_{\boldsymbol{\lambda}}$ in (11.15) and (11.16) respectively, are both row vectors. The differentiation of quadratic and bilinear forms with respect to vectors (MacFarlane 1970) is required for the *LQP* problem (Section 1.4).
17. The Lie bracket $[\boldsymbol{f}, \boldsymbol{g}]$ is sometimes defined as $\boldsymbol{f_x g} - \boldsymbol{g_x f}$, i.e. the negative of the expression given in (11.17). Another name for Lie bracket is *Jacobi bracket*.
18. 0 represents the null vector field i.e. $0(x)$ is the zero vector in the tangent space $T_x(M)$ for all $x \in M$.
19. Double suffix summation convention, i.e. any repeated suffix is to be summed over that suffix ($i=1, \ldots, n$ in this case).
20. Note that the control u plays the role of a parameter in this analysis.
21. For an infinite set D of vector fields, D is called *involutive* if $[\boldsymbol{f}, \boldsymbol{g}] \in D$ whenever $\boldsymbol{f}$ and $\boldsymbol{g}$ belong to D.
22. $L(\boldsymbol{f}, \boldsymbol{g}_1, \ldots, \boldsymbol{g}_m)$ could easily be of *infinite* dimension.
23. By definition $L(\boldsymbol{f}, \boldsymbol{g}_1, \ldots, \boldsymbol{g}_m)(\boldsymbol{x})$ is always a *finite*-dimensional vector space with dimension not greater than n.

SOLUTIONS TO THE EXERCISES

Chapter 1

1. Let $x_1 = p$, $x_2 = \dot{p}$, $x_3 = q$, and $x_4 = \dot{q}$, so that $\dot{x}_1 = x_2$ and $\dot{x}_3 = x_4$. Then the given equations are

$$\dot{x}_2 + 3x_2 + 9x_3 = 0, \qquad \dot{x}_4 + 2x_4 + x_1 = \cos t.$$

Thus,

$$\begin{aligned} \dot{x}_1 &= x_2, \\ \dot{x}_2 &= -3x_2 - 9x_3, \\ \dot{x}_3 &= x_4, \\ \dot{x}_4 &= -x_1 - 2x_4 + \cos t, \end{aligned}$$

and

$$\boldsymbol{A} = \begin{bmatrix} 0 & 1 & 0 & 0 \\ 0 & -3 & -9 & 0 \\ 0 & 0 & 0 & 1 \\ -1 & 0 & 0 & -2 \end{bmatrix}, \qquad \boldsymbol{b} = \begin{bmatrix} 0 \\ 0 \\ 0 \\ 1 \end{bmatrix}.$$

Let $y_1 = p + q = x_1 + x_3$ and $y_2 = \dot{p} - \dot{q} = x_2 - x_4$. Then

$$\boldsymbol{y} = \begin{bmatrix} y_1 \\ y_2 \end{bmatrix} = \begin{bmatrix} 1 & 0 & 1 & 0 \\ 0 & 1 & 0 & -1 \end{bmatrix} \begin{bmatrix} x_1 \\ x_2 \\ x_3 \\ x_4 \end{bmatrix} = \boldsymbol{Cx},$$

so $\boldsymbol{C} = \begin{bmatrix} 1 & 0 & 1 & 0 \\ 0 & 1 & 0 & -1 \end{bmatrix}$.

2. Taking Laplace transforms with zero initial conditions, we get

$$(s^2 + 4s + 2)\bar{y} = (s + 2)\bar{u},$$

and the transfer function g is given by

$$g(s) = \frac{\bar{y}}{\bar{u}} = \frac{s + 2}{s^2 + 4s + 2}.$$

3. The input sequence is given by $u(-5)=-6$, $u(-4)=-5$, $u(-3)=-4$, $u(-2)=-3$, $u(-1)=-2$, $u(0)=-1$. The z-transform of this sequence is

$$\bar{u}(z)=\sum_{t=-5}^{0} u(t)z^{-t}=-6z^5-5z^4-4z^3-3z^2-2z-1.$$

4. $\dfrac{dx_1}{dx_2}=\dfrac{\dot{x}_1}{\dot{x}_2}=-\dfrac{x_2}{x_1}$. Thus $\int x_1 dx_1=-\int x_2 dx_2$, i.e. $\frac{1}{2}x_1^2=-\frac{1}{2}x_2^2+C$ (where C is constant of integration). At $t=0$, it is given that $x_1=0$ and $x_2=1$ so $C=\frac{1}{2}$. Hence, $x_1^2+x_2^2=1$ which is the unit circle in $\mathbb{R}^2$, with centre the origin and radius 1. The system traverses this circle in a clockwise direction, since $\dot{x}_1>0$ (i.e. x_1 increasing with time) when $x_2>0$, and $\dot{x}_1<0$ when $x_2<0$.

5. Let $\dot{x}_{n+1}=L(\boldsymbol{x}, \boldsymbol{u})$ and $x_{n+1}(t_0)=0$. Integrating with respect to time between t_0 and t_f yields

$$x_{n+1}(t_f)=\int_{t_0}^{t_f} L(\boldsymbol{x}, \boldsymbol{u})dt.$$

6. Let $x_3=t$. Then $\dot{x}_3=1$, and $x_3(0)=0$, and the system becomes

$$\begin{aligned} \dot{x}_1&=3x_2-x_3x_1, & x_1(0)&=1\\ \dot{x}_2&=-x_1+e^{x_3}, & x_2(0)&=-1\\ \dot{x}_3&=1, & x_3(0)&=0. \end{aligned}$$

Chapter 2

1. Suppose that $A\in \mathrm{So}(n, \mathbb{R})$. Then A is an $n\times n$ orthogonal matrix with real components and will belong to $\mathrm{G}\ell(n, \mathbb{R})$, the set of all $n\times n$ real nonsingular matrices, if A is a nonsingular matrix. But A is an orthogonal matrix so by definition $AA^{\mathrm{T}}=I=A^{\mathrm{T}}A$. Thus $A^{\mathrm{T}}=A^{-1}$ and so the inverse matrix A^{-1} exists. It follows that A is nonsingular and so $A\in \mathrm{G}\ell(n, \mathbb{R})$. Hence

$$A\in \mathrm{So}(n, \mathbb{R}) \quad\Rightarrow\quad A\in \mathrm{G}\ell(n, \mathbb{R})$$

i.e. $\mathrm{So}(n, \mathbb{R})\subseteq \mathrm{G}\ell(n, \mathbb{R})$. But $\mathrm{So}(n, \mathbb{R})\neq \mathrm{G}\ell(n, \mathbb{R})$ because $\mathrm{G}\ell(n, \mathbb{R})$ contains nonsingular matrices which are not orthogonal. For example, with $n=2$, we have

$$A=\begin{bmatrix} 1 & 0\\ 0 & 2 \end{bmatrix}\in \mathrm{G}\ell(2, \mathbb{R}) \quad \text{but} \quad A^{-1}=\begin{bmatrix} 1 & 0\\ 0 & \frac{1}{2} \end{bmatrix}\neq A^{\mathrm{T}}.$$

Thus, $\mathrm{So}(n, \mathbb{R})\subset \mathrm{G}\ell(n, \mathbb{R})$.

2. The relation $\sim$ is (i) reflexive because $\boldsymbol{P}=\mathbf{I}_n^{-1}\boldsymbol{P}\mathbf{I}_n$ for each $\boldsymbol{P}\in \mathbb{R}^{n\times n}$. (ii) To show symmetry, observe first that, for any $\boldsymbol{P}, \boldsymbol{Q}\in \mathbb{R}^{n\times n}$ and for some nonsingular matrix $\boldsymbol{T}$, the condition $\boldsymbol{P}\sim \boldsymbol{Q}$ implies that $\boldsymbol{Q}=\boldsymbol{T}^{-1}\boldsymbol{P}\boldsymbol{T}$. Premultiply by $\boldsymbol{T}$ and postmultiply by $\boldsymbol{T}^{-1}$ in the last equation to give $\boldsymbol{P}=\boldsymbol{T}\boldsymbol{Q}\boldsymbol{T}^{-1}$. Writing $\boldsymbol{U}=\boldsymbol{T}^{-1}\in \mathbb{R}^{n\times n}$, we obtain $\boldsymbol{P}=\boldsymbol{U}^{-1}\boldsymbol{Q}\boldsymbol{U}$, i.e. $\boldsymbol{Q}\sim \boldsymbol{P}$. (iii) The relation is transitive because, for any $\boldsymbol{P}, \boldsymbol{Q}, \boldsymbol{R}\in \mathbb{R}^{n\times n}$, we have $\boldsymbol{P}\sim \boldsymbol{Q}\Rightarrow \boldsymbol{Q}=\boldsymbol{T}^{-1}\boldsymbol{P}\boldsymbol{T}$ and $\boldsymbol{Q}\sim \boldsymbol{R}\Rightarrow \boldsymbol{R}=\boldsymbol{S}^{-1}\boldsymbol{Q}\boldsymbol{S}$ for nonsingular matrices

$\boldsymbol{T}, \boldsymbol{S} \in \mathbb{R}^{n \times n}$. Thus,

$$\boldsymbol{R} = \boldsymbol{S}^{-1}(\boldsymbol{T}^{-1}\boldsymbol{P}\boldsymbol{T})\boldsymbol{S} = (\boldsymbol{S}^{-1}\boldsymbol{T}^{-1})\boldsymbol{P}(\boldsymbol{T}\boldsymbol{S})$$
$$= (\boldsymbol{T}\boldsymbol{S})^{-1}\boldsymbol{P}(\boldsymbol{T}\boldsymbol{S}).$$

Writing $\boldsymbol{M} = \boldsymbol{T}\boldsymbol{S} \in \mathbb{R}^{n \times n}$ yields $\boldsymbol{R} = \boldsymbol{M}^{-1}\boldsymbol{P}\boldsymbol{M}$ and so $\boldsymbol{P} \sim \boldsymbol{R}$.

3. Clearly, $\sim$ is reflexive because $ab = ba$, so that $(a, b) \sim (a, b)$, and $\sim$ is symmetric because $(a, b) \sim (c, d) \Rightarrow ad = bc \Rightarrow da = cb \Rightarrow cb = da \Rightarrow (c, d) \sim (a, b)$. To show transitivity, suppose that $(a, b) \sim (c, d)$ and $(c, d) \sim (p, q)$. Then $ad = bc$ and $cq = dp$. It follows that $adcq = bcdp$ or, what is the same thing, $cd(aq - bp) = 0$ (note the use of commutative, associative, and distributive laws in $\mathbb{Z}$). Since $d \neq 0$ by definition, either $c = 0$ or $aq - bp = 0$. If $c = 0$, then $ad = bc$ and $cq = dp$ give $a = 0$ and $p = 0$, in which case $aq - bp = 0$ is also true. Hence, $aq = bp$ in all cases and this implies that $(a, b) \sim (p, q)$, so $\sim$ is transitive.

4. (i) $\boldsymbol{x} = (x_1, \ldots, x_n) \in \mathbb{R}^n$; clearly $\boldsymbol{x} \leqslant \boldsymbol{x}$ since $x_i = x_i$ $(i = 1, \ldots, n)$. (ii) If $\boldsymbol{x}, \boldsymbol{y} \in \mathbb{R}^n$, with $\boldsymbol{x} \leqslant \boldsymbol{y}$ and $\boldsymbol{y} \leqslant \boldsymbol{x}$, then $x_i \leqslant y_i$ and $y_i \leqslant x_i$, in which case $x_i = y_i$ $(i = 1, \ldots, n)$, i.e. $\boldsymbol{x} = \boldsymbol{y}$. (iii) If $\boldsymbol{x}, \boldsymbol{y}, \boldsymbol{z} \in \mathbb{R}^n$, with $\boldsymbol{x} \leqslant \boldsymbol{y}$ and $\boldsymbol{y} \leqslant \boldsymbol{z}$, then $x_i \leqslant y_i$ and $y_i \leqslant z_i$ $(i = 1, \ldots, n)$. Therefore $x_i \leqslant z_i$ $(i = 1, \ldots, n)$ and this implies $\boldsymbol{x} \leqslant \boldsymbol{z}$.

5. $a + d = b + c$ iff $b - a = d - c$ $(= s$, say).

Hence $(a, b) \sim (c, d)$ iff both points in $\mathbb{R}^2$ lie on the same straight line of gradient 1, i.e. the line with equation $y - x = s$. The equivalence classes are these straight lines, corresponding to each $s \in \mathbb{R}$, and the union of them is the whole space $\mathbb{R}^2$.

6. $[0]_{\text{mod } 2} = \{ \ldots, -6, -4, -2, 0, 2, 4, 6, \ldots\}$
$= 2\mathbb{Z}$, the set of all even integers.

$[1]_{\text{mod } 2} = \{ \ldots, -7, -5, -3, -1, 1, 3, 5, 7, \ldots\}$
$= 2\mathbb{Z} + 1$, the set of all odd integers.

Chapter 3

1. $x^2 - 3x + 2 = (x - 1)(x - 2)$, so f is not defined at the points $x = 1$ and $x = 2$. The largest domain for f is therefore $\mathbb{R} \setminus \{1, 2\}$.

2. $f(t) = t^2 - 1$. $f(3) = (3)^2 - 1 = 8$. From Fig. S3.1, it is clear that $[-1, 2] \xrightarrow{f} [-1, 3]$. Similarly, $(-\infty, \infty) \xrightarrow{f} [-1, \infty)$ so $\operatorname{im} f = [-1, \infty)$. f is not surjective because for example, there is no $t \in \mathbb{R}$ such that $f(t) = -2$.

3. We have $\operatorname{im} f = [-1, 1] \subset \mathbb{R}$, so f is not surjective. Also, f is not 1–1 because, for example, $x = 0$ and $x = \pi$ both give $\sin x = 0$ (see Fig. S3.2). Therefore f is not injective. Choose the codomain of f as $[-1, 1]$. Redefine f as a function $f: \mathbb{R} \to [-1, 1]$ in which case f is surjective. Now restrict the domain of f to $[-\frac{1}{2}\pi, \frac{1}{2}\pi] \subset \mathbb{R}$. Define a function $g: [-\frac{1}{2}\pi, \frac{1}{2}\pi] \to [-1, 1]$ by $g(x) = \sin x$. The $g = f \restriction [-\frac{1}{2}\pi, \frac{1}{2}\pi]$ is the bijective restriction of f to $[-\frac{1}{2}\pi, \frac{1}{2}\pi]$.

4. $\{\det \boldsymbol{A} : \boldsymbol{A} \in \mathbb{R}^{n \times n}\} = \mathbb{R}$ for all $\boldsymbol{A} \in \mathbb{R}^{n \times n}$, so $\varphi: \mathbb{R}^{n \times n} \to \mathbb{R}$ is surjective, φ is not injective because many different $n \times n$ matrices in $\mathbb{R}^{n \times n}$ have the same determinant.

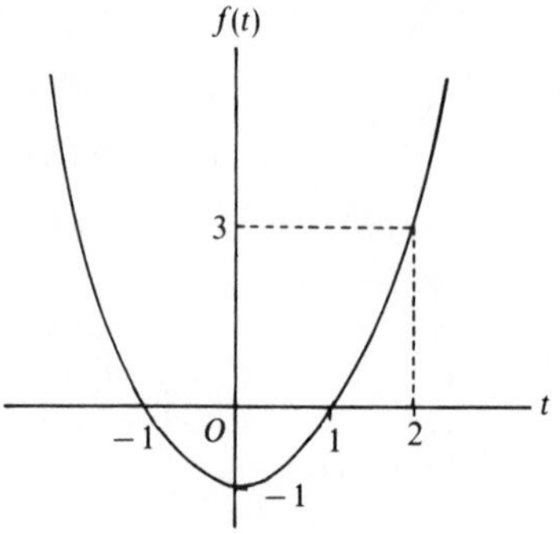

Fig. S3.1

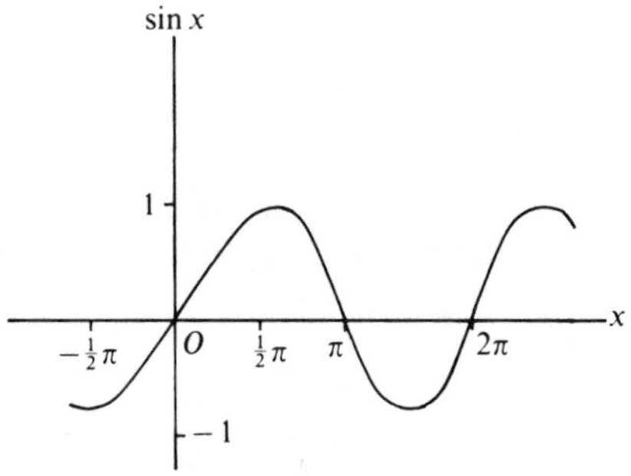

Fig. S3.2

$\varphi(\boldsymbol{A})=0 \Rightarrow \det \boldsymbol{A}=0$ and $\boldsymbol{A}$ is a singular matrix. If $\boldsymbol{B}\in\mathbb{R}^{n\times n}$ is a nonsingular matrix, then $\varphi(\boldsymbol{B})\in\mathbb{R}\setminus\{0\}$. Thus,

$$\varphi^{-1}(\mathbb{R}\setminus\{0\})=G\ell(n, \mathbb{R}).$$

5. (i) Obviously $f(x)=f(x)$, so that $x\sim x$ for all $x\in X$ (reflexive).
   (ii) $x\sim y\Rightarrow f(x)=f(y)$, i.e $f(y)=f(x)$ and so $y\sim x$ (symmetric).
   (iii) $x\sim y$ and $y\sim z$ together imply that $f(x)=f(y)$ and $f(y)=f(z)$. Thus, $f(x)=f(z)$, i.e. $x\sim z$ (transitive).

6. Suppose f is injective and that $y\in f(A)\cap f(X\setminus A)$. Then $y=f(a)$ and $y=f(b)$ for some $a\in A$ and $b\in X\setminus A$ (i.e. $b\notin A$). Since f is injective, $f(a)=f(b)\Rightarrow a=b$, but this contradicts the fact that $a\in A$ with $b\notin A$. Therefore

$$f(A)\cap f(X\setminus A)=\varnothing.$$

Now assume that $f(A)\cap f(X\setminus A)=\varnothing$ for all $A\subseteq X$. Let $x,y\in X$ with $x\neq y$. Since $f(\{x\})\cap f(X\setminus\{x\})=\varnothing$ (taking $A=\{x\}$), we have $f(x)\neq f(y)$. Thus, $x\neq y\Rightarrow f(x)\neq f(y)$ and so f is injective.

7. $[x]=[y] \Rightarrow \dot{x}(0)=\dot{y}(0)$, so α is single-valued and therefore a mapping. Also, $[x]\neq[y] \Rightarrow \dot{x}(0)\neq\dot{y}(0)$ so α is injective. Finally, for any $c\in\mathbb{R}$, there exists x such that $\dot{x}(0)=c$. Thus, α is also surjective and therefore bijective.

8. The set S of rational numbers in $[0\,,\,1]$ can be arranged in a sequence, grouping by common denominators in the following way:

$$(0),\ (1),\ (\tfrac{1}{2}),\ (\tfrac{1}{3},\tfrac{2}{3}),\ (\tfrac{1}{4},\tfrac{2}{4},\tfrac{3}{4}),\ (\tfrac{1}{5},\tfrac{2}{5},\tfrac{3}{5},\tfrac{4}{5}),\ \dots\ .$$

Any rational number that is repeated, such as $\frac{2}{4}$, $\frac{3}{6}$, etc., may be deleted from the sequence. Thus, S is countable.

Similarly, the set of rational numbers in any closed interval $[n\,,\,n+1]$ $(n\in\mathbb{Z})$ is countable. Denoting the rational number h/k by (h, k) the nonnegative rational numbers belonging to these closed intervals can be displayed in the following way:

$[0\,,\,1]$: $(1, 1), (1, 2), (1, 3), \dots, (1, i), \dots$

$[1\,,\,2]$: $(2, 1), (2, 2), \dots$

$[2\,,\,3]$: $(3, 1), \dots$

$\vdots$

$[i-1\,,\,i]$: $(i, 1), \dots$

Thus, one is led to a zig-zag count which counts (with possible repeats) all the non-negative rational numbers. Therefore, the union of these countable sets is countable, and similarly for the union of all the countable sets $[-1\,,\,0]$, $[-2\,,\,1]$, ... Thus, $\mathbb{Q}$ is countable.

9.
$$\begin{aligned}(g\circ f)(t)&=g\big(f(t)\big) && \text{(by the definition of } g\circ f)\\ &=g(t-1) && \text{(by the definition of } f)\\ &=\sin(t-1) && \text{(by the definition of } g).\end{aligned}$$

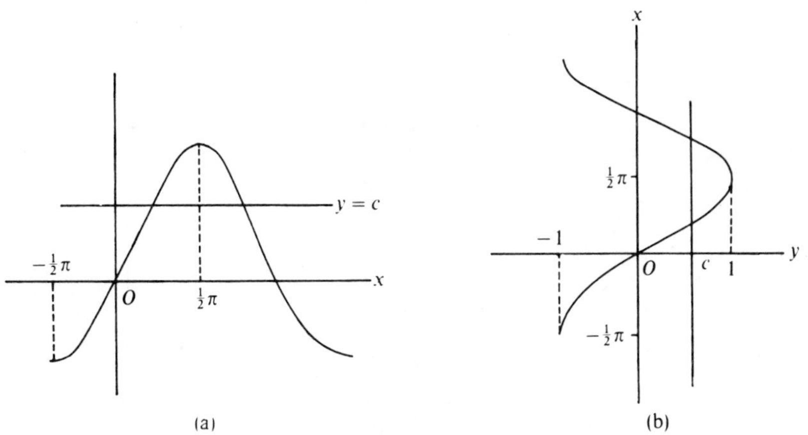

Fig. S3.3

Thus, $g \circ f$ is a mapping from $\mathbb{R}$ to $\mathbb{R}$ given by $t \mapsto \sin(t-1)$. Similarly, $(f \circ g)(t) = f(g(t)) = f(\sin t) = \sin t - 1$, so that $f \circ g : \mathbb{R} \to \mathbb{R}$ is defined by $t \mapsto \sin t - 1$.

Note that $gf \neq fg$ because, for example, $gf(\pi) \neq fg(\pi)$. Composition of maps is not commutative.

10. The line $y = c$ (with $0 < c < 1$) cuts the graph of f in an infinite number of points (Fig. S3.3(a)). The inverse of Fig. S3.3(a) is the graph shown in Fig. S3.3(b) which does not represent a function since any y in the interval $[-1, 1]$ (e.g. $y = c$) has an infinite number of images x. Thus, f^{-1} does not exist. However, by restricting the function f to the domain $[-\frac{1}{2}\pi, \frac{1}{2}\pi]$ in which the line $y = c$ cuts the graph in only one point then the bijective restriction $g = f \upharpoonright [-\pi, \pi]$ is obtained from f as in Exercise 3. Since g is a bijective mapping the inverse function g^{-1} does exist.

Chapter 4

1. For $\boldsymbol{A}, \boldsymbol{B} \in \mathbb{R}^{n \times n}$, we have $\boldsymbol{AB} \in \mathbb{R}^{n \times n}$ and $\boldsymbol{BA} \in \mathbb{R}^{n \times n}$, so $(\mathbb{R}^{n \times n}, \cdot)$ is closed. Matrix multiplication is associative, i.e. $\boldsymbol{A}(\boldsymbol{BC}) = (\boldsymbol{AB})\boldsymbol{C}$ for all $\boldsymbol{A}, \boldsymbol{B}, \boldsymbol{C} \in \mathbb{R}^{n \times n}$. Thus, $(\mathbb{R}^{n \times n}, \cdot)$ is a semigroup. If $\boldsymbol{A} \in \mathbb{R}^{n \times n}$, then $\boldsymbol{A}\mathbf{I}_n = \boldsymbol{A} = \mathbf{I}_n\boldsymbol{A}$, so $\mathbf{I}_n$ is the identity in $\mathbb{R}^{n \times n}$.

2. Let F be the set of all mappings $f : X \to X$, with $X \neq \varnothing$. If $f, g \in F$, then $g \circ f : X \to X$ is an element of F, so F is closed under composition.

$$h \circ (g \circ f) = (h \circ g) \circ f \quad \forall f, g, h \in F, \quad \text{(associativity)}.$$

Identity map $\mathrm{id}_X : X \to X$ defined by $\mathrm{id}_X(x) = x$ for $x \in X$.

3. (i) $\mathrm{G}\ell(n, \mathbb{R}) \subset \mathbb{R}^{n \times n}$ so that $(\mathrm{G}\ell(n, \mathbb{R}), \cdot)$ is associative by Exercise 1. $\mathrm{G}\ell(n, \mathbb{R})$ is closed with respect to matrix multiplication because, if $\boldsymbol{A}, \boldsymbol{B} \in \mathrm{G}\ell(n, \mathbb{R})$, then $\det \boldsymbol{A} \neq 0$ and $\det \boldsymbol{B} \neq 0$. The products $\boldsymbol{AB}$ and $\boldsymbol{BA}$ are real $n \times n$ matrices with nonzero determinants, e.g. $\det \boldsymbol{AB} = \det \boldsymbol{A} \det \boldsymbol{B} \neq 0$. Therefore $\boldsymbol{AB}, \boldsymbol{BA} \in \mathrm{G}\ell(n, \mathbb{R})$. The matrix $\mathbf{I}_n$ belongs to $\mathrm{G}\ell(n, \mathbb{R})$ and is thus the identity in $\mathrm{G}\ell(n, \mathbb{R})$. Now $\det \boldsymbol{A} \neq 0$ for all $\boldsymbol{A} \in \mathrm{G}\ell(n, \mathbb{R})$ and so the real $n \times n$ inverse matrix $\boldsymbol{A}^{-1}$ exists. Furthermore,[18] $\det \boldsymbol{A}^{-1} = 1/\det \boldsymbol{A} \neq 0$. Thus, $\boldsymbol{A}^{-1} \in \mathrm{G}\ell(n, \mathbb{R})$ and $(\mathrm{G}\ell(n, \mathbb{R}), \cdot)$ is a group.

(ii)
$$\begin{aligned} \boldsymbol{A} + \boldsymbol{B} &\in \mathbb{R}^{m \times n} && \text{(closed under addition)}, \\ \boldsymbol{A} + (\boldsymbol{B} + \boldsymbol{C}) &= (\boldsymbol{A} + \boldsymbol{B}) + \boldsymbol{C} && \text{(associativity)}, \\ \boldsymbol{A} + \boldsymbol{B} &= \boldsymbol{B} + \boldsymbol{A} && \text{(commutativity)}, \end{aligned}$$

for all $\boldsymbol{A}, \boldsymbol{B}, \boldsymbol{C} \in \mathbb{R}^{m \times n}$. The identity element is $\mathbf{0} \in \mathbb{R}^{m \times n}$. For each $\boldsymbol{A} \in \mathbb{R}^{m \times n}$, there is a unique inverse $-\boldsymbol{A} \in \mathbb{R}^{m \times n}$ such that

$$\boldsymbol{A} = \begin{bmatrix} a_{11} & \cdots & a_{1n} \\ a_{21} & \cdots & a_{2n} \\ \vdots & & \vdots \\ a_{m1} & \cdots & a_{mn} \end{bmatrix} \Rightarrow -\boldsymbol{A} = \begin{bmatrix} -a_{11} & \cdots & -a_{1n} \\ -a_{21} & \cdots & -a_{2n} \\ \vdots & & \vdots \\ -a_{m1} & \cdots & -a_{mn} \end{bmatrix},$$

and so $\boldsymbol{A} + (-\boldsymbol{A}) = \mathbf{0}$ and $(-\boldsymbol{A}) + \boldsymbol{A} = \mathbf{0}$.

4. $G \subset F$ with F as in Exercise 2. Because F is a semigroup, so is G. Let ι be the identity map, so that $\iota(x)=x$ for all $x \in X$. Then $\iota \in G$. Also, $g\iota(x)=g(\iota(x))=g(x)$ and $g(x)=\iota(g(x))=\iota \circ g(x)$ for all $g \in G$ and all $x \in X$, so that $\iota \circ g=g=g \circ \iota$ and ι is the identity in G. Also, g^{-1} exists because g is bijective, so

$$g \circ g^{-1}=\iota=g^{-1} \circ g.$$

Thus, G under composition is a group.

5. (i) Exactly as in Example 4.9, e.g.

$$\frac{p_1}{q_1}+\frac{p_2}{q_2}=\frac{p_1 q_2+p_2 q_1}{q_1 q_2} \in \mathbb{Q} \quad \text{(closure)}$$

for all p_1/q_1 and p_2/q_2 in $\mathbb{Q}$ ($q_1 q_2 \neq 0$ because $q_1 \neq 0$ and $q_2 \neq 0$). $0 \in \mathbb{Q}$ and $-q \in \mathbb{Q}$ for all $q \in \mathbb{Q}$.

(ii) Zero must be excluded from $\mathbb{Q}$ otherwise some elements in $\mathbb{Q}$ do not have inverses. For example, $p/q \in \mathbb{Q} (p=0, q \neq 0)$ but $q/p=q/0$ is not defined. For all p_1/q_1 and p_2/q_2 in $\mathbb{Q}$, we have $(p_1/q_1) \cdot (p_2/q_2) \in \mathbb{Q}$ (closure). The identity is $1=1/1 \in \mathbb{Q}$. If $p/q \in \mathbb{Q} \backslash \{0\}$, then $p \neq 0$ and $q \neq 0$, so that the inverse to p/q is q/p which exists $((p/q) \cdot (q/p)=1)$. The rational numbers are commutative and so $(\mathbb{Q} \backslash \{0\}, \cdot)$ is an abelian group.

(iii) Matrix multiplication is not commutative. For example, in the case of $n=2$, the matrices

$$\boldsymbol{A}=\begin{bmatrix} 2 & 3 \\ 1 & 4 \end{bmatrix} \quad \text{and} \quad \boldsymbol{B}=\begin{bmatrix} 1 & 2 \\ 1 & 3 \end{bmatrix}$$

belong to $G\ell(2, \mathbb{R})$, but $\boldsymbol{AB} \neq \boldsymbol{BA}$.

6. If $x(t)=e^{at}x_0$, then $\dot{x}(t)=a(e^{at}x_0)=ax(t)$ and $x(0)=e^0 x_0=x_0$. Proof that $\{e^{at}: t \in \mathbb{R}\}$ is an abelian group under multiplication of operators follows closely that for the group $\{e^t: t \in \mathbb{R}\}$ in the text. The identity is $e^{a0}=e^0=1$ and the inverse of any element e^{at} is e^{-at}.

7. From Table 4.1 given in Example 4.2, $(S, \cdot)$ is closed. Matrix multiplication is associative. The identity in $(S, \cdot)$ is $\boldsymbol{I}=\mathbf{I}_2$, and inverses of elements in S are $\boldsymbol{I}_2^{-1}=\boldsymbol{I}_2$, $\boldsymbol{J}^{-1}=-\boldsymbol{J}$, $\boldsymbol{K}^{-1}=-\boldsymbol{K}$, $\boldsymbol{L}^{-1}=-\boldsymbol{L}$, $(-\boldsymbol{I})^{-1}=-\boldsymbol{I}$, $(-\boldsymbol{J})^{-1}=\boldsymbol{J}$, $(-\boldsymbol{K})^{-1}=\boldsymbol{K}$, $(-\boldsymbol{L})^{-1}=\boldsymbol{L}$. Thus, $(S, \cdot)$ is a group, but not an abelian group because the commutative law is not valid in $(S, \cdot)$, e.g. $\boldsymbol{KJ}=-\boldsymbol{L}$ and $\boldsymbol{JK}=\boldsymbol{L}$, so $\boldsymbol{KJ} \neq \boldsymbol{JK}$.

The order of S is the number of different elements in S, i.e. $|S|=8$ (see Section 2.1). The group S is called the *quaternion group*.

8. $\mathrm{So}(n, \mathbb{R})$ is closed with respect to multiplication because, if $\boldsymbol{A}, \boldsymbol{B} \in \mathrm{So}(n, \mathbb{R})$, then $\boldsymbol{AB}$ is an $n \times n$ matrix and $(\boldsymbol{AB})^{\mathsf{T}}=\boldsymbol{B}^{\mathsf{T}}\boldsymbol{A}^{\mathsf{T}}=\boldsymbol{B}^{-1}\boldsymbol{A}^{-1}=(\boldsymbol{AB})^{-1}$, i.e. $\boldsymbol{AB}$ is orthogonal. Thus, $\boldsymbol{AB}$ is an $n \times n$ orthogonal matrix and so $\boldsymbol{AB} \in \mathrm{So}(n, \mathbb{R})$. Matrix multiplication is associative. The set $\mathrm{So}(n, \mathbb{R})$ contains the identity $\mathbf{I}_n$ with respect to multiplication $(\mathbf{I}_n^{\mathsf{T}}=\mathbf{I}_n^{-1}$ so $\mathbf{I}_n \in \mathrm{So}(n, \mathbb{R}))$. For $\boldsymbol{A} \in \mathrm{So}(n, \mathbb{R})$, the matrix $\boldsymbol{A}$ has an inverse in $\mathrm{So}(n, \mathbb{R})$ with respect to matrix multiplication because $\boldsymbol{A}^{\mathsf{T}}$ exists and $\boldsymbol{A}^{\mathsf{T}}=\boldsymbol{A}^{-1}$. The inverse $\boldsymbol{A}^{-1}$ belongs to $\mathrm{So}(n, \mathbb{R})$ because

$$(\boldsymbol{A}^{-1})^{\mathsf{T}}=(\boldsymbol{A}^{\mathsf{T}})^{\mathsf{T}}=\boldsymbol{A}=(\boldsymbol{A}^{-1})^{-1} \quad \forall\, \boldsymbol{A} \in \mathrm{So}(n, \mathbb{R}).$$

Therefore, A^{-1} is an $n\times n$ matrix and orthogonal so that $A^{-1}\in \mathrm{So}(n,\mathbb{R})$. Note that $\mathrm{So}(n,\mathbb{R})$ is a subgroup of $\mathrm{G}\ell(n,\mathbb{R})$.

9. $(H,\cdot)$ is a subgroup with identity 1 and inverses $1^{-1}=1, (-1)^{-1}=-1$. Multiplication tables are given below.

H:

$\cdot$	1	−1
1	1	−1
−1	−1	1

G:

$\cdot$	1	−1	j	−j
1	1	−1	j	−j
−1	−1	1	−j	j
j	j	−j	−1	1
−j	−j	j	1	−1

$G/H=\{H, \mathrm{j}H\}$ with $\mathrm{j}H=\{\mathrm{j}, -\mathrm{j}\}$.

$$(\mathrm{j}H)(\mathrm{j}H)=-H=\{-1,1\}=H.$$

Then there is a map $\varphi: H\to G/H$ defined by $\varphi(1)=H$ and $\varphi(-1)=\mathrm{j}H$ which is an isomorphism from H to G/H, i.e. $G/H\simeq H$.

10.
$$\begin{bmatrix} a & -b \\ b & a \end{bmatrix}=\begin{bmatrix} a & 0 \\ 0 & a \end{bmatrix}+\begin{bmatrix} 0 & -b \\ b & 0 \end{bmatrix}=a\boldsymbol{I}+b\boldsymbol{J},$$

where

$$\boldsymbol{I}=\begin{bmatrix} 1 & 0 \\ 0 & 1 \end{bmatrix} \quad\text{and}\quad \boldsymbol{J}=\begin{bmatrix} 0 & -1 \\ 1 & 0 \end{bmatrix}.$$

Correspondence between $a\boldsymbol{I}+b\boldsymbol{J}$ and the complex number $a+b\mathrm{j}$ ($\mathrm{j}=\sqrt{(-1)}$) is 1–1 and onto, i.e. bijective. The mapping $\theta: S\to\mathbb{C}$ between the set $S=\{a\boldsymbol{I}+b\boldsymbol{J}: a,b\in\mathbb{R}\}$ and $\mathbb{C}$ defined by $\theta(a\boldsymbol{I}+b\boldsymbol{J})=a+b\mathrm{j}$ preserves addition and multiplication because

$$(a\boldsymbol{I}+b\boldsymbol{J})+(a'\boldsymbol{I}+b'\boldsymbol{J})=(a+a')\boldsymbol{I}+(b+b')\boldsymbol{J},$$

$$(a+b\mathrm{j})+(a'+b'\mathrm{j})=(a+a')+(b+b')\mathrm{j},$$

i.e. $\theta(\boldsymbol{C})+\theta(\boldsymbol{C}')=\theta(\boldsymbol{C}+\boldsymbol{C}')$, where $\boldsymbol{C}=a\boldsymbol{I}+b\boldsymbol{J}\in S$ and $\boldsymbol{C}'=a'\boldsymbol{I}+b'\boldsymbol{J}\in S$, and $\boldsymbol{J}^2=-\boldsymbol{I}$, while

$$(a\boldsymbol{I}+b\boldsymbol{J})(a'\boldsymbol{I}+b'\boldsymbol{J})=(aa'-bb')\boldsymbol{I}+(ab'+a'b)\boldsymbol{J},$$

$$(a+b\mathrm{j})(a'+b'\mathrm{j})=(aa'-bb')+(ab'+a'b)\mathrm{j},$$

i.e. $\theta(\boldsymbol{C})\theta(\boldsymbol{C}')=\theta(\boldsymbol{C}\boldsymbol{C}')$. Thus, $S\cong\mathbb{C}$.

11.
$$\theta(t_1+t_2)=\begin{bmatrix} \mathrm{e}^{a(t_1+t_2)} & 0 \\ 0 & \mathrm{e}^{a(t_1+t_2)} \end{bmatrix},$$

$$\theta(t_1)\theta(t_2)=\begin{bmatrix} \mathrm{e}^{at_1} & 0 \\ 0 & \mathrm{e}^{at_1} \end{bmatrix}\begin{bmatrix} \mathrm{e}^{at_2} & 0 \\ 0 & \mathrm{e}^{at_2} \end{bmatrix}$$

$$=\begin{bmatrix} \mathrm{e}^{a(t_1+t_2)} & 0 \\ 0 & \mathrm{e}^{a(t_1+t_2)} \end{bmatrix},$$

so that $\theta(t_1+t_2)=\theta(t_1)\theta(t_2)$ and θ is a homomorphism. Note the operation is addition of real numbers in $\mathbb{R}$ but multiplication of matrices in $G\ell(2, \mathbb{R})$. The map θ is actually a monomorphism if $a \neq 0$, because it is a 1–1 (but not an onto) map.

12. If e_1 and e_2 are the identities of G_1 and G_2 respectively, then (4.20) gives $\theta(e_1)=e_2$, whence $e_1 \in \ker \theta$, and so the kernel of θ is non-empty. If $x,y \in \ker \theta$, so that $\theta(x)=e_2$ and $\theta(y)=e_2$, then

$$\begin{aligned}\theta(x^{-1}y) &= \theta(x^{-1})\theta(y) && \text{(by (4.19))}\\ &= [\theta(x)]^{-1}\theta(y) && \text{(by (4.21))}\\ &= (e_2)^{-1}e_2 = e_2.\end{aligned}$$

Hence $x^{-1}y \in \ker \theta$ and so $x,y \in \ker \theta \Rightarrow x^{-1}y \in \ker \theta$. It follows from (4.22) that $\ker \theta$ is a subgroup of G_1.

13. In the definition of the product of two subsets

$$XY \triangleq \{xy : x \in X,\ y \in Y\},$$

the subsets X and Y are clearly non-empty (see also Note 4.11).

Since the product XY of any two subsets X and Y of a group G is a subset of G, the set $\mathscr{P}(G)$ is closed under the type of multiplication envisaged. For associativity, it is required to prove that $X(YZ)=(XY)Z$ for any non-empty subsets X, Y, Z of G. To do this, it must be proved that $g \in X(YZ) \Rightarrow g \in (XY)Z$, so that $X(YZ) \subseteq (XY)Z$, and also that $g \in (XY)Z \Rightarrow g \in X(YZ)$, so that $(XY)Z \subseteq X(YZ)$ (see (2.1)). Both cases can be dealt with together in the following way:

$$\begin{aligned} g \in X(YZ) &\Leftrightarrow g = xs \text{ for some } x \in X \text{ and } s \in YZ\\ &\Leftrightarrow g = x(yz) \text{ for some } x \in X,\ y \in Y, \text{ and } z \in Z\\ &\Leftrightarrow g = (xy)z \quad \text{(associativity in } G)\\ &\Leftrightarrow g = tz \text{ for some } t \in XY\\ &\Leftrightarrow g \in (XY)Z.\end{aligned}$$

Since each step is reversible, the above statements can be read either from top to bottom to give $g \in X(YZ) \Rightarrow g \in (XY)Z$ or from bottom to top to give $g \in X(YZ) \Leftarrow g \in (XY)Z$. Thus, $X(YZ)=(XY)Z$, so the associative law holds together with closure. Hence, $\mathscr{P}(G)$ is a semigroup with respect to the set multiplication of Section 4.5. In particular, brackets can be dropped from $X(YZ)$ and $(XY)Z$, since XYZ is the same set irrespective of the order in which multiplication is carried out.

14. Since H is a subgroup, it is closed under multiplication so that $HH \subseteq H$. Also, with 1 as the identity element of H (and of G), we have $1h \in HH$ for any $h \in H$, i.e. $h \in HH$, so that $H \subseteq HH$. It follows that $HH = H$.

15. It is shown in the text that $\mathscr{K}$ is a semigroup. Now $H = 1H \in \mathscr{K}$ and $K = gH \in \mathscr{K}$, so

$$KH = (gH)H = g(HH) = gH = K.$$

Thus, the coset H itself is the identity element in $\mathscr{K}$. Finally, for any coset $K = gH \in \mathscr{K}$

there exists a coset $g^{-1}H \in \mathscr{K}$ since $g^{-1} \in G$. Furthermore,

$$(g^{-1}H)(gH) = (g^{-1}g)H = 1H = H,$$

$$(gH)(g^{-1}H) = (gg^{-1})H = 1H = H.$$

Since H is the identity in $\mathscr{K}$, the coset $g^{-1}H$ is the inverse of gH in $\mathscr{K}$.

16. $\nu(g_1g_2) = (g_1g_2)H = (g_1H)(g_2H) = \nu(g_1)\nu(g_2)$, so ν is a homomorphism. Also, if $gH \in G/H$, then $gH = \nu(g)$, so that ν is surjective. Thus, ν is an epimorphism. Again, $g \in \ker \nu$ iff $\nu(g) = H$, the identity element in G/H, i.e. iff $gH = H$.

Now, if $gH = H$, then $g1 \in H$, i.e. $g \in H$. On the other hand, if $g \in H$, then $gH = H$. Thus $gH = H$ iff $g \in H$.

It has now been shown that $g \in \ker \nu \Leftrightarrow g \in H$ and so $\ker \nu = H$.

17. Let $H = \ker \theta$ and consider the map $\varphi: G_1/H \to G_2$ defined for all $g \in G_1$ by $\varphi(gH) = \theta(g)$. To justify this definition, it is necessary to demonstrate that using a different label for the element $gH \in G_1/H$, say kH, still gives $\varphi(kH) = \theta(g)$ (this is the difficulty posed by equivalence classes and cosets having many possible labels as discussed in Section 4.5). For $g, k \in G_1$, we have

$$\begin{aligned} gH = kH \quad &\Leftrightarrow \quad g^{-1}(gH) = g^{-1}(kH) \\ &\Leftrightarrow \quad H = (g^{-1}k)H \\ &\Leftrightarrow \quad g^{-1}k \in H \quad \text{(see Solution 16 above).} \end{aligned}$$

But $H = \ker \theta$, so that

$$\begin{aligned} gH = kH \quad &\Leftrightarrow \quad g^{-1}k \in \ker \theta \\ &\Leftrightarrow \quad \theta(g^{-1}k) = e_2 \quad \text{(the identity in } G_2\text{)} \\ &\Leftrightarrow \quad \theta(g^{-1})\theta(k) = e_2 \quad \text{(since } \theta \text{ is a homomorphism)} \\ &\Leftrightarrow \quad (\theta(g))^{-1}\theta(k) = e_2 \quad \text{(by (4.21))} \\ &\Leftrightarrow \quad \theta(g) = \theta(k). \end{aligned}$$

The definition of map φ given above is therefore justified. Also, the result

$$\theta(g) = \theta(k) \quad \Rightarrow \quad gH = kH$$

means that, by the definition of φ, we have

$$\varphi(gH) = \varphi(kH) \quad \Rightarrow \quad gH = kH,$$

i.e. φ is an injective mapping. Also, φ is a homomorphism because, if $g, k \in G_1$, then

$$\begin{aligned} \varphi\big((gH)(kH)\big) &= \varphi\big((gk)H\big) \quad \text{(by (4.24))} \\ &= \theta(gk) \quad \text{(by the definition of } \varphi\text{)} \\ &= \theta(g)\theta(k) \quad \text{(since } \theta \text{ is a homomorphism)} \\ &= \varphi(gH)\varphi(kH) \quad \text{(by the definition of } \varphi\text{).} \end{aligned}$$

Finally, the image set of φ is equal to the image set of θ, because

$$\operatorname{im} \varphi = \{\varphi(gH) : g \in G_1\} = \{\theta(g) : g \in G_1\} = \operatorname{im} \theta.$$

Thus, φ is a monomorphism whose image set is equal to im θ. Taking the codomain of φ as im $\varphi=\text{im}\,\theta$ then makes φ a bijective mapping and hence an isomorphism from G_1/H to im θ, i.e.

$$\text{im}\,\theta \cong G_1/\ker\theta.$$

18. $\ker\theta=\{t\in\mathbb{R}:\theta(t)=1\}$.

$$\theta(t)=1 \quad\Rightarrow\quad \cos 2\pi t+\mathrm{j}\sin 2\pi t=1$$
$$\Rightarrow\quad \cos 2\pi t=1 \text{ and } \sin 2\pi t=0,$$

i.e. $t=k\in\mathbb{Z}$. Hence $\ker\theta=\mathbb{Z}$. Now $z=\theta(t)=\cos 2\pi t+\mathrm{j}\sin 2\pi t=e^{\mathrm{j}2\pi t}$ satisfies $|z|=1$ with arg z arbitrary so that im $\theta=G$. By the first isomorphism theorem, $G\cong\mathbb{R}/\mathbb{Z}$.

Chapter 5

1. (i) $3\mathbb{Z}=\{a\in\mathbb{Z}:3|a\}=\{0, \pm 3, \pm 6, \pm 9, \ldots\}$. It is easy to show that this set of integers satisfies the ring axioms and that multiplication of integers is commutative. However, $1\notin 3\mathbb{Z}$, so the ring does not have an identity.

(ii) $\mathbb{Z}_2=\{0, 1\}$, Example 5.6 shows that $\mathbb{Z}_2$ is a commutative ring, with $1\cdot 1=1$ and $0\cdot 1=0$, so 1 is the nonzero identity.

2. $\mathbb{Z}_6=\{0, 1, 2, 3, 4, 5\}$. We have $2\cdot 3=6=0$ and $3\cdot 4=12=0$. Thus, $\mathbb{Z}_6$ contains zero-divisors such as 2, 3, and 4 and therefore is not an integral domain.

3. (i) First note that $x[y+(-y)]=x0$. To prove that $x0=0$, note that $0+0=0$, so that $x(0+0)=x0$. By the distributive law, $x0+x0=x0$. Thus, $x0=x0+(-x0)=0$. Then

$$x[y+(-y)]=0,$$

and again the distributive law gives $xy+x(-y)=0$, whence

$$x(-y)=-(xy).$$

(ii) $ax=ay$ gives $ax+(-ay)=0$, and so $ax+a(-y)=0$, by (i); hence $a(x-y)=0$, by the distributive law.

4. $S+T=4\mathbb{Z}+6\mathbb{Z}=\{0, \pm 4, \pm 8, \pm 12, \ldots\}+\{0, \pm 6, \pm 12, \pm 18, \ldots\}$

$$=\{0, \pm 2, \pm 4, \pm 6, \ldots\}=2\mathbb{Z}.$$

$$ST=4\mathbb{Z}.\,6\mathbb{Z}=\{0, \pm 24, \pm 48, \pm 72, \ldots\}=24\mathbb{Z}.$$

Note that $4\mathbb{Z}$ and $6\mathbb{Z}$ are subsets of the ring $\mathbb{Z}$. On the other hand, $\mathbb{Z}_4+\mathbb{Z}_6$ has no meaning because the rings $\mathbb{Z}_4$ and $\mathbb{Z}_6$ are not subsets of some larger ring. Note also that $S\cap T=12\mathbb{Z}$.

5. Tables for S are given below, and it is easy to show that S satisfies all the properties

+	0	a
0	0	a
a	a	0

	0	a
0	0	0
a	0	a

of a ring. However, the identity in the subring S is a whereas the identity in the ring R is c. The fact that the identity of a subring may be different from the identity of the parent ring is an awkward fact of life.

6. Let $p=(p_i)$, $q=(q_i)$, $r=(r_i)$ all belonging to $\mathbb{R}[z]$, where $p_i,q_i,r_i \in \mathbb{R}$. Then

$$q+r=(q_i+r_i) \quad \text{and} \quad p+(q+r)=(p_i+(q_i+r_i)).$$

Similarly,

$$p+q=(p_i+q_i) \quad \text{and} \quad (p+q)+r=((p_i+q_i)+r_i).$$

Since addition in $\mathbb{R}$ is associative,

$$p+(q+r)=(p+q)+r,$$

i.e. addition in $\mathbb{R}[z]$ is associative.

7. Let p, q, and r be as given in Solution 6. Using (5.4), we get

$$pq=(s_i) \text{ where } s_i=\sum_{k+j=i} p_k q_j.$$

Then

$$(pq)r=(t_i) \text{ where } t_i=\sum_{k+j=i} s_k r_j=\sum_{k+j=i}\left(\sum_{h+l=k} p_h q_l\right) r_j$$

$$=\sum_{h+l+j=i} p_h q_l r_j. \qquad (*)$$

Similarly,

$$qr=(u_i) \text{ where } u_i=\sum_{k+j=i} q_k r_j$$

and

$$p(qr)=(v_i) \text{ where } v_i=\sum_{k+j=i} p_k u_j=\sum_{k+j=i} p_k\left(\sum_{h+l=j} q_h r_l\right)$$

$$=\sum_{k+h+l=i} p_k q_h r_l. \qquad (**)$$

Since k, h, l, and j are dummy variables, the expressions in (*) and (**) are equal, i.e. $p(qr)=(pq)r$ and multiplication in $\mathbb{R}[z]$ is associative.

8. It is a lengthy but straightforward process to verify that R is a noncommutative ring with a 1. Direct multiplication shows that every nonzero element of R has a multiplicative inverse

$$\frac{1}{|a|^2+|b|^2}\begin{bmatrix} \bar{a} & -b \\ \bar{b} & a \end{bmatrix}.$$

Thus, R is a noncommutative division ring.

9. A field F is a commutative ring with identity. Let $a,b \in F$ with $a \neq 0$, and suppose that $ab=0$. Then $(1/a)ab=(1/a)0=0$; also $(1/a)ab=[(1/a)a]b=1b=b$. Hence, $b=0$ and so F has no zero-divisors, i.e. F is an integral domain.

An integral domain D is a commutative ring with identity and contains no zero-divisors. However, not all nonzero elements of D need have multiplicative-inverses in which case D is not a field. A case in point is $\mathbb{Z}$ which is a ring by Example 5.1. Since multiplication of integers is commutative, $\mathbb{Z}$ is a commutative ring and $\mathbb{Z}$ has identity 1. For all nonzero $a,b \in \mathbb{Z}$, we have $ab \neq 0$, so $\mathbb{Z}$ has no zero-divisors. Hence $\mathbb{Z}$ is an integral domain. But, for example, $2 \in \mathbb{Z}$ does not have a multiplicative inverse, because $\frac{1}{2} \notin \mathbb{Z}$. Thus, $\mathbb{Z}$ is not a field.

Chapter 6

1. From the definitions of addition and scalar multiplication, it is clear that $f+g$ and αf belong to C[0, 1]. The null function on C[0, 1] is the zero element 0, i.e. $0(t)=0$ for $t \in [0, 1]$. With $\alpha = -1$, scalar multiplication of f by -1 yields

$$(-1f)(t)=(-f)(t)=-f(t),$$
$$[f+(-f)](t)=f(t)-f(t)=0=0(t);$$

thus $f+(-f)=0$, and so $-f$ is the additive inverse of f. For all $f,g \in$ C[0, 1], we have $f+g=g+f$, so that C[0, 1] is an additive abelian group. It is then a simple matter to verify properties (6.3) with $m,n \in \mathbb{R}$ and $u,v \in$ C[0, 1].

2. Solutions of $\ddot{x}+x=0$ are of the form $A \cos t + B \sin t$, where A and B are arbitrary constants. It is easy to check that the set of all such solutions is an additive abelian group. Scalar multiplication of the above solution by $\alpha \in \mathbb{R}$ is given by $(\alpha A)\cos t + (\alpha B)\sin t$, and (6.3) is easily verified.

3.
$$\begin{bmatrix} 1 & 3 & 1 \\ 2 & -1 & -5 \\ -3 & 2 & 8 \\ 4 & 1 & -7 \end{bmatrix} \begin{bmatrix} c_1 \\ c_2 \\ c_3 \end{bmatrix} = \begin{bmatrix} 0 \\ 0 \\ 0 \\ 0 \end{bmatrix}$$

$$\Rightarrow \begin{bmatrix} 1 & 3 & 1 \\ 2 & -1 & -5 \\ -3 & 2 & 8 \end{bmatrix} \begin{bmatrix} c_1 \\ c_2 \\ c_3 \end{bmatrix} = \begin{bmatrix} 0 \\ 0 \\ 0 \end{bmatrix}.$$

The determinant of the 3×3 matrix is zero, so there exists a nontrivial solution to the set of three homogeneous equations. That is, not all of c_1, c_2, c_3 are zero and so $\boldsymbol{x}_1, \boldsymbol{x}_2, \boldsymbol{x}_3$ are linearly dependent. The original 4×3 matrix has rank 2, and partitioning and block multiplication gives

$$\begin{bmatrix} 1 & 3 \\ 2 & -1 \end{bmatrix} \begin{bmatrix} c_1 \\ c_2 \end{bmatrix} + \begin{bmatrix} 1 \\ -5 \end{bmatrix} c_3 = \begin{bmatrix} 0 \\ 0 \end{bmatrix}.$$

Then
$$\begin{bmatrix} c_1 \\ c_2 \end{bmatrix} = \begin{bmatrix} 1 & 3 \\ 2 & -1 \end{bmatrix}^{-1} \begin{bmatrix} -1 \\ 5 \end{bmatrix} c_3$$
$$= -\frac{1}{7}\begin{bmatrix} -1 & -3 \\ -2 & 1 \end{bmatrix} \begin{bmatrix} -1 \\ 5 \end{bmatrix} c_3 = \begin{bmatrix} 2 \\ -1 \end{bmatrix} c_3.$$

c_3 is arbitrary, so choose $c_3 = 1$. Then $c_1 = 2$, $c_2 = -1$, and so

$$2\boldsymbol{x}_1 - \boldsymbol{x}_2 + \boldsymbol{x}_3 = \boldsymbol{0}.$$

To conclude, $\{\boldsymbol{x}_1, \boldsymbol{x}_2\}$ is a linearly independent set, and $\boldsymbol{x}_3 = \boldsymbol{x}_2 - 2\boldsymbol{x}_1$.

4. $[\boldsymbol{e}_1, \boldsymbol{e}_2, \boldsymbol{e}_3]\boldsymbol{x} = [\boldsymbol{x}_1, \boldsymbol{x}_2, \boldsymbol{x}_3]\boldsymbol{x}$, i.e.

$$\boldsymbol{I}\boldsymbol{x} = \boldsymbol{x} = \begin{bmatrix} 2 & 1 & 1 \\ -1 & 2 & -1 \\ 3 & -1 & -1 \end{bmatrix} \begin{bmatrix} 1 \\ 2 \\ 3 \end{bmatrix} = \begin{bmatrix} 7 \\ 0 \\ -2 \end{bmatrix}.$$

5. Let the bases W and Z be given in terms of column vectors by $W = \{\boldsymbol{w}_1, \boldsymbol{w}_2, \boldsymbol{w}_3\}$ and $Z = \{\boldsymbol{z}_1, \boldsymbol{z}_2, \boldsymbol{z}_3\}$ and let any vector $\boldsymbol{y}$ be given relative to W and Z by

$$\boldsymbol{y} = y_1\boldsymbol{w}_1 + y_2\boldsymbol{w}_2 + y_3\boldsymbol{w}_3 = \tilde{y}_1\boldsymbol{z}_1 + \tilde{y}_2\boldsymbol{z}_2 + \tilde{y}_3\boldsymbol{z}_3.$$

Represent these two expressions by writing

$\boldsymbol{y}_W = (y_1, y_2, y_3)$ and $\boldsymbol{y}_Z = (\tilde{y}_1, \tilde{y}_2, \tilde{y}_3)$ $(\in \mathbb{R}^3)$. Denote the matrix $[\boldsymbol{w}_1, \boldsymbol{w}_2, \boldsymbol{w}_3]$ by $\boldsymbol{W}$ and $[\boldsymbol{z}_1, \boldsymbol{z}_2, \boldsymbol{z}_3]$ by $\boldsymbol{Z}$; then

$$\boldsymbol{W}\boldsymbol{y}_W = \boldsymbol{Z}\boldsymbol{y}_Z.$$

It follows that

$$\boldsymbol{y}_W = \boldsymbol{W}^{-1}\boldsymbol{Z}\boldsymbol{y}_Z,$$

where $\boldsymbol{W}^{-1}$ exists because the vectors $\boldsymbol{w}_1, \boldsymbol{w}_2, \boldsymbol{w}_3$ are basis vectors and therefore linearly independent. Writing $\boldsymbol{P} = \boldsymbol{W}^{-1}\boldsymbol{Z}$ so that $\boldsymbol{y}_W = \boldsymbol{P}\boldsymbol{y}_Z$, we obtain

$$\boldsymbol{Z} = \begin{bmatrix} 1 & 1 & 1 \\ 0 & 0 & 1 \\ 0 & 1 & 1 \end{bmatrix}, \quad \boldsymbol{W} = \begin{bmatrix} 1 & 2 & 1 \\ 1 & 2 & 2 \\ 2 & 1 & 2 \end{bmatrix},$$

$$\boldsymbol{W}^{-1} = \tfrac{1}{3}\begin{bmatrix} 2 & -3 & 2 \\ 2 & 0 & -1 \\ -3 & 3 & 0 \end{bmatrix}, \quad \boldsymbol{P} = \tfrac{1}{3}\begin{bmatrix} 2 & 4 & 1 \\ 2 & 1 & 1 \\ -3 & -3 & 0 \end{bmatrix},$$

$$\boldsymbol{P}^{-1} = \begin{bmatrix} -1 & 1 & -1 \\ 1 & -1 & 0 \\ 1 & 2 & 2 \end{bmatrix}.$$

Now, $y_W = Py_Z$, $x_W = Px_Z$, and $y_Z = Ax_Z$. Therefore $x_Z = P^{-1}x_W$ and $Py_Z = PAx_Z$. It follows that

$$y_W = Py_Z = PAP^{-1}x_W = Bx_W,$$

where

$$B = PAP^{-1} = \tfrac{1}{3}\begin{bmatrix} 8 & 22 & 7 \\ 5 & 16 & 7 \\ -9 & -18 & -9 \end{bmatrix}.$$

6. In (6.15), addition is pointwise and multiplication is map composition. Also, $\alpha+\beta$ and $\alpha\circ\beta$ belong to End M, because

$$\begin{aligned}(\alpha+\beta)(m_1+m_2) &= \alpha(m_1+m_2)+\beta(m_1+m_2)\\ &= \alpha(m_1)+\alpha(m_2)+\beta(m_1)+\beta(m_2)\\ &= \alpha(m_1)+\beta(m_1)+\alpha(m_2)+\beta(m_2) \quad \text{(since } M \text{ is abelian)}\\ &= (\alpha+\beta)(m_1)+(\alpha+\beta)(m_2),\end{aligned}$$

so that $\alpha+\beta\in \text{End } M$, and

$$\begin{aligned}\alpha\circ\beta(m_1+m_2) &= \alpha\big(\beta(m_1+m_2)\big)\\ &= \alpha\big(\beta(m_1)+\beta(m_2)\big),\end{aligned}$$

and this equals $\alpha\circ\beta(m_1)+\alpha\circ\beta(m_2)$ (without using commutativity), so that $\alpha\circ\beta\in\text{End } M$.

Thus, (6.15) gives closed binary operations on End M. The distributive laws follow similarly by the definition, e.g.

$$\begin{aligned}[(\alpha+\beta)\circ\gamma](m) &= (\alpha+\beta)\big(\gamma(m)\big)\\ &= \alpha\big(\gamma(m)\big)+\beta\big(\gamma(m)\big)\\ &= \alpha\circ\gamma(m)+\beta\circ\gamma(m)\\ &= (\alpha\circ\gamma+\beta\circ\gamma)(m).\end{aligned}$$

7. If $p,q\in\mathbb{R}[z]$ and $S,S_1,S_2\in\mathbb{R}^{m\times n}[z]$, then $p(z)S\in\mathbb{R}^{m\times n}[z]$ and

(i) $p(z)(S_1+S_2)=p(z)S_1+p(z)S_2$,

(ii) $[p(z)+q(z)]S=p(z)S+q(z)S$,

(iii) $p(z)[q(z)S]=[p(z)q(z)]S$,

(iv) $1S=S$.

Thus, (6.13) is satisfied and, since $\mathbb{R}^{m\times n}[z]$ is an additive abelian group and $\mathbb{R}[z]$ is a ring with a 1, then $\mathbb{R}^{m\times n}[z]$ is an $\mathbb{R}[z]$-module. Also $\mathbb{R}^{n\times n}[z]$ is an abelian group with respect to addition and a semigroup with respect to multiplication. Furthermore, (5.1) is satisfied in $\mathbb{R}^{n\times n}[z]$, so that this vector space is a ring, clearly noncommutative for $n>1$ since matrix multiplication is not commutative. Any polynomial matrix $S\in\mathbb{R}^{m\times n}[z]$ can be written as a matrix polynomial

$$S = A_0 + A_1 z + A_2 z^2 + \cdots + A_k z^k$$

for some $k \in \mathbb{N} \cup \{0\}$, where $A_0, ..., A_k \in \mathbb{R}^{m \times n}$. Thus,

$$\begin{bmatrix} 1+3z & 2+z^2 \\ z^2 & 1+z^2 \end{bmatrix} = \begin{bmatrix} 1 & 2 \\ 0 & 1 \end{bmatrix} + \begin{bmatrix} 3 & 0 \\ 0 & 0 \end{bmatrix} z + \begin{bmatrix} 0 & 1 \\ 1 & 1 \end{bmatrix} z^2.$$

8. Equations (6.13) give the conditions to be checked.

(i) $r(m_1+m_2)=rm_1+rm_2$: let $f \in F[x]$ and $v_1, v_2 \in V$. Then, by definition,

$$\begin{aligned} f(v_1+v_2) &= f(\alpha)(v_1+v_2) \\ &= f(\alpha)(v_1)+f(\alpha)(v_2) \quad \text{(linearity of } f(\alpha)) \\ &= f(v_1)+f(v_2) \quad \text{(by definition).} \end{aligned}$$

(ii) $(r_1+r_2)m=r_1m+r_2m$: let $v \in V$ and $f, g \in F[x]$, where

$$f=a_0+a_1x+\cdots+a_nx^n, \qquad g=b_0+b_1x+\cdots+b_nx^n$$

(not all the coefficients a_i and b_i in f and g have to be nonzero) so that

$$f+g=(a_0+b_0)+(a_1+b_1)x+\cdots+(a_n+b_n)x^n.$$

Then

$$\begin{aligned} (f+g)(v) &= (a_0+b_0)v+(a_1+b_1)\alpha(v)+\cdots+(a_n+b_n)\alpha^n(v) \\ &= [a_0v+a_1\alpha(v)+\cdots+a_n\alpha^n(v)]+[b_0v+b_1\alpha(v)+\cdots+b_n\alpha^n(v)] \\ &= f(v)+g(v) \qquad \text{(by definition).} \end{aligned}$$

(iii) $(r_1r_2)m = r_1(r_2m)$: again using f and g from (ii),

$$fg = \sum_{k=0}^{2n} (a_0b_k + a_1b_{k-1} + \cdots + a_kb_0)x^k$$

by (5.4); then, by definition,

$$\begin{aligned} (fg)v &= (a_0b_0)v + (a_0b_1 + a_1b_0)\alpha(v) + \cdots + (a_0b_{2n} + a_1b_{2n-1} + \cdots + a_{2n}b_0)\alpha^{2n}(v) \\ &= (a_0\iota + a_1\alpha + \cdots + a_n\alpha^n)[b_0v + b_1\alpha(v) + \cdots + b_n\alpha^n(v)] \\ &= f(\alpha)[g(\alpha)(v)] = f(gv). \end{aligned}$$

(iv) $1m = m$: $1 = 1x^0 \in F[x]$, and $1v = v$.

9. Let $z_1 = a + \mathrm{j}b$ and $z_2 = c + \mathrm{j}d$. Then

$$z_1z_2 = (ac - bd) + \mathrm{j}(ad + bc), \qquad \overline{z_1z_2} = (ac - bd) - \mathrm{j}(ad + bc).$$

$\varphi\colon \mathbb{C} \to \mathbb{C}$ is defined by $z \mapsto \bar{z}$, so that $\varphi(z_1z_2) = \overline{z_1z_2}$.

Now $\varphi(z_1) = a - \mathrm{j}b$ and $\varphi(z_2) = c - \mathrm{j}d$. Therefore

$$\varphi(z_1)\varphi(z_2) = (ac + bd) - \mathrm{j}(ad + bc).$$

Thus, $\varphi(z_1z_2) = \varphi(z_1)\varphi(z_2)$. Similarly $\varphi(z_1 + z_2)$ is easily seen to be $\varphi(z_1) + \varphi(z_2)$. Therefore φ is a ring homomorphism. However,

$$\begin{aligned} z_1\varphi(z_2) &= (a + \mathrm{j}b)(c - \mathrm{j}d) \\ &= (ac + bd) + \mathrm{j}(bc - ad) \neq \varphi(z_1z_2), \end{aligned}$$

so φ is not a $\mathbb{C}$-homomorphism. A specific counterexample is

$$\mathrm{j}\varphi(\mathrm{j}) = \mathrm{j}(-\mathrm{j}) = 1, \qquad \varphi(\mathrm{j}\cdot\mathrm{j}) = \varphi(-1) = -1.$$

10. $o(2) = \{n \in \mathbb{Z} : n\cdot 2 = 0 \pmod 4\}$. But $2n = 0 \pmod 4$ iff $4|2n$ or $2|n$, so $n \in \{0, \pm 2, \pm 4, ...\} = 2\mathbb{Z}$. Thus, $o(2) = 2\mathbb{Z}$. Also, $o(3) = \{n \in \mathbb{Z} : n\cdot 3 = 0 \pmod 4\}$. Clearly, $3n = 0 \pmod 4$ iff $4|3n$. This happens only when $3n = 4a$ or $n = 4a/3$ for some $a \in \mathbb{Z}$. Thus, $n \in \mathbb{Z}$ only when $3|a$, so $a \in 3\mathbb{Z}$, whence $n \in 4\mathbb{Z}$. It follows that $o(3) = 4\mathbb{Z}$.

11. Since T is the set of torsion elements in M, we have $0 \in T$. Using the properties of the integral domain R, we see that, if $t_1,t_2 \in T$, then $r_1t_1 = 0 = r_2t_2$ for some nonzero elements $r_1,r_2 \in R$. Then

$$\begin{aligned} r_1r_2(t_1 - t_2) &= (r_2r_1)t_1 - (r_1r_2)t_2 \\ &= r_2(r_1t_1) - r_1(r_2t_2) \\ &= 0. \end{aligned}$$

Since R is an integral domain, $r_1r_2 \neq 0$ and so $t_1 - t_2 \in T$. Furthermore,

$$r_1(rt_1) = r(r_1t_1) = r0 = 0 \quad \forall\, r \in R,$$

and, again because $r_1 \neq 0$, it follows that $rt_1 \in T$. Thus, we have shown that (i) $0 \in T$, (ii) $t_1,t_2 \in T$ implies $t_1 - t_2 \in T$, and (iii) $t_1 \in T$ and $r \in R$ together imply $rt_1 \in T$, so T is a submodule of M.

Let $T + m \in M/T$ and suppose there is a nonzero $r \in R$ such that

$$r(T + m) = T.$$

Then $rm \in T$ and it follows that there is a nonzero $r_1 \in R$ such that $r_1(rm) = 0$, i.e.

$$(r_1r)m = 0.$$

Now, $r_1r \neq 0$ so that $m \in T$. Therefore, $T + m = T$ and M/T is torsion-free.

12. The linearity of f and (6.20) give

$$\boldsymbol{f}(\boldsymbol{u} \circ \boldsymbol{v}) = \boldsymbol{f}(z^{1+\deg \boldsymbol{v}}\boldsymbol{u}) + \boldsymbol{f}(\boldsymbol{v}) = z^{1+\deg \boldsymbol{v}}\boldsymbol{f}(\boldsymbol{u}) + \boldsymbol{f}(\boldsymbol{v}).$$

Thus (4.26) $\Rightarrow$ (6.19). From the definitions, it is clear that (6.19) $\Rightarrow$ (4.26). Hence $[\boldsymbol{u}] = (\boldsymbol{u})_f$ (Kalman et al 1969).

Chapter 7

1. (a) Properties (7.3) must be checked. (i) $\rho(\boldsymbol{x}, \boldsymbol{y}) \geqslant 0$ because the modulus of a real number is non-negative. Property (ii) is obvious since $|x_i - y_i| = |y_i - x_i|$ $(i = 1, 2)$. (iii)

$$\begin{aligned} \rho(\boldsymbol{x}, \boldsymbol{y}) &= |x_1 - y_1| + |x_2 - y_2| \\ &= |(x_1 - z_1) + (z_1 - y_1)| + |(x_2 - z_2) + (z_2 - y_2)| \\ &\leqslant |x_1 - z_1| + |z_1 - y_1| + |x_2 - z_2| + |z_2 - y_2| \text{ by (7.1)}, \\ &= \rho(\boldsymbol{x}, \boldsymbol{z}) + \rho(\boldsymbol{z}, \boldsymbol{y}). \quad \forall\, \boldsymbol{x},\boldsymbol{y},\boldsymbol{z} \in \mathbb{R}^2 \quad (\text{with } \boldsymbol{z} = (z_1, z_2)). \end{aligned}$$

(b) Again in (7.3), (i) and (ii) follow immediately from the definition of ρ.
(iii) $\rho(a, b) \leqslant 1$ for all $a,b \in A$. For $a,b,c \in A$ with $a = b = c$, we have

$$\rho(a, b) = 0 = \rho(a, c) + \rho(c, b),$$

and (iii) is satisfied with equality. If a, b, c are not all equal, then $\rho(a, c) + \rho(c, b) \geqslant 1$ so that

$$\rho(a, b) \leqslant \rho(a, c) + \rho(c, b).$$

Thus, (iii) is verified for all $a,b,c \in A$.

2. Check properties (7.3). (i) $\rho(p, q) \geqslant 0$. If $\rho(p, q) = 0$, then $\rho_X(x_1, x_2) = 0 = \rho_Y(y_1, y_2)$, so that $x_1 = x_2$ and $y_1 = y_2$. Hence $p = q$.
(ii) $\rho_X(x_2, x_1) = \rho_X(x_1, x_2)$ and $\rho_Y(y_2, y_1) = \rho_Y(y_1, y_2)$,

so that

$$\rho(q, p) = \max\{\rho_X(x_2, x_1), \rho_Y(y_2, y_1)\} = \rho(p, q).$$

(iii) Let $r = (x_3, y_3) \in X \times Y$, so that

$$\rho(p, r) = \max\{\rho_X(x_1, x_3), \rho_Y(y_1, y_3)\}.$$

Clearly $\rho(p, r)$ is equal to either $\rho_X(x_1, x_3)$ or $\rho_Y(y_1, y_3)$ (or both if $\rho_X(x_1, x_3) = \rho_Y(y_1, y_3)$). Assume that $\rho_X(x_1, x_3) \geqslant \rho_Y(y_1, y_3)$. Then

$$\rho(p, r) = \rho_X(x_1, x_3) \leqslant \rho_X(x_1, x_2) + \rho_X(x_2, x_3)$$
$$\leqslant \rho(p, q) + \rho(q, r).$$

The last step is valid because

$$\rho(p, q) = \max\{\rho_X(x_1, x_2), \rho_Y(y_1, y_2)\},$$
$$\rho(q, r) = \max\{\rho_X(x_2, x_3), \rho_Y(y_2, y_3)\},$$

and hence

$$\rho_X(x_1, x_2) \leqslant \rho(p, q) \text{ and } \rho_X(x_2, x_3) \leqslant \rho(q, r).$$

The same result is obtained if $\rho_Y(y_1, y_3) > \rho_X(x_1, x_3)$.

3. Check (7.3). (i) and (ii) obviously hold. For (iii), suppose $x,y,z \in A$. Then

$$\frac{\rho(x, y)}{1 + \rho(x, y) + \rho(y, z)} \leqslant \rho'(x, y),$$

$$\frac{\rho(y, z)}{1 + \rho(x, y) + \rho(y, z)} \leqslant \rho'(y, z).$$

But $\rho(x, z) \leqslant \rho(x, y) + \rho(y, z)$ since ρ is a metric. Therefore [7]

$$\rho'(x, z) = \frac{\rho(x, z)}{1 + \rho(x, z)} \leqslant \frac{\rho(x, y) + \rho(y, z)}{1 + \rho(x, y) + \rho(y, z)}$$

$$= \frac{\rho(x, y)}{1 + \rho(x, y) + \rho(y, z)} + \frac{\rho(y, z)}{1 + \rho(x, y) + \rho(y, z)}$$

$$\leqslant \rho'(x, y) + \rho'(y, z).$$

4. If $\|x + y\| = 0$, then obviously $\|x + y\| \leqslant \|x\| + \|y\|$, since $\|x\|$ and $\|y\|$ are both non-negative. Suppose $\|x + y\| > 0$; then

$$\begin{aligned}\|x + y\|^2 &= \sum_{i=1}^{n} (x_i + y_i)^2 = \sum_i |x_i + y_i|\,|x_i + y_i| \\ &\leqslant \sum_i |x_i + y_i|(|x_i| + |y_i|) \quad \text{(by (7.1))} \\ &= \sum_i |x_i + y_i|\,|x_i| + \sum_i |x_i + y_i|\,|y_i| \\ &\leqslant \|x + y\|\,\|x\| + \|x + y\|\,\|y\| \quad \text{(by the Schwartz inequality)} \\ &= \|x + y\|(\|x\| + \|y\|).\end{aligned}$$

Thus, dividing by the positive real number $\|x + y\|$, we get

$$\|x + y\| \leqslant \|x\| + \|y\|.$$

This result is a special case ($p = 2$) of the *Minkowski inequality*

$$\left(\sum_i |x_i + y_i|^p\right)^{1/p} \leqslant \left(\sum_i |x_i|^p\right)^{1/p} + \left(\sum_i |y_i|^p\right)^{1/p} \qquad (1 \leqslant p < \infty).$$

5. Check (7.7). (i) $\|x\| \geqslant 0$ since a sum of squares is non-negative. Furthermore, $\|x\| = 0$ iff $x_i = 0$ for $i = 1, \ldots, n$, i.e. $\|x\| = 0$ iff $x = \mathbf{0}$.

(ii) $\lambda x = \lambda(x_1, x_2, \ldots, x_n) = (\lambda x_1, \lambda x_2, \ldots, \lambda x_n)$ for all $\lambda \in \mathbb{R}$. Hence

$$\|\lambda x\| = \sqrt{\sum_{i=1}^{n} (\lambda x_i)^2} = \sqrt{\sum_i \lambda^2 x_i^2} = |\lambda| \sqrt{\sum_i x_i^2} = |\lambda|\,\|x\|.$$

(iii) $\|x + y\| \leqslant \|x\| + \|y\|$, for all $x, y \in \mathbb{R}^n$, by the Minkowski inequality (see Solution 4 above).

6. $$\begin{aligned}\|x\| - \|y\| &= \|(x - y) + y\| - \|y\| \\ &\leqslant \|x - y\| + \|y\| - \|y\| \quad \text{by (7.7 (iii))} \\ &= \|x - y\|.\end{aligned}$$

Chapter 8

1. Let $s_k = (1 + 1/k)^k$. With $k, r \geqslant 2$ the $(r + 1)$th term in the Binomial expansion of s_k is

$$\frac{k(k-1)(k-2)\cdots(k-r+1)}{r!}\left(\frac{1}{k}\right)^r,$$

i.e.

$$\frac{1}{r!}\left(1 - \frac{1}{k}\right)\left(1 - \frac{2}{k}\right)\cdots\left(1 - \frac{r-1}{k}\right).$$

The rth term in the expansion of s_{k+1} is similarly

$$\frac{1}{r!}\left(1-\frac{1}{k+1}\right)\left(1-\frac{2}{k+1}\right)\cdots\left(1-\frac{r-1}{k+1}\right).$$

So, clearly, each term in the expansion of s_{k+1} is not less than the corresponding one in the expansion of s_k. Furthermore,

$$s_{k+1}=\left(1+\frac{1}{k+1}\right)^{k+1}$$

contains one more positive term in its expansion than s_k, so that $s_{k+1} > s_k$. The sequence (s_k) is therefore an increasing one. It is also bounded above because

$$\begin{aligned} s_k=\left(1+\frac{1}{k}\right)^k &= 1+k\left(\frac{1}{k}\right)+\frac{k(k-1)}{2!}\left(\frac{1}{k}\right)^2+\cdots+\frac{k(k-1)\cdots 3\cdot 2\cdot 1}{k!}\left(\frac{1}{k}\right)^k \\ &< 1+1+\frac{1}{2!}+\cdots+\frac{1}{k!} \\ &\leqslant 1+1+(\tfrac{1}{2})+\cdots+(\tfrac{1}{2})^{k-1} \\ &< 3 \end{aligned}$$

since the series $1+\frac{1}{2}+\cdots+(\frac{1}{2})^{k-1}$ is a geometric progression, with common ratio $\frac{1}{2}$, first term 1, and k terms in all; so it has a sum

$$\frac{1-(\frac{1}{2})^k}{1-\frac{1}{2}}=2[1-(\tfrac{1}{2})^k]=2-(\tfrac{1}{2})^{k-1}<2.$$

Now the axiom of completeness guarantees that $(1+1/k)^k$ converges to a real number e such that $2 < \mathrm{e} < 3$.

2. For $t\in(-1, 1)$, it is clear that $x^{(k)}(t)\to 1$ as $k\to\infty$ since $t^{2k}\to 0$ as $k\to\infty$. Also, if $|t|=1$, then $x^{(k)}(t)\to\frac{1}{2}$ as $k\to\infty$. Further, if $t\in\mathbb{R}\backslash[-1, 1]$, then $x^{(k)}(t)\to 0$ as $k\to\infty$. Thus, the sequence $(x^{(k)}(\bullet))$ of continuous functions has a limit which is discontinuous and so, by Note 10.3, the convergence of $(x^{(k)}(t))$ is nonuniform in t.

3. If $f=g$, then $\rho(f, g)=0$. It is obvious that $\rho(f, g)\geqslant 0$. If $\rho(f, g)=0$, then $f_1=g_1$ and $f_2=g_2$, so that $f=g$. The expression for $\rho(f, g)$ involves $[g_1(x)-f_1(x)]^2$ and $[g_2(x)-f_2(x)]^2$ and so is no different from $\rho(f, g)$. Finally, by Minkowski's inequality (Solution 7.4) with $p=2$, for $f, g, h\in\mathscr{B}(\mathbb{R}^2)$, we have

$$\begin{aligned} \rho(f, h)+\rho(h, g) &\geqslant \{[f_1(x)-h_1(x)]^2+[f_2(x)-h_2(x)]^2\}^{\frac{1}{2}} \\ &\quad +\{[h_1(x)-g_1(x)]^2+[h_2(x)-g_2(x)]^2\}^{\frac{1}{2}} \\ &\geqslant \{[f_1(x)-g_1(x)]^2+[f_2(x)-g_2(x)]^2\}^{\frac{1}{2}}. \end{aligned}$$

Taking the supremum of the last expression yields

$$\rho(f, g)\leqslant\rho(f, h)+\rho(h, g).$$

This exercise can be associated with a system of differential equations $\dot{x}_1=f(x_1, x_2)$, $\dot{x}_2=g(x_1, x_2)$ (Chillingworth 1976).

4. If $f = 0$, then $\|f\| = 0$. If $\|f\| = 0$, then $\|f(x)\| = 0$ and so $f(x) = 0$.

$$\|\lambda f\| = \sup \frac{\|\lambda f(x)\|}{\|x\|} = \sup \frac{|\lambda| \, \|f(x)\|}{\|x\|} = |\lambda| \sup \frac{\|f(x)\|}{\|x\|} = |\lambda| \, \|f\|.$$

$$\frac{\|f(x)\|}{\|x\|} + \frac{\|g(x)\|}{\|x\|} = \frac{\|f(x)\| + \|g(x)\|}{\|x\|} \geqslant \frac{\|f(x) + g(x)\|}{\|x\|} \quad \forall f,g.$$

$$\|f\| + \|g\| = \sup \frac{\|f(x)\|}{\|x\|} + \sup \frac{\|g(x)\|}{\|x\|} = F + G, \text{ say. Then}$$

$$\frac{\|f(x)\|}{\|x\|} + \frac{\|g(x)\|}{\|x\|} \leqslant F + G.$$

Hence

$$\sup \left\{ \frac{\|f(x)\|}{\|x\|} + \frac{\|g(x)\|}{\|x\|} \right\} \leqslant F + G.$$

But

$$\|f + g\| = \sup \frac{\|f(x) + g(x)\|}{\|x\|}$$

and so $\|f + g\| \leqslant \|f\| + \|g\|$. Assume $x \neq 0$, write $\|x\| = a$, and let $y = (1/a)x$. Then

$$\|y\| = \left\| \frac{1}{a} x \right\| = \frac{1}{a} \|x\| = 1.$$

$$\begin{aligned} \|f\| &= \sup_{\substack{x \in A \\ x \neq 0}} \frac{1}{a} \|f(x)\| = \sup_{\substack{x \in A \\ x \neq 0}} \left\| \frac{1}{a} f(x) \right\| \\ &= \sup_{\substack{x \in A \\ x \neq 0}} \left\| f\left(\frac{1}{a} x\right) \right\| \text{ since } f \text{ is linear,} \\ &= \sup_{\substack{y \in A \\ \|y\| = 1}} \|f(y)\|. \end{aligned}$$

5. First note that a real continuous function on a closed interval is bounded (Section 8.2). Therefore $\|x\|$ is well-defined. Properties (7.7) must be checked. First, $|x(t)| \geqslant 0$ for $t \in [0, 1]$, so that $\|x\| \geqslant 0$. Furthermore, $\|x\| = 0$ iff $|x(t)| = 0$ for all $t \in [0, 1]$, i.e. $\|x\| = 0$ iff $x = 0$, the null function on $[0, 1]$. Thus, (7.7(i)) is satisfied. Let $\lambda \in \mathbb{R}$. Then $\|\lambda x\| = \max |(\lambda x)(t)| = \max |\lambda x(t)| = \max |\lambda| |x(t)| = |\lambda| \max |x(t)| = |\lambda| \, \|x\|$, so (7.7(ii)) is valid. Finally, if $x_1, x_2 \in C[0, 1]$, then

$$\begin{aligned} \|x_1 + x_2\| &= \max |(x_1 + x_2)(t)| = \max |x_1(t) + x_2(t)| \\ &\leqslant \max \{|x_1(t)| + |x_2(t)|\} \\ &\leqslant \max |x_1(t)| + \max |x_2(t)| = \|x_1\| + \|x_2\|. \end{aligned}$$

Let $(x^{(k)})$ be any Cauchy sequence in C[0, 1]. Then, given $\varepsilon > 0$, there exists N such that

$$\rho(x^{(m)}, x^{(n)}) = \max_{t \in [0,1]} |x^{(m)}(t) - x^{(n)}(t)| < \varepsilon \quad \forall\, m,n > N. \tag{*}$$

Hence, for any $t' \in [0, 1]$ and any $m,n > N$, we obtain

$$|x^{(m)}(t') - x^{(n)}(t')| < \varepsilon.$$

Thus, $\left(x^{(1)}(t'), x^{(2)}(t'), \ldots\right)$ is a Cauchy sequence in $\mathbb{R}$. But $\mathbb{R}$ is complete, by the Cauchy convergence test, so that this last sequence converges to some $x(t') \in \mathbb{R}$. Each $t \in [0, 1]$ may be associated with a unique $x(t) \in \mathbb{R}$, defining a function x on $[0, 1]$.

Let $n \to \infty$ in (*) so that

$$\max_{t \in [0,1]} |x^{(m)}(t) - x(t)| \leqslant \varepsilon \quad \forall\, m > N.$$

It follows that

$$|x^{(m)}(t) - x(t)| \leqslant \varepsilon \quad \forall\, t \in [0, 1] \quad \forall\, m > N.$$

Thus, $(x^{(m)}(t))$ converges to $x(t)$ uniformly on $[0, 1]$. The $x^{(m)}$'s are continuous on $[0, 1]$ and the convergence is uniform, so the limit function x is continuous on $[0, 1]$. Thus, $x \in C[0, 1]$ and $x^{(m)} \to x$ as $m \to \infty$. This shows that C[0, 1] is complete.

The normed space $(C[0, 1], \|\bullet\|)$ with the norm defined by (7.10) is not complete (Kreyszig 1978).

6. Consider the functions $x^{(m)}$ represented in Fig. S8.1(a). Here $\rho(x^{(m)}, x^{(n)})$ represents the area of the triangle ABC in Fig. S8.1(b), and

$$\rho(x^{(m)}, x^{(n)}) < \varepsilon$$

whenever $m,n > 1/\varepsilon$, for any given $\varepsilon > 0$.

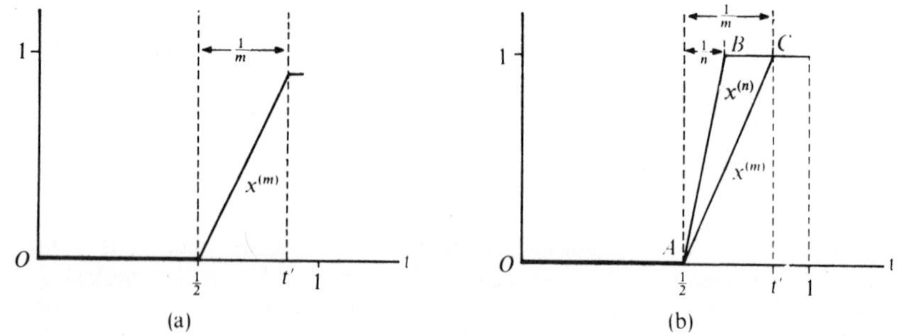

Fig. S8.1

It turns out that this Cauchy sequence does not converge. To see this, assume that $(x^{(m)})$ converges to x in C[0, 1] so that $\rho(x^{(m)}, x) \to 0$. Now,

$$x^{(m)}(t) = \begin{cases} 0 & \text{for } t \in [0, \tfrac{1}{2}], \\ 1 & \text{for } t \in [t', 1], \end{cases}$$

where $t' = \frac{1}{2} + 1/m$. Hence

$$\rho(x^{(m)}, x) = \int_0^1 |x^{(m)}(t) - x(t)|\,dt$$

$$= \int_0^{\frac{1}{2}} |x(t)|\,dt + \int_{\frac{1}{2}}^{t'} |x^{(m)}(t) - x(t)|\,dt + \int_{t'}^{1} |1 - x(t)|\,dt.$$

which is nonnegative since each integrand is nonnegative. So each term tends to zero as $m \to \infty$. This implies that $x(t) = 0$ for $t \in [0, \frac{1}{2}]$ and $x(t) = 1$ for $t \in (\frac{1}{2}, 1)$, which is a contradiction to $x \in C[0, 1]$. Thus, the sequence $(x^{(m)})$ does not converge to a limit in $C[0, 1]$, and so $(C[0, 1], \rho)$ is not complete.

Chapter 9

1. (i) $B(0, \frac{1}{4}) = \{x \in \mathbb{R}: |x| < \frac{1}{4}\}$ which is just the open interval $(-\frac{1}{4}, \frac{1}{4})$.

(ii) $B(0, \frac{1}{4}) = \{x \in [0, 1]: |x| < \frac{1}{4}\} = [0, \frac{1}{4})$. Notice that the centre of the ball is the real number 0, and this lies on the edge of the interval $[0, \frac{1}{4})$!

(iii) $B(0, 1) = \{(x_1, x_2) \in \mathbb{R}^2: |x_1| + |x_2| < 1\}$ which is the interior of the square shown in Fig. S9.1.

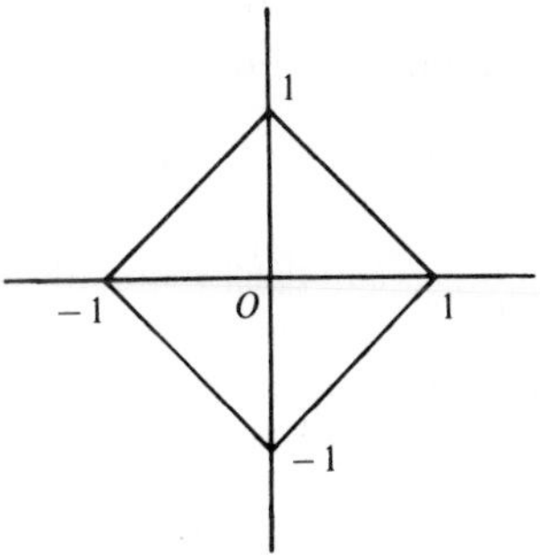

Fig. S9.1

2. $B(\frac{1}{16}, \frac{3}{16}) = \{x \in [0, 1]: |x - \frac{1}{16}| < \frac{3}{16}\}$. Now $|x - \frac{1}{16}| < \frac{3}{16}$ can be written as $-\frac{3}{16} < x - \frac{1}{16} < \frac{3}{16}$. Then $-\frac{3}{16} < x - \frac{1}{16} \Rightarrow x > -\frac{1}{8}$ and $x - \frac{1}{16} < \frac{3}{16} \Rightarrow x < \frac{1}{4}$. Thus, $|x - \frac{1}{16}| < \frac{3}{16}$ is equivalent to $-\frac{1}{8} < x < \frac{1}{4}$. But $x \in [0, 1]$, so that $B(\frac{1}{16}, \frac{3}{16}) = [0, \frac{1}{4})$, which is the interval also represented by $B(0, \frac{1}{4})$ in Exercise 1(ii) above. This shows that a ball representing a given interval is not necessarily unique.

3. $[0, \frac{1}{2})$ is open in $[0, 1]$ but not open in $\mathbb{R}$. This shows that whether a set S is open in a space X depends on X.

4. If $x \in S$ then $x \in S_i$ for some i. Then, since S_i is an open set, there is an $\varepsilon > 0$ such that

$$B(x, \varepsilon) \subseteq S_i \subseteq S.$$

Thus, $x \in \operatorname{int} S$ and so every x in S belongs to $\operatorname{int} S$, i.e. $\operatorname{int} S = S$ and so S is open.

5. It suffices to show that, if U and V are open sets in a metric space X, then $U \cap V$ is open in X. The result is obvious if $U \cap V = \emptyset$ [1] so assume $U \cap V \neq \emptyset$. If $x \in U \cap V$, then $x \in U$ and so

$$B(x, \varepsilon_1) \subseteq U$$

for some $\varepsilon_1 > 0$. Similarly, $x \in V$ and so

$$B(x, \varepsilon_2) \subseteq V$$

for some $\varepsilon_2 > 0$. Now take $\varepsilon = \min\{\varepsilon_1, \varepsilon_2\}$. It follows that $B(x, \varepsilon) \subseteq U \cap V$ and so $U \cap V$ is open.

The result is always true for finite intersections because, if $\varepsilon_i > 0$ $(i = 1, \ldots, n)$, then $\min\{\varepsilon_1, \ldots \varepsilon_n\} > 0$. For the infinite case, $\inf_i \varepsilon_i$ could be zero. For example, $X = \mathbb{R}^2$, $U_n = B(0, 1 + 1/n)$ $(n = 1, 2, \ldots)$. Each U_n is an open set but

$$\bigcap_{n=1}^{\infty} U_n = \{x \in \mathbb{R}^2 : \rho(x, 0) \leqslant 1\}$$

is not open.

6. Given any open interval $T_1 \subset \mathbb{R}$ and a real number $x \in T_1$, then there exists an open interval E such that $x \in E \subset T_1$. Hence, the set $\mathscr{B}$ of all open intervals forms a base for the usual topology on $\mathbb{R}$ (Example 9.15).

7. Let $x_1, x_2 \in S$ so that $c(x_1) \leqslant 0$ and $c(x_2) \leqslant 0$. Since c is convex in X, we have

$$c[(1-\alpha)x_1 + \alpha x_2] \leqslant (1-\alpha)c(x_1) + \alpha c(x_2)$$

$(0 \leqslant \alpha \leqslant 1)$ which is $\leqslant 0$, from the first result. Hence

$$(1-\alpha)x_1 + \alpha x_2 \in S,$$

and so S is convex.

8. The set $\mathbb{Q}$ of rational numbers is countable (Solution 3.8) and so can be arranged in terms of a sequence $(s_n : n = 1, 2, \ldots)$. For each element s_n of this sequence, there is an open ball $B_n = B(s_n, \varepsilon/2^n)$ which is the interval $(s_n - \varepsilon/2^n, s_n + \varepsilon/2^n)$, with ε arbitrarily small. The sequence $(B_1, B_2, \ldots)$ of open intervals has a total length of

$$\varepsilon[1 + \tfrac{1}{2} + \tfrac{1}{4} + \cdots + (\tfrac{1}{2})^n + \cdots] = 2\varepsilon.$$

Thus, all rational numbers have been covered by a sequence of open intervals with a total length that can be made arbitrarily small. This is why the rationals are sometimes described as sparse in $\mathbb{R}$.

9. The usual topology on $\mathbb{R}^2$ is the collection of sets formed by the unions of open sets [12] $\{x \in \mathbb{R}^2 : \|x - a\| < \delta\}$, where here the norm is taken from the usual metric on $\mathbb{R}^2$ (Example 7.1). Similarly for $\mathbb{R}^n$, with $n \in \mathbb{N}$.

10. The set of all complex numbers whose real and imaginary parts are rational numbers form a countable dense subset of $\mathbb{C}$. Thus, $\mathbb{C}$ is separable.

Chapter 10

1. Suppose we require $\delta \leqslant 1$. Then $|t - t_0| < \delta \Rightarrow |t - t_0| < 1$. Therefore

$$|t| = |(t - t_0) + t_0| \leqslant |t - t_0| + |t_0| < 1 + |t_0|,$$

so that $|t - t_0| < 1 \Rightarrow |t| < 1 + |t_0|$. Hence

$$\begin{aligned}|x(t) - x(t_0)| &= |t^2 - t_0^2| = |(t + t_0)(t - t_0)| = |t + t_0||t - t_0| \\ &\leqslant (|t| + |t_0|)|t - t_0| \\ &< (1 + 2|t_0|)|t - t_0|\end{aligned}$$

for any $t_0 \in \mathbb{R}$. Thus, by choosing $\delta > 0$ such that $\delta \leqslant \min\{1, \varepsilon/(1 + 2|t_0|)\}$, it is guaranteed that $|x(t) - x(t_0)| < \varepsilon$. This is because, if $\varepsilon/(1 + 2|t_0|) \geqslant 1$, then $\delta \leqslant 1$ and $1 + 2|t_0| \leqslant \varepsilon$ so that $|x(t) - x(t_0)| < \varepsilon$ whereas, if $\varepsilon/(1 + 2|t_0|) < 1$, then $\delta \leqslant \varepsilon/(1 + 2|t_0|)$ and again $|x(t) - x(t_0)| < (1 + 2|t_0|)\delta \leqslant \varepsilon$.

Therefore x is continuous at all points $t_0 \in \mathbb{R}$, and so is continuous on $\mathbb{R}$. However, with $\delta \leqslant \min\{1, \varepsilon/(1 + 2|t_0|)\}$, it is clear that δ depends on both ε and t_0, and so x is not uniformly continuous on $\mathbb{R}$.

2.
$$u(t) - u(t_0) = \frac{1}{1 + t^2} - \frac{1}{1 + t_0^2} = \frac{(t_0 - t)(t_0 + t)}{(1 + t^2)(1 + t_0^2)}$$

$$(|t|, |t_0| \leqslant 1).$$

Then

$$|u(t) - u(t_0)| = \frac{|t_0 - t||t_0 + t|}{|1 + t^2||1 + t_0^2|} \leqslant |t_0 - t||t_0 + t|.$$

But $|t_0 + t| \leqslant |t_0| + |t| \leqslant 2$, so that

$$|u(t) - u(t_0)| \leqslant 2|t_0 - t|.$$

Given that $|u(t) - u(t_0)| < \varepsilon$, choose $\delta = \frac{1}{2}\varepsilon$, so that

$$|t_0 - t| < \delta \quad \Rightarrow \quad |u(t) - u(t_0)| < 2\delta = \varepsilon.$$

Then, given any ε, there exists $\delta = \frac{1}{2}\varepsilon$ such that

$$|t_0 - t| = |t - t_0| < \delta \quad \Rightarrow \quad |u(t) - u(t_0)| < \varepsilon.$$

Hence u is continuous at a point $t_0 \in [-1, 1]$. Because the choice of δ does not depend upon t_0, the function u is uniformly continuous on $[-1, 1]$.

3. By Example 10.6, the function u given by $u(t) = t \sin 1/t$ is continuous at $t = 0$, and so it follows immediately that $t^2 \sin 1/t$ has the same property since the product of two continuous functions is continuous.

4. Approaching the origin along the parabola $x_2^2 = x_1$, we find $f(x_1, x_2) = x_1^2/(2x_1^2) = \frac{1}{2}$. But, at the origin, $f(x_1, x_2) = 0$ and so f is discontinuous there.

5. X is the unit circle excluding the point $(0, -1)$. We can represent f as the stereographic projection of X from $(0, -1)$ onto the x_1 axis, just as in Example 10.11. The function f is bijective and continuous. The inverse function $f^{-1}: \mathbb{R} \to X$ given by

$$f^{-1}(t) = \left(\frac{2t}{1 + t^2}, \frac{1 - t^2}{1 + t^2}\right)$$

is also continuous, so that f is a homeomorphism.

6. f is bijective and continuous. The inverse function $f^{-1}: \mathbb{R} \to (-1, 1)$ given by $t = 2\pi^{-1} \arctan f(t)$ is also continuous. Hence f is a homeomorphism and $(-1, 1)$ and $\mathbb{R}$ are homeomorphic.

7. $\lim_{h\to 0}\frac{u(h)-u(0)}{h}=\lim_{h\to 0}\sin\frac{1}{h}$. Neither the left-hand limit ($h\to 0^-$) nor the right-hand limit ($h\to 0^+$) exists, since $\sin 1/h$ oscillates between 1 and -1 an infinite number of times as $h\to 0$ from either direction. Since the limits do not exist, the function is not differentiable at $t=0$.

8. Choose standard bases $\{\boldsymbol{e}_1,\dots,\boldsymbol{e}_n\}$ for $\mathbb{R}^n$ and $\{\varepsilon_1,\dots,\varepsilon_m\}$ for $\mathbb{R}^m$, and let $\boldsymbol{h}=h_1\boldsymbol{e}_1+\cdots+h_n\boldsymbol{e}_n$. Then

$$\mathrm{D}\boldsymbol{f}(\boldsymbol{x})\boldsymbol{h}=h_1\mathrm{D}\boldsymbol{f}(\boldsymbol{x})\boldsymbol{e}_1+\cdots+h_n\mathrm{D}\boldsymbol{f}(\boldsymbol{x})\boldsymbol{e}_n\quad\forall\,\boldsymbol{x}\in X.$$

Write $\mathrm{D}\boldsymbol{f}(\boldsymbol{x})\boldsymbol{e}_i=\boldsymbol{a}_i\in\mathbb{R}^m$ $(i=1,\dots,n)$ so that

$$\begin{aligned}\mathrm{D}\boldsymbol{f}(\boldsymbol{x})\boldsymbol{h}&=h_1\boldsymbol{a}_1+\cdots+h_n\boldsymbol{a}_n\\&=h_1(k_{11}\boldsymbol{\varepsilon}_1+\cdots+k_{m1}\boldsymbol{\varepsilon}_m)+\cdots+h_n(k_{1n}\boldsymbol{\varepsilon}_1+\cdots+k_{mn}\boldsymbol{\varepsilon}_m)\\&=(h_1k_{11}+\cdots+h_nk_{1n})\boldsymbol{\varepsilon}_1+\cdots+(h_1k_{m1}+\cdots+h_nk_{mn})\boldsymbol{\varepsilon}_m\\&=\lambda_1\boldsymbol{\varepsilon}_1+\cdots+\lambda_m\boldsymbol{\varepsilon}_m,\end{aligned}$$

where $\lambda_j=h_1k_{j1}+\cdots+h_nk_{jn}$ $(j=1,\dots,m)$, i.e.

$$\begin{bmatrix}\lambda_1\\ \vdots\\ \lambda_m\end{bmatrix}=\begin{bmatrix}k_{11}&\cdots&k_{1n}\\ \vdots&&\vdots\\ k_{m1}&\cdots&k_{mn}\end{bmatrix}\begin{bmatrix}h_1\\ \vdots\\ h_n\end{bmatrix}.$$

As in Example 10.20, it can be shown that $k_{ji}=\partial f_j/\partial x_i$. Hence, $\mathrm{D}\boldsymbol{f}(\boldsymbol{x})$ is represented by the matrix $[\partial f_j/\partial x_i]$, i.e the Jacobian matrix

$$\begin{bmatrix}\partial f_1/\partial x_1&\cdots&\partial f_1/\partial x_n\\ \vdots&&\vdots\\ \partial f_m/\partial x_1&\cdots&\partial f_m/\partial x_n\end{bmatrix},$$

often denoted by $\dfrac{\partial(f_1,\dots,f_m)}{\partial(x_1,\dots,x_n)}$.

9. Using Exercise 8 with $f_1(x_1,x_2)=x_1^2+x_2^2$ and $f_2(x_1,x_2)=2x_1x_2$, we have

$$\mathrm{D}\boldsymbol{f}(\boldsymbol{x})=\begin{bmatrix}2x_1&2x_2\\2x_2&2x_1\end{bmatrix}.$$

Now, $\det \mathrm{D}\boldsymbol{f}(\boldsymbol{x})=4(x_1^2-x_2^2)\neq 0$ unless $x_1=\pm x_2$. Thus, the rank of $\mathrm{D}\boldsymbol{f}(\boldsymbol{x})$ is 2 unless $x_1=\pm x_2$, in which case the rank is 1 except at the origin where the rank is zero.

10. By Example 10.22,

$$\frac{\partial g}{\partial x}=\frac{\partial g}{\partial u}\frac{\partial u}{\partial x}+\frac{\partial g}{\partial v}\frac{\partial v}{\partial x}=2u\cdot 2x+(-2v)\cdot 2y=4x(x^2-3y^2),$$

$$\frac{\partial g}{\partial y}=\frac{\partial g}{\partial u}\frac{\partial u}{\partial y}+\frac{\partial g}{\partial v}\frac{\partial v}{\partial y}=2u\cdot(-2y)+(-2v)\cdot 2x=4y(y^2-3x^2).$$

11. f is differentiable on (a, b) and therefore continuous. For any $t_0 \in (a, b)$,

$$\dot{f}(t_0) = \lim_{t \to t_0} \frac{f(t) - f(t_0)}{t - t_0} > 0.$$

This implies that $f(t) \to f(t_0)$ as $t \to t_0$. With $x = f(t)$ and $x_0 = f(t_0)$, the limit

$$\lim_{x \to x_0} \frac{t - t_0}{f(t) - f(t_0)}$$

exists. But this is just the definition for the derivative $(f^{-1})^*$ of the inverse function given by $f^{-1}(x) = t$, namely

$$\lim_{x \to x_0} \frac{f^{-1}(x) - f^{-1}(x_0)}{x - x_0}.$$

Note that $\dot{f}(t) \cdot (f^{-1})^*(x) = 1$ or, in other words, $\mathrm{d}f/\mathrm{d}t = 1/(\mathrm{d}t/\mathrm{d}f)$.

12. $x_1\dot{x}_1 + x_2\dot{x}_2 = 0$ so that $x_1^2 + x_2^2 =$ constant. From the initial conditions, the constant is 1, so the solution of the given differential equations is the unit circle $x_1^2 + x_2^2 = 1$ in $\mathbb{R}^2$, with centre the origin. This circle is the image of the path φ given by $\varphi(t) = \big(x_1(t), x_2(t)\big)$, with

$$x_1(t) = \cos t + \sin t,$$

$$x_2(t) = \cos t - \sin t.$$

At time $t = \frac{1}{4}\pi$, we have $\dot{\varphi}(\frac{1}{4}\pi) = \begin{bmatrix} \dot{x}_1 \\ \dot{x}_2 \end{bmatrix} = \begin{bmatrix} \frac{1}{2}\sqrt{2} \\ -\frac{1}{2}\sqrt{2} \end{bmatrix}$, which is a unit vector radiating from the origin of state space $\mathbb{R}^2$ at an angle $-\frac{1}{4}\pi$ to the positive x_1 axis. Translation of this vector to the point $\varphi(\frac{1}{4}\pi)$ gives the velocity $\dot{\varphi}(\frac{1}{4}\pi)$ at this point. Remember that $\dot{\varphi}(\frac{1}{4}\pi)$ is the same thing as $\mathsf{D}\varphi(\frac{1}{4}\pi) \cdot 1$.

13. $\boldsymbol{f}$ moves each point $(x, y) \in \mathbb{R}^2$ by t units in the x direction. The inverse mapping takes each point $(x + t, y)$ back to (x, y). Thus, the inverse mapping $\boldsymbol{f}_t^{-1} : \mathbb{R}^2 \to \mathbb{R}^2$ is given by $(x, y) \mapsto (x - t, y) = \boldsymbol{f}\big(-t, (x, y)\big)$, denoted by $\boldsymbol{f}_{-t}$. Both $\boldsymbol{f}$ and $\boldsymbol{f}^{-1}$ are homeomorphisms and C^∞ mappings. Hence, for fixed t, the function $\boldsymbol{f}_t$ is a diffeomorphism. Note that $\boldsymbol{f}$ is not a diffeomorphism because it is not bijective.

14. Choose any norm, say the Euclidean norms on $\mathbb{R}^n$ and $\mathbb{R}^m$ and the cubic norm $\|\boldsymbol{A}\| = \max_{i,j} |a_{ij}|$ on $\mathbb{R}^{m \times n}$. For any $\boldsymbol{x}_0 \in \mathbb{R}^n$ and $\boldsymbol{x} \in \mathbb{R}^n$, we have

$$\|\boldsymbol{A}\boldsymbol{x} - \boldsymbol{A}\boldsymbol{x}_0\| = \|\boldsymbol{A}(\boldsymbol{x} - \boldsymbol{x}_0)\| \leqslant \|\boldsymbol{A}\| \, \|\boldsymbol{x} - \boldsymbol{x}_0\| \text{ (by (8.14))}$$

i.e. $\|\boldsymbol{A}\boldsymbol{x} - \boldsymbol{A}\boldsymbol{x}_0\| \leqslant (\max_{i,j} |a_{ij}|) \, \|\boldsymbol{x} - \boldsymbol{x}_0\|$. Thus, given $\varepsilon > 0$, by choosing $\delta = \varepsilon/(\max_{i,j} |a_{ij}|)$, we obtain

$$\|\boldsymbol{x} - \boldsymbol{x}_0\| < \delta \;\Rightarrow\; (\max_{i,j} |a_{ij}|) \, \|\boldsymbol{x} - \boldsymbol{x}_0\| < \varepsilon,$$

i.e.

$$\|\boldsymbol{x} - \boldsymbol{x}_0\| < \delta \;\Rightarrow\; \|\boldsymbol{A}\boldsymbol{x} - \boldsymbol{A}\boldsymbol{x}_0\| < \varepsilon,$$

whence $\boldsymbol{f}$ is continuous at $\boldsymbol{x}_0$. Since $\boldsymbol{x}_0$ is arbitrary in $\mathbb{R}^n$, it follows that $\boldsymbol{f}$ is continuous on $\mathbb{R}^n$.

15. $\boldsymbol{f}$ is linear if

$$\boldsymbol{f}(\alpha \boldsymbol{a} + \beta \boldsymbol{b}) = \alpha \boldsymbol{f}(\boldsymbol{a}) + \beta \boldsymbol{f}(\boldsymbol{b}) \quad \forall\ \boldsymbol{a},\boldsymbol{b} \in X \quad \forall\ \alpha,\beta \in \mathbb{R}.$$

Let $\boldsymbol{a} = (a_1, a_2, ...)$ and $\boldsymbol{b} = (b_1, b_2, ...)$. Then

$$\alpha \boldsymbol{a} + \beta \boldsymbol{b} = (\alpha a_1 + \beta b_1, \alpha a_2 + \beta b_2, ...)$$

and $\boldsymbol{f}(\alpha \boldsymbol{a} + \beta \boldsymbol{b}) = (\alpha a_2 + \beta b_2, \alpha a_3 + \beta b_3, ...)$. On the other hand,

$$\begin{aligned} \alpha \boldsymbol{f}(\boldsymbol{a}) + \beta \boldsymbol{f}(\boldsymbol{b}) &= (\alpha a_2, \alpha a_3, ...) + (\beta b_2, \beta b_3, ...) \\ &= (\alpha a_2 + \beta b_2, \alpha a_3 + \beta b_3, ...) \\ &= \boldsymbol{f}(\alpha \boldsymbol{a} + \beta \boldsymbol{b}), \end{aligned}$$

and the result follows.

Now let $\boldsymbol{a}^{(n)} = \big(0, 0, ..., (n-1)!, 0, 0, ...\big)$, where the term $(n-1)!$ occurs in the nth position. Then

$$\|\boldsymbol{a}^{(n)}\| = \frac{|a_n|}{n!} = \frac{(n-1)!}{n!} = \frac{1}{n}.$$

However, $\boldsymbol{f}(\boldsymbol{a}^{(n)}) = \big(0, 0, ..., (n-1)!, 0, 0, ...\big)$, where the term $(n-1)!$ now occurs in the $(n-1)$th position. So now

$$\|\boldsymbol{f}(\boldsymbol{a}^{(n)})\| = \frac{|a_{n-1}|}{(n-1)!} = \frac{(n-1)!}{(n-1)!} = 1.$$

Hence $\|\boldsymbol{f}(\boldsymbol{a}^{(n)}) - \boldsymbol{f}(\boldsymbol{0})\| = \|\boldsymbol{f}(\boldsymbol{a}^{(n)})\| = 1$, no matter how small the quantity $\|\boldsymbol{a}^{(n)} - \boldsymbol{0}\|$ is made; thus, no matter how small δ is chosen, the condition $\|\boldsymbol{a}^{(n)} - \boldsymbol{0}\| < \delta$ does not imply $|\boldsymbol{f}(\boldsymbol{a}^{(n)}) - \boldsymbol{f}(\boldsymbol{0})| < \varepsilon$, for any given $\varepsilon \leqslant 1$. So f is not continuous at the point $(0, 0, ...)$ (see p. 206).

16. $f(x, u) = x \cos xu$, so $f(1, \frac{1}{2}\pi) = \cos \frac{1}{2}\pi = 0$ and

$$\begin{aligned} f_x &= \cos xu - xu \sin xu, \qquad & f_x(1, \tfrac{1}{2}\pi) &= -\tfrac{1}{2}\pi \neq 0, \\ f_u &= -x^2 \sin xu, & f_u(1, \tfrac{1}{2}\pi) &= -1 \neq 0. \end{aligned}$$

By the implicit-function theorem, $x \cos xu = 0$ has a unique solution $u = g(x)$ in a neighbourhood of $(1, \frac{1}{2}\pi)$. The inverse function g^{-1} also exists in some neighbourhood of that point. By the same theorem,

$$\frac{\mathrm{d}u}{\mathrm{d}x} = -\frac{f_x}{f_u} = \frac{-xu \sin xu + \cos xu}{x^2 \sin xu}.$$

When $x^2 \sin xu \neq 0$ (i.e. $xu \neq k\pi$, for $k \in \mathbb{Z}$), $\mathrm{d}u/\mathrm{d}x = 0$ when

$$xu \sin xu = \cos xu,$$

i.e. when $\tan xu = 1/xu$.

Chapter 11

1. If r is the distance between the two particles, then the gravitational force $\boldsymbol{F}$ is from m_2 to m_1 with magnitude

$$|\boldsymbol{F}| = Gm_1m_2/r^2$$

where G is the gravitational constant. Choosing cartesian coordinates (x, y, z) for m_2, we have

$$r^2 = x^2 + y^2 + z^2$$

and $\boldsymbol{r} = x\boldsymbol{i} + y\boldsymbol{j} + z\boldsymbol{k}$. The unit vector in direction from the free to the fixed particle is $-\boldsymbol{r}/r$ (minus sign since direction is towards the origin). Then

$$\boldsymbol{F} = |\boldsymbol{F}|(-\boldsymbol{r}/r) = -\frac{Gm_1m_2}{r^3}\boldsymbol{r} = -\frac{Gm_1m_2}{r^3}(x\boldsymbol{i} + y\boldsymbol{j} + z\boldsymbol{k}),$$

and this vector function describes the gravitational force acting on m_2. The vector field then assigns to each point $\boldsymbol{p} \in \mathbb{R}^3\backslash\{0\}$ a vector

$$X_{\boldsymbol{p}} = -\frac{Gm_1m_2}{r^3}\left(x\frac{\partial}{\partial x} + y\frac{\partial}{\partial y} + z\frac{\partial}{\partial z}\right) \in \mathrm{T}_{\boldsymbol{p}}(\mathbb{R}^3).$$

2. If X and Y are vector fields on $\mathbb{R}^n$, then addition is defined by $(X + Y)(\boldsymbol{x}) = X(\boldsymbol{x}) + Y(\boldsymbol{x}) \in \mathrm{T}_{\boldsymbol{x}}(\mathbb{R}^n)$, and this addition is clearly associative and commutative. The identity element is the null vector field 0 (Note 11.18) and the inverse of any X is $-X$ (cf. Example 4.9).

3. $X(\boldsymbol{p}) = (1, 0, 0)$ so that $X_{\boldsymbol{p}}f = X(\boldsymbol{p})\cdot(\nabla f)_{\boldsymbol{p}} = (\partial f/\partial x)_{\boldsymbol{p}}$. Similarly, for the directions (0, 1, 0) and (0, 0,1), the directional derivatives are $(\partial f/\partial y)_{\boldsymbol{p}}$ and $(\partial f/\partial z)_{\boldsymbol{p}}$ respectively.

4. Since $\boldsymbol{x} = (2, 1, 3)$, we have $X(\boldsymbol{x}) = (1, 0, -2)$, i.e. $X(\boldsymbol{x})$ is vector $\boldsymbol{i}_1 - 2\boldsymbol{i}_3$ in $\mathrm{T}_{\boldsymbol{x}}(\mathbb{R}^3)$. By direct calculation, $(\nabla f)_{\boldsymbol{x}} = (4, 2, 6)$. The directional derivative of f at $\boldsymbol{x}$ in direction of $X(\boldsymbol{x})$ is

$$X_{\boldsymbol{x}}f = X(\boldsymbol{x})\cdot(\nabla f)_{\boldsymbol{x}} = -8.$$

5. The space E^2 together with a metric is a metric space which is Hausdorff and has a countable basis of open sets. An orthogonal set of axes enables a homeomorphism to be defined as $\varphi\colon \mathrm{E}^2 \to \mathbb{R}^2$. Then E^2 is covered by a single chart (E^2, φ) with $\varphi(E^2) = \mathbb{R}^2$. Hence, E^2 is a topological manifold.

6. Any rigid motion is made up of a rotation about the origin of $\mathbb{R}^3$ followed by a translation. So a rigid motion can be defined as $T\colon \mathbb{R}^3 \to \mathbb{R}^3$ given by $T\boldsymbol{x} = \boldsymbol{A}\boldsymbol{x} + \boldsymbol{b}$, with $\boldsymbol{A} \in \mathrm{O}(3)$ and $\boldsymbol{b} \in \mathbb{R}^3$. With the usual metric ρ for $\mathbb{R}^3$, rigidity is ensured by $\rho(T\boldsymbol{x}, T\boldsymbol{y}) = \rho(\boldsymbol{x}, \boldsymbol{y})$ (an isometry). The translation takes the origin to $\boldsymbol{b}$, so the composition of two rigid motions gives another rigid motion, and composition is the group operation. Any rigid motion is a diffeomorphism. For any $\boldsymbol{A},\boldsymbol{B} \in \mathrm{O}(3)$, the mappings $(\boldsymbol{A}, \boldsymbol{B}) \mapsto \boldsymbol{AB}$ and $\boldsymbol{A} \mapsto \boldsymbol{A}^{-1}$ are both C^∞. Assign to each rigid motion T the pair $(\boldsymbol{A}, \boldsymbol{b}) \in \mathrm{O}(3) \times \mathbb{R}^3$. The group's manifold structure is then obtained from this 1–1 correspondence, and the group is a Lie group.

7. With $\boldsymbol{a}$ fixed, to each $t \in \mathbb{R}$ there is associated a translation $\varphi_t\colon \mathbb{R}^3 \to \mathbb{R}^3$ given by $\boldsymbol{x} \mapsto \boldsymbol{x} + t\boldsymbol{a}$. This is a C^∞ action on M, and the orbits of $\boldsymbol{x}$ are straight lines parallel to vector $\boldsymbol{a}$.

8. If $\boldsymbol{a},\boldsymbol{b},\boldsymbol{c} \in \mathbb{R}^3$, then $\boldsymbol{a} \times (\boldsymbol{b} \times \boldsymbol{c}) = (\boldsymbol{a}\cdot\boldsymbol{c})\boldsymbol{b} - (\boldsymbol{a}\cdot\boldsymbol{b})\boldsymbol{c}$ (Kreyszig 1979). Similarly, $(\boldsymbol{a} \times \boldsymbol{b}) \times \boldsymbol{c} = -\boldsymbol{c} \times (\boldsymbol{a} \times \boldsymbol{b}) = -(\boldsymbol{c}\cdot\boldsymbol{b})\boldsymbol{a} + (\boldsymbol{c}\cdot\boldsymbol{a})\boldsymbol{b}$. But the first vector lies in the plane of $\boldsymbol{b}$ and $\boldsymbol{c}$, whereas the second vector lies in the plane of $\boldsymbol{b}$ and $\boldsymbol{a}$. Thus, in general, these two vectors are unequal and the operation $\times$ is nonassociative. In fact,

$$\boldsymbol{a} \times (\boldsymbol{b} \times \boldsymbol{c}) + \boldsymbol{b} \times (\boldsymbol{c} \times \boldsymbol{a}) + \boldsymbol{c} \times (\boldsymbol{a} \times \boldsymbol{b}) = \boldsymbol{0}$$

(cf. (11.23)).

9. $$\begin{aligned}[(f+g), f+g] &= [(f+g), f] + [(f+g), g] \quad \text{(by (11.22))}\\ &= [f,f] + [g,f] + [f,g] + [g,g] \quad \text{(again by (11.22))}\\ &= [g,f] + [f,g] \quad \text{(by (11.19)).}\end{aligned}$$

But $[(f+g), (f+g)] = 0$, by (11.19), so that $[g,f] + [f,g] = 0$; whence $[g,f] = -[f,g]$.

10. $$\begin{aligned}[g_3, g_1](x) &= (g_1)_x g_3 - (g_3)_x g_1\\ &= \begin{bmatrix} 0 & 0 & 0\\ 0 & 0 & 1\\ 0 & -1 & 0\end{bmatrix}\begin{bmatrix} x_2\\ -x_1\\ 0\end{bmatrix} - \begin{bmatrix} 0 & 1 & 0\\ -1 & 0 & 0\\ 0 & 0 & 0\end{bmatrix}\begin{bmatrix} 0\\ x_3\\ -x_2\end{bmatrix}\\ &= \begin{bmatrix} 0\\ 0\\ x_1\end{bmatrix} - \begin{bmatrix} x_3\\ 0\\ 0\end{bmatrix} = \begin{bmatrix} -x_3\\ 0\\ x_1\end{bmatrix} = g_2(x).\end{aligned}$$

11.

$$f = \begin{bmatrix} 0\\ 0\\ x_1x_2\end{bmatrix}, \quad g_1 = \begin{bmatrix} 1\\ 0\\ 0\end{bmatrix}, \quad g_2 = \begin{bmatrix} 0\\ 1\\ 0\end{bmatrix},$$

$$[f, g_1] = -\begin{bmatrix} 0 & 0 & 0\\ 0 & 0 & 0\\ x_2 & x_1 & 0\end{bmatrix}\begin{bmatrix} 1\\ 0\\ 0\end{bmatrix} = \begin{bmatrix} 0\\ 0\\ -x_2\end{bmatrix},$$

$$[f, g_2] = -\begin{bmatrix} 0 & 0 & 0\\ 0 & 0 & 0\\ x_2 & x_1 & 0\end{bmatrix}\begin{bmatrix} 0\\ 1\\ 0\end{bmatrix} = \begin{bmatrix} 0\\ 0\\ -x_1\end{bmatrix},$$

$$[g_2, [f, g_1]] = \begin{bmatrix} 0 & 0 & 0\\ 0 & 0 & 0\\ 0 & -1 & 0\end{bmatrix}\begin{bmatrix} 0\\ 1\\ 0\end{bmatrix} = \begin{bmatrix} 0\\ 0\\ -1\end{bmatrix},$$

$$[g_1, [f, g_2]] = \begin{bmatrix} 0 & 0 & 0 \\ 0 & 0 & 0 \\ -1 & 0 & 0 \end{bmatrix} \begin{bmatrix} 1 \\ 0 \\ 0 \end{bmatrix} = \begin{bmatrix} 0 \\ 0 \\ -1 \end{bmatrix}.$$

12. The set $\mathbb{R}^{2\times 2}$ is a vector space of dimension 4 (Examples 6.2 and 6.10). For all $\boldsymbol{A},\boldsymbol{B},\boldsymbol{C} \in \mathbb{R}^{2\times 2}$, we have

$$\boldsymbol{A}(\boldsymbol{B}+\boldsymbol{C}) = \boldsymbol{AB} + \boldsymbol{AC}, \qquad (\boldsymbol{A}+\boldsymbol{B})\boldsymbol{C} = \boldsymbol{AC} + \boldsymbol{BC},$$

and $(\alpha \boldsymbol{A})\boldsymbol{B} = \alpha(\boldsymbol{AB}) = \boldsymbol{A}(\alpha \boldsymbol{B})$ for all $\alpha \in \mathbb{R}$.

In this linear algebra, multiplication is associative but not commutative. Since $\mathbb{R}^2$ is an additive abelian group, with multiplication associative and distributive over addition, it is a ring (Section 5.1).

REFERENCES

Abraham, R. and Marsden, J. E. (1978). Foundations of mechanics (2nd edn). Benjamin/Cummings, New York.

Allwright, J. C. (1982). LQP: dominant output feedbacks. *IEEE Transactions on Automatic Control*, **27**, 915–21.

Apostol, T. M. (1974). *Mathematical analysis* (2nd edn). Addison-Wesley, Reading, Massachusetts.

Arnold, V. I. (1978). *Mathematical methods of classical mechanics*. Springer-Verlag, New York.

Athans, M. and Falb, P. L. (1966). *Optimal control: an introduction to the theory and its applications*. McGraw-Hill, New York.

Banks, S. P. (1988). *Mathematical theories of nonlinear systems*. Prentice Hall, London.

Barnett, S. and Cameron, R. G. (1985). *Introduction to mathematical control theory* (2nd edn). Clarendon Press, Oxford.

Barra, G. de (1981). *Measure theory and integration*. Ellis Horwood, Chichester.

Bell, D. J. and Jacobson, D. H. (1975). *Singular optimal control problems*. Academic Press, London.

Bell, D. J. (1984). Lie brackets and singular optimal control. *IMA Journal of Mathematical Control and Information*, **1**, 83–94.

Boothby, W. M. (1975). *An introduction to differentiable manifolds and Riemannian geometry*. Academic Press, New York.

Brockett, R. W. (1972). System theory on group manifolds and coset spaces. *SIAM Journal on Control*, **10**, 265–84.

Brockett, R. W. (1973). Lie algebras and Lie groups in control theory. In *Geometric methods in system theory* (ed. D. Q. Mayne and R. W. Brockett) pp. 43–82. Reidel, Dordrecht, Holland.

Brockett, R. W. (1976). Nonlinear systems and differential geometry. *Proceedings of the IEEE*, **64**, 61–72.

Chillingworth, D. R. J. (1976). *Differential topology with a view to applications*. Pitman, London.

Citron, S. J. (1969). *Elements of optimal control*. Holt, Rinehart and Winston, New York.

Coddington, E. A. and Levinson, N. (1955). *Theory of ordinary differential equations*. McGraw-Hill, New York.

Courant, R. (1936). *Differential and integral calculus*, Vol. 2. Blackie, London.

Courant, R. and Robbins, H. (1941). *What is mathematics*? Oxford University Press, Oxford.

Fliess, M. (1986). Some remarks on nonlinear invertibility and dynamic state-feedback. In *Theory and applications of nonlinear control systems* (ed. C. I. Byrnes and A. Lindquist), pp. 115–21. North-Holland, Amsterdam.

Fuller A. T. (1963). Study of an optimum nonlinear control system. *Journal of Electronics and Control*, **15**, 63–71.

Gabasov, R. and Kirillova, F. M. (1972). High order necessary conditions for optimality. *SIAM Journal on Control*, **10**, 127–68.

Goldstein, H. (1950). *Classical mechanics.* Addison-Wesley, Reading, Massachusetts.

Graham, A. (1981). *Kronecker products and matrix calculus with applications.* Ellis Horwood, Chichester.

Green, D. H. (1986). *Modern logic design.* Addison-Wesley, Wokingham, England.

Halmos, P. R. (1974). *Naive set theory.* Springer-Verlag, New York.

Hartley, B. and Hawkes, T. O. (1970). *Rings, modules and linear algebra.* Chapman and Hall, London.

Hermes, H. and Haynes, G. (1963). On the nonlinear control problem with control appearing linearly. *SIAM Journal on Control*, **1**, 85–108.

Hermes, H. and LaSalle, J. P. (1969). *Functional analysis and time optimal control.* Academic Press, New York.

Hill, R. (1986). *A first course in coding theory.* Clarendon Press, Oxford.

Hirsch, M. W. and Smale, S. (1974). *Differential equations, dynamical systems and linear algebra.* Academic Press, New York.

Hirschorn, R. M. (1979). Invertibility of nonlinear control systems. *SIAM Journal on Control and Optimization.* **17**, 289–97.

Horn, R. A. and Johnson, C. R. (1985). *Matrix analysis.* Cambridge University Press, Cambridge.

Isidori, A. (1985). *Nonlinear control systems: an introduction.* Springer-Verlag, Berlin.

Jenner, W. E. (1963). *Rudiments of algebraic geometry.* Oxford University Press, New York.

Kalman, R. E., Falb, P. L., and Arbib, M. A. (1969). *Topics in mathematical system theory.* McGraw-Hill, New York.

Krasovskii, N. N. (1963). *Stability of motion.* Stanford University Press, Stanford, California.

Krener, A. J. (1974). A generalization of Chow's theorem and the bang-bang theorem to nonlinear control problems. *SIAM Journal on Control*, **12**, 43–52.

Krener, A. J. (1977). The high order maximal principle and its application to singular extremals. *SIAM Journal on Control and Optimization*, **15**, 256–93.

Kreyszig, E. (1978). *Introductory functional analysis with applications.* Wiley, New York.

Kreyszig, E. (1979). *Advanced engineering mathematics* (4th edn). Wiley, New York.

Lambert, J. D. (1973). *Computational methods in ordinary differential equations.* Wiley, London.

Lee, E. B. and Markus, L. (1967). *Foundations of optimal control theory.* Wiley, New York.

Lefschetz, S. (1963). *Differential equations: geometric theory* (2nd edn). Interscience, New York.

Lobry, C. (1970). Controlabilite des systemes non lineaires. *SIAM Journal on Control*, **8**, 573–605.

Luenberger, D. G. (1969). *Optimization by vector space methods.* Wiley, New York.

MacFarlane, A. G. J. (1970). *Dynamical system models.* Harrap, London.

MacLane, S. and Birkoff, G. (1979). *Algebra* (2nd edn). Macmillan, New York.

McDanell, J. P. and Powers, W. F. (1971). Necessary conditions for joining optimal singular and nonsingular subarcs. *SIAM Journal on Control*, **9**, pp. 161–73.

Nicholson, H. (ed.) (1980). *Modelling of dynamical systems*, Vol. 1 (Vol. 2, 1981). Peter Peregrinus, Stevenage, U.K..

Northcott, D. G. (1984). *Multilinear algebra.* Cambridge University Press, Cambridge.

Nowinski, J. L. (1981). Applications of functional analysis in engineering. *International Journal of Engineering Science*, **19**, 1377–90.

Ogata, K. (1987). *Discrete-time control systems.* Prentice Hall, Englewood Cliffs, N.J..

Parzynski, W. R. and Zipse, P. W. (1987). *Introduction to mathematical analysis.* McGraw-Hill, Singapore.

Patterson, E. M. and Rutherford, D. E. (1965). *Elementary abstract algebra.* Oliver and Boyd, Edinburgh.

Pommaret, J. F. (1983). *Differential Galois theory.* Gordon and Breach, New York.

Pontryagin, L. S., Boltyanskii, V. G., Gamkrelidze, R. V., and Mishchenko, E. F. (1962). *The mathematical theory of optimal processes.* Interscience, New York.

Rosenbrock, H. H. (1970a). *State-space and multivariable theory.* Nelson, London.

Rosenbrock, H. H. (1970b). Further properties of minimal indices. *Electronics Letters*, **6**, 450.

Rosenbrock, H. H. and Storey, C. (1970). *Mathematics of dynamical systems.* Nelson, London.

Silverman, L. M. (1969). Inversion of multivariable linear systems. *IEEE Transactions on Automatic Control*, **14**, 270–76.

Simmons, G. F. (1963). *Introduction to topology and modern analysis.* McGraw-Hill, New York.

Spivak, M. (1965). *Calculus on manifolds.* Benjamin, New York.

Stewart, I. (1973). *Galois theory.* Chapman and Hall, London.

Strang, G. (1980). *Linear algebra and its applications* (2nd edn). Academic Press, New York.

Sussmann, H. J. and Jurdjevic, V. (1972). Controllability of nonlinear systems. *Journal of Differential Equations*, **12**, 95–116.

Tropper, M. (1969). *Linear algebra.* Nelson.

Vaisman, I. (1984). *A first course in differential geometry.* Marcel Dekker, New York.

Walsh, G. R. (1975). *Methods of optimization.* Wiley, London.

Weir, A. J. (1973). *Lebesgue integration and measure.* Cambridge University Press, London.

Whitelaw, T. A. (1978). *An introduction to abstract algebra.* Blackie, Glasgow.

Wolfe, M. A. (1978). *Numerical methods for unconstrained optimization.* Van Nostrand Reinhold, New York.

INDEX

Notation in index: e, exercise; n, note; s, solution